Lecture Notes in Electrical Engineering

Volume 453

"Lecture Notes in Electrical Engineering (LNEE)" is a book series which reports the latest research and developments in Electrical Engineering, namely:

- Communication, Networks, and Information Theory
- Computer Engineering
- Signal, Image, Speech and Information Processing
- Circuits and Systems
- Bioengineering

LNEE publishes authored monographs and contributed volumes which present cutting edge research information as well as new perspectives on classical fields, while maintaining Springer's high standards of academic excellence. Also considered for publication are lecture materials, proceedings, and other related materials of exceptionally high quality and interest. The subject matter should be original and timely, reporting the latest research and developments in all areas of electrical engineering.

The audience for the books in LNEE consists of advanced level students, researchers, and industry professionals working at the forefront of their fields. Much like Springer's other Lecture Notes series, LNEE will be distributed through Springer's print and electronic publishing channels.

More information about this series at http://www.springer.com/series/7818

Vijay Nath
Editor

Proceedings of the International Conference on Microelectronics, Computing & Communication Systems

MCCS 2015

Springer

Editor
Vijay Nath
Department of Electronics &
Communication Engineering
Birla Institute of Technology, Mesra
Ranchi, Jharkhand
India

ISSN 1876-1100 ISSN 1876-1119 (electronic)
Lecture Notes in Electrical Engineering
ISBN 978-981-10-5564-5 ISBN 978-981-10-5565-2 (eBook)
https://doi.org/10.1007/978-981-10-5565-2

Library of Congress Control Number: 2017955665

Printed on acid-free paper

This Springer imprint is published by Springer Nature
The registered company is Springer Nature Singapore Pte Ltd.
The registered company address is: 152 Beach Road, #21-01/04 Gateway East, Singapore 189721, Singapore

Preface

In the direction of the cashless world, computing systems play a major role in terms of reliability, robustness, correctness and performance. This conference gives the ideas how to work in electronic media safely and securely. Second point, the manufacturing companies contribute a major role in the development of the country. However, it is facing several challenges such as rapid product development, flexibility, low to medium volume, transportation and low cost. Many advanced/unconventional technologies/tools/software are being developed worldwide to face these challenges. Among these technologies, IC design and manufacturing has become more popular due to the ability of precise work. For the research, development, sharing knowledge and exchange ideas in current trends, the first International Conference on Microelectronics, Computing & Communication Systems (MCCS-2015) was organized by Indian Society of VLSI Education (ISVE), Ranchi, at Advanced Regional Telecom Training Centre (ARTTC) near Jumar River, Hazaribagh Road, Ranchi, from 14 to 15 November 2015. In this conference, around 150 papers were received and 50 chapters in pedagogy (Washington Accord) in which 33 reviewed, registered and presented papers/chapters were accepted for publication in conference proceeding of Springer book series Lecture Notes in Electrical Engineering. Pedagogy pattern of chapters provides new ideas to learner to enhance their knowledge and scope of employability. Since this platform provides outcome-based learning, this is beneficial to students, researchers, professors and industrial people to recognize or evaluate the value of his/her current job. All the papers/chapters have been blind-reviewed by three expert reviewers, and detailed comments were passed to the concerned authors with decisions. All the presentation sessions were reviewed by six-member expert committee.

The conference on "Investigations on the Logic Performance of Hybrid CMOSFETs Comprising p-Ge/n-InGaAs MOSFETs with Barrier Layers" began with welcome address by Dr. Vijay Nath, General Chair of the conference, and a keynote address by Prof. Abhijt Biswas, Institute of Radio Physics & Electronics, Kolkata University.

The theme of the first session was Recent Trends in Microelectronics, device & circuits, and the session featured talks by Prof. A.A. Khan, former VC, Ranchi University. Future Scopes of Communication System was presented by Sh. Prasad Vijay Bhushan Pandey, Chairman ISVE & DGM, ETR BSNL, Ranchi. The session on Signal Processing consisted of talks by Prof. J.K. Mandal, Kalyani University, and Prof. D. Acharya, President, ISTM, Kolkata. Dr. Soma Mandal, Institute of Radio Physics & Electronics, Kolkata University, delivered her lecture on "Electrical Equivalent Model for Gene Regulatory System".

The Spokesperson for the session on Telecommunication Systems & Switching was Sh. Prasad Vijay Bhushan Pandey who was holding the position of DGM ETR BSNL Jharkhand Ranchi Circle and chairman ISVE Ranchi. The spokespersons for the session on Electronics System Design and Manufacturing (ESDM) were Prof. Vijay Nath, BIT, Mesra, Ranchi, and Prof. S.P. Tiwari, IIT Jodhpur. The spokesperson for Biomedical Instrumentation was Prof. Anand Kumar Thakur, SSMC Ranchi University. The session on Environmental Science and Engineering consisted of talks by Prof. K.K. Khatua, NIT Rourkela. This conference included oral session, poster sessions, tutorials, invited talks, keynote address by renowned scientists, professors and industries related to the theme of the conference. In this conference, original theoretical, practical, experimental simulations, development, applications, measurement- and testing-based papers were invited on wide areas from electrical, electronics, computer, communication, information technology, biomedical instrumentations, aerospace applications and environmental science and engineering, etc. The main aim of the conference was to bring together scientists, researchers, engineers, professors, industries that exchange and share their knowledge, experiences, technological developments and researches in current trends.

The sub-area of the conference covered were microelectronic devices, MEMS, VLSI design, IC technology, IC fabrication and testing, VLSI signal processing, VLSI for wireless communication & bioengineering, VLSI for electronic system design and manufacturing, image processing, digital signal processing, embedded system, robotics, electric power system, hybrid vehicles, renewable energy, green energy, cloud computing, algorithm development and implementation, computer networks, ICT applications, computer architecture, information security, data mining, mobile communication and computing, ad hoc network, wireless sensor network, EMI and EMC, satellite communication, fibre optics communication and optical networks, quantum dots, telemedicine, RFID and telemetry systems, aerospace, and environmental science and engineering, etc.

Totally, 33 papers represent in this volume the cover theme of the conference, i.e. design, simulation, verification, implementation and applications of micro- and nanoelectronics, computing and communication systems. In the first session, **Microelectronics, device & circuits**, papers were presented under the chairmanship of Dr. A.A. Khan and session chair persons were Dr. K.K. Senapati, Dr. Aminul Islam, Dr. J.K. Mandal, Dr. Soma Berman, Dr. P.R. Thakura and Dr. Abhijit Biswas. Shahiruddin et al. presented their paper on Single-Mode Negative Dispersion Hexagonal Photonic Crystal Fiber. Rifaqat Ali et al. described A Secure

Three-Factor Remote User Authentication Scheme Using Elliptic Curve Cryptosystem. S. Selvi et al. demonstrated the Implementation of Fingerprint-Based Biometric System and Its Integration with HRMS Application at RDCIS, SAIL. Rasika Dhavse et al. described Fabrication and Investigation of low voltage programmable flash memory gate stack. Sushma Kamlu et al. defined an Effective Method for Maintenance Scheduling of Vehicles Using Neural Network. Susmita Mandal et al. described her view on universally verifiable certificate less signcryption scheme for MANET. Abhijit Biswas et al. presented paper on impact of sidewall spacer layers on the analog/RF performance of nanoscale double gate junctionless transistors. R.C. Barik et al. described A Novel Data Encryption Approach in the Grid-Structured Binary Image. Bhattu.Hari Prasad Naik et al. explained the Analysis of Electromagnetic Wave Using Explicit FDTD in TM Mode with Extrapolation.

In the second session, **VLSI Signal Processing**, papers were presented under the chairmanship of Dr. Abhijit Biswas and session chair persons were Dr. J.K. Mandal, Dr. Soma Berman, Dr. P.R. Thakura, Prof. D. Acharjee and Dr. N. Chattoraj. S.S. Panigrahi et al. described A DEA-Based Evolutionary Computation Model for Stock Market Forecasting. Monalisa Dutta et al. demonstrated the Electrical Equivalent Model for Gene Regulatory System. Sneha Jain et al. described the Colour Image Segmentation Techniques: A Survey. Parivesh Pandey et al. presented his survey on Wireless Image Sensor Networks: A Review. Suprojit Nandy et al. demonstrated their strategy on Design of a Low-Cost Heart Rate Monitoring System.

In the third session, **ICT**, papers were presented under the chairmanship of Dr. J. K. Mandal and session chair persons were Dr. R.K. Lal, Dr. Soma Berman, Dr. Abhijit Biswas, Dr. N. Chattoraj and Dr. S.K. Mahapatra. A. Uma et al. demonstrated their approach for Design of DA-Based FIR Filter Architectures Using LUT Reduction Techniques. K. Rajalakshmi et al. described fractional delay FIR filter architecture using numeric strength reduction techniques. Yogesh Kumar Sharma et al. presented their work on Lifetime Enhancement of WSN Based on Modified Heterogeneous Leach Protocol. N. Chattoraj et al. described Modeling and Investigation of Electrothermally Actuated Micro-gripper. Vijay Nath et al. described an ultra-low-power Internet-controlled home automation. Kamalini Devi et al. described their work on Depth-Averaged Velocity Distribution for symmetrical and asymmetrical compound channels. Vijay Nath et al. demonstrated their design work for A 0.533-dB Noise Figure, 7-mW Narrowband Low-Noise Amplifier for Global Positioning System Application. Deepak Prasad et al. described the Design of Ultra-Low-Power CMOS Class E Power Amplifier. Shaligram Prajapat et al. described the mechanism of cryptic mining for automatic variable key-based cryptosystem.

In the fourth session, **Hybrid Electronics & Space Engineering**, papers were presented under the chairmanship of Dr. K.K. Khatua and session chair persons were Dr. A.K. Tiwary, Dr. Sukalyan Chakraborty, Prof. Shahiruddin, Prof. D. Acharjee, Dr. Abhijit Biswas and Dr. P.R. Thakura. Abha Sharma et al. explained the Improved Clustering for Categorical Data with Genetic Algorithm.

Niranjan Raj et al. described the Balanced Wrapper Design to Test the Embedded Core Partitioned into Multiple Layer for 3D SoC Targeting Power and Number of TSVs. Abhijit Biswas et al. described the Investigations on the Logic Performance of Hybrid CMOSFETs Comprising p-Ge/ n-InGaAs MOSFETs with Barrier Layers. Tara Prasanna Dash et al. showed the Design and Simulation of Strained-Si/SiGe Channel p-MOSFETs. Bhabnai Shankar Das et al. demonstrated the application of Lateral Distribution Method and Modified-Lateral Distribution Method to compound channel having converging floodplain. Tara Prasanna Dash et al. defined the Silicon–Germanium Channel heterostructure p-MOSFETs. Vijay Nath et al. demonstrated the strategy of Digital Hardware Design and IC Technology in Pedagogy. Ranjan Mishra et al. defined the Antenna Path Loss Propagation Model in the Dehradun Valley at 1800 MHz in L-Band. Rajesh Kumar Lal et al. described technique for Reduction of Dark Current in QWIP. Md Maqubool Hosain showed the Design of Circular Disc Monopole Antenna for UWB application.

Authors and editors have taken utmost care in presenting the information and acknowledging the original sources whenever necessary. Editors express their gratitude towards the authors, organizers of IC-MCCS and staff of Springer (India) for publication of this research book/proceeding possible. Readers are requested to provide their valuable feedback on the quality of presentation and inadvertent error or omission of information if any. We expect that the book will be welcomed by students as well as practising engineers/researchers/professors.

Ranchi, India Vijay Nath

Acknowledgements

We extend our thanks to all the authors for contributing to this book/proceeding by sharing their valuable research findings. We specially thank a number of reviewers for promptly reviewing the papers submitted to the conference. We are grateful to the volunteers, invited speakers, session chairs, sponsors, subcommittee members and members of International Advisory Committee, National Advisory Committee, Technical Programme Committee, Joint Secretary and Scientific Advisory Committee for successful conduct of the conference. The editors express their heartfelt gratitude towards Smt. Srimati Dagur, President, IETE, New Delhi, Sh. Sanjay Kumar Jha, Executive Engineer, Govt of Jharkhand & Chairman of IETE Ranchi; Sh. Prasad Vijay Bhushan Pandey, DGM ETR, Jharkhand Circle, Ranchi, & Chairman ISVE Ranchi; Prof. A.A. Khan, former VC, Ranchi University; Prof. M.K. Mishra, VC, BIT, Mesra; Dr. Labh Singh, CGM, BSNL, Ranchi; Prof. R.K. Pandey, VC, Ranchi University; Prof. P.K. Barhai, former VC, BIT, Mesra; Sh. R. Mishra, former CMD HEC Ranchi; Dr. Abhijit Biswas, Professor, Kolkata University; Dr. J.K. Mandal, Professor, Kalyani University; Prof. D. Acharjee, President, ISTM Kolkata; Dr. Vibha Rani Gupta, Professor, BIT, Mesra; Dr. B.K. Mishra, Professor, BIT, Mesra; Dr. V.K. Jha, BIT, Mesra; Sh. Ajay Kumar, AGM (admin) ARTTC BSNL, Ranchi & Secretary IETE, Ranchi; Dr. P.R. Thakura, Executive member of ISVE & Professor of BIT, Mesra, Ranchi; Dr. Anand Kumar Thakur Treasurer, IETE Ranchi, for their endless support, encouragement, motivation to organize such prestigious event that paved the way for this book on Microelectronics, Computing & Communication Systems (MCCS). At last, we express our sincere gratitude towards the staff of Springer who helped in publishing this book.

Contents

About the Editor

Dr. Vijay Nath received his bachelor's degree in Physics and master's degree in Electronics from DDU Gorakhpur University, India, in 1998 and 2001, respectively. He received a PGDCN from MMMUT (MMMEC), Gorakhpur (gold medallist), in 1999. He received his Ph.D. in VLSI Design & Technology from Dr. RML Avadh University, Faizabad, in association with CEERI, Pilani, in 2008. He was a member of the faculty in the Department of Electronics, DDU Gorakhpur University, Gorakhpur (2002–2006). In 2006, he joined as a faculty in the Department of Electronics and Communication Engineering, BIT, Mesra, Ranchi, India. Currently, he is Professor-In-Charge of VLSI Design Lab, Department of ECE, BIT, Mesra, Ranchi. His research interests include low-power VLSI circuits, mixed CMOS VLSI circuits, MEMS and NEMS sensors, CMOS signal-processing circuits, ASICs, embedded system designs, intelligent instrumentations, smart cardiac pacemaker and early-stage detection of cancer. He has to his credit more than 100 publications in international journals and conferences. He is a member of several reputed professional and academic bodies including IETE, ISVE and IEEE. He has completed several R&D projects of Government of India funded by DST, DRDO, MHRD and MoCIT. He has ongoing project of Government of India funded by RESPOND, ISRO. He has developed VLSI Design course in pedagogy (e-learning) funded by MHRD on the pattern of Washington Accord.

Single Mode Negative Dispersion Hexagonal Photonic Crystal Fiber

Shahiruddin, Akash Kumar and Dharmendra K. Singh

Abstract A photonic crystal fiber (PCF) with circular air holes having low dispersion and low confinement loss is analyzed. By deliberate selection of dimensions of air holes and spacing between air holes, it is possible to obtain the two required properties of solid core PCF at wide wavelength range that is negative dispersion and low confinement loss which is of the order of 10^{-7} dB/m. At 1.55 μm wavelength with common pitch (Λ), the simulated results have been observed at different diameters. The intended design finds applications in communication fields.

Keywords Air fill fraction · Confinement loss · Dispersion · Finite element method · Photonic crystal fiber

1 Introduction

Photonic crystal fiber (PCF) with periodic arrangement of air holes consecutively along the longitudinal direction shows a lot of interesting distinctiveness unachievable by conventional optical fiber. By alteration of the air holes or with special transverse structure design, the dispersion can be modified to compliment the weakness of optical system [1]. PCF is divided into two types of fibers. The first one is index-guiding PCF which guides light by total internal reflection (TIR) between a solid core and cladding region with multiple air holes [2, 3]. On the other hand, the second one is photonic band gap (PBG) effect at the operating wavelength to guide light in a low index core region [4, 5]. As we know that the

Shahiruddin (✉) · A. Kumar
Department of Electronics and Communication Engineering,
Birla Institute of Technology, Patna Campus, Patna, India
e-mail: shahir@bitmesra.ac.in

D. K. Singh
Department of Electronics and Communication Engineering,
National Institute of Technology, Patna, India
e-mail: dksingh@nitp.ac.in

V. Nath (ed.), *Proceedings of the International Conference on Microelectronics, Computing & Communication Systems*, Lecture Notes in Electrical Engineering 453,
https://doi.org/10.1007/978-981-10-5565-2_1

Table 1 Dispersion and confinement loss at 1.55 μM

Author	Simulated results	
	Dispersion	Confinement loss
Hsu et al. [7]	$D = -51{,}625$ ps/km nm	$L_c = 6.54 \times 10^{-4}$ db/km
Medjouria et al. [8]	$D = 18.43$ ps/km nm	$L_c = 0.5 \times 10^{-5}$ db/km
Medjouria et al. [9]	$D = 18$ ps/km nm	$L_c = 7.64 \times 10^{-3}$ db/km
Seifouria et al. [10]	$D = -2450$ ps/km nm	$L_c = 0.013$ db/m
Medjouria et al. [11]	$D = 62$ ps/km nm for triangular lattice $D = 50$ ps/km nm for sq. lattice	$L_{c=}$ 3.5×10^{-4} db/km for triangular lattice $L_c = 2 \times 10^{-2}$ db/km for sq. lattice

confinement loss can be reduced by introducing multiple air holes as much as possible, but the number of air holes in the cladding area is limited. The finite element method approach is used.

The dispersion must be compensated in the long-distance optical data transmission system to suppress the broadening of pulse. One way to realize this is to use the dispersion compensating fibers (DCFs) having large negative dispersion [6]. In PCF, the dispersion can be controlled with unique freedom. The dispersion of the air holes near the PCF core affects the dispersion characteristics. When the circular air holes are replaced with several elliptical air holes near the core is carried out to obtain better dispersion characteristic properties. As yet, the proposed PCF's dispersion and confinement loss values in different papers are shown in Table 1. In this work, we explore the impending in the hexagonal lattice of PCF to provide a new method to flexibly control the dispersion. Zero or negative dispersion is achieved in this paper. The confinement loss characteristic of PCF is achieved at 10^{-7} dB/m. Moreover, we discuss the influence of the diameter of air hole rings on dispersion as well as confinement loss. The design of structure and their parameters are calculated by RSoft FEMSIM.

2 Designed Structure

Figure1a shows geometric structure of the proposed hexagonal PCF. The air hole diameters are represented by d and d_1, and the pitch is labeled as Λ. The rings are formed of hollow air channels running along the entire length of the fiber. The core is surrounded by a cladding of effective refractive index, n_{eff}. The wavelength window of interest here is 0.8–2.0 μm. PCFs are designed for variable air fill fraction, d/Λ, where d is the air hole diameters and Λ the hole-to-hole spacing. The core carries the light signal which is characterized by its diameter, and the core material is of higher refractive index. The cladding is made of one or more layers of

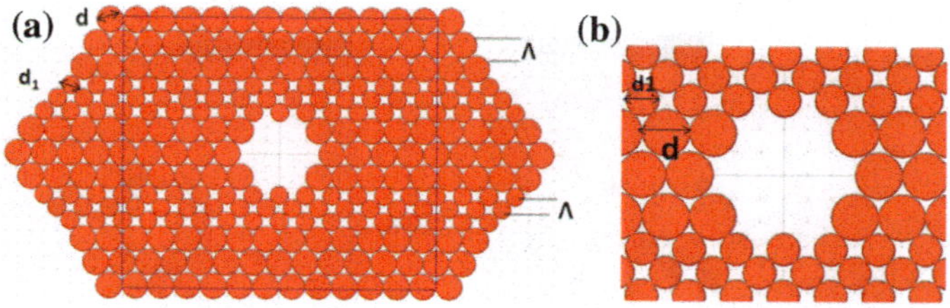

Fig. 1 **a** Cross section of proposed PCF and air hole arrangement, **b** close look of core region

materials of lower refractive index which keeps the light in the core. It can be obtained as:

$$\mathbf{N_A} = \sqrt{n_{\text{core}}^2 - n_{\text{cladding}}^2} \tag{1}$$

Initially, the diameters of the proposed PCF are d = 2 μm and d_1 = 1.4 μm. The distance between the centers of the neighboring air holes is Λ = 2 μm. Figure 1b explores the core region with two different diameters of air holes used in PCF.

Figure 2 depicts the light is confined in the core region. The diameter will change to observe the light confinement at different air fill fraction. In Fig. 2a, change the diameter of air hole d = 1.9 μm, d_1 = 1.3 μm, and constant pitch (Λ) = 2 μm in all cases. Figure 2b explores with change in diameter d = 2 μm, d_1 = 1.4 μm. The diameter d = 2.1 μm, d_1 = 1.5 μm is considered in Fig. 2c, Fig. 2d depicts light which is confined after change in air hole diameter d = 2.2 μm, d_1 = 1.6 μm. Table 2 shows the constant values of Λ and air fill fraction of different diameter of solid core PCF.

3 Numerical Results and Discussion

Figure 3 shows wavelength dependence of refractive index of proposed PCF for different air filling conditions. In this figure, we have changed the diameter of the air holes in the PCF to observe the change of refractive index and plot it by using MATLAB. The light is confined in the core with β. The light will propagate constantly along the core and it should not propagate along the cladding. The refractive index n is defined as:

$$\beta = nk_0 \tag{2}$$

where k_0 is the free-space propagation constant.

At a proposed wavelength the refractive index defined by value β of the respective material.

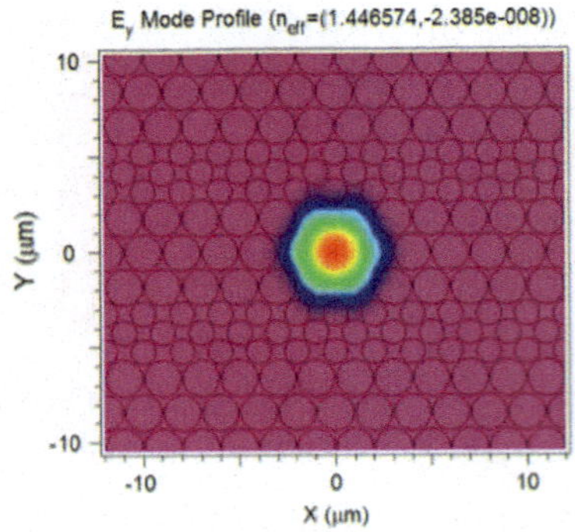

(a) With parameter d=1.9μm, d_l=1.3μm

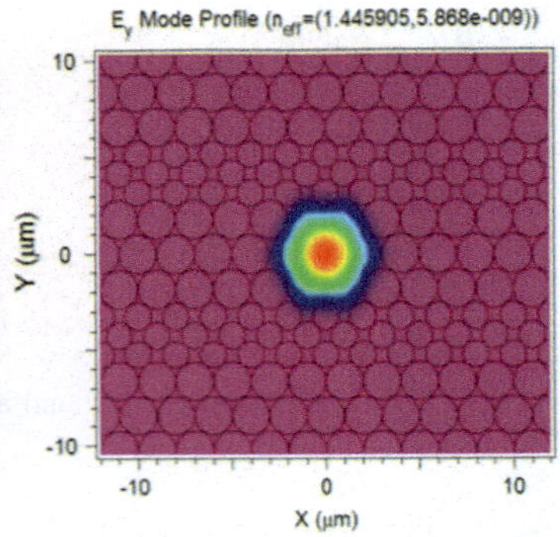

(b) With parameter d=2μm, d_l=1.4μm

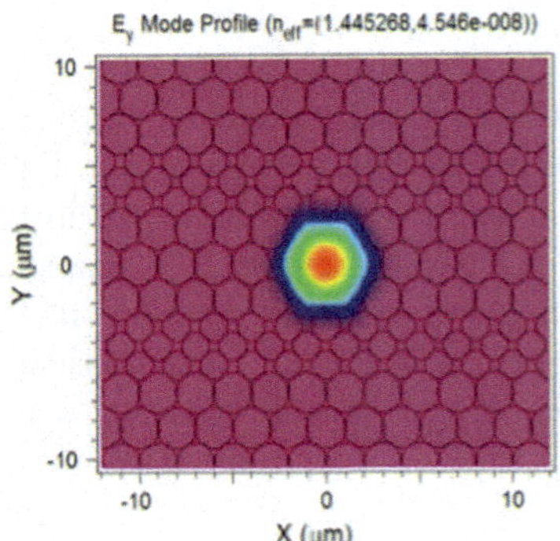

(c) With parameter d=2.1μm, d_l=1.5μm

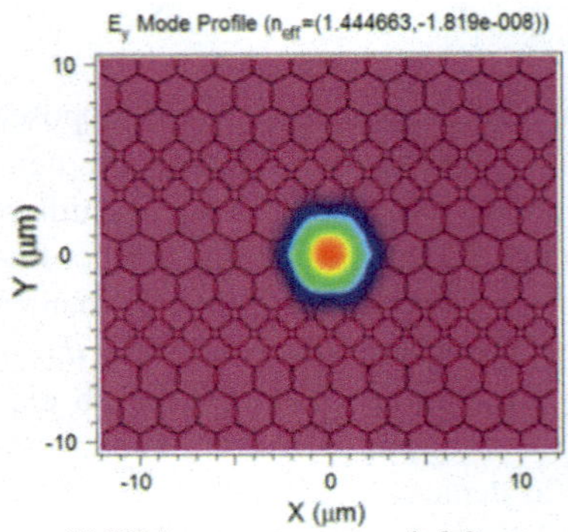

(d) With parameter d=2.2μm, d_l=1.6μm

Fig. 2 Field intensity distribution of PCF with constant Λ = 2 μm

Table 2 Different values of pitch and air filling fraction of solid core PCF

Λ (μm)	d/Λ (μm)	d_l/Λ (μm)
2	0.95	0.65
2	1	0.7
2	1.05	0.75
2	1.1	0.8

The effective index model will be used for an investigation of bending loss and dispersion properties.

Chromatic dispersion or total dispersion $D\ (\lambda)$ is a sum of both waveguide dispersion $Dw\ (\lambda)$ and material dispersion $Dm\ (\lambda)$, i.e.,

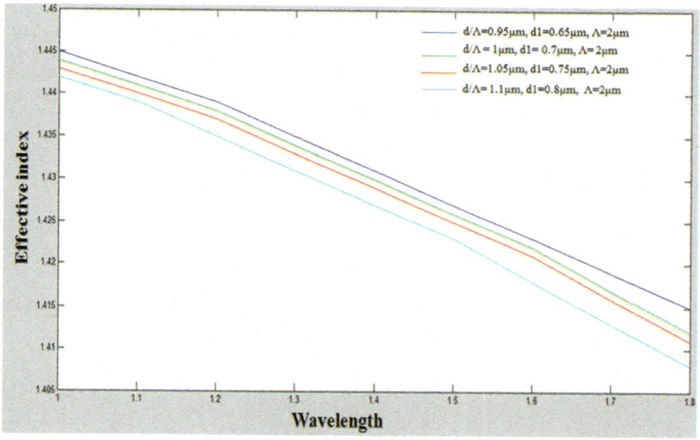

Fig. 3 Refractive index of variable diameter of PCF

$$D(\lambda) = Dw(\lambda) + Dm(\lambda) \tag{3}$$

Waveguide dispersion Dw (λ) occurs due to wavelength dependence of propagation constant of propagating mode, and material dispersion Dm (λ) occurs due to wavelength dependence of refractive index of material.

The dispersion (D) is proportional to the second derivative of the n_{eff}, with respect to the wavelength (λ) and is calculated by:

$$D(\lambda) = -\frac{\lambda}{c}\frac{d^2\mathrm{Re}[n_{\mathrm{eff}}]}{d\lambda^2} \tag{4}$$

where D is total dispersion of the PCF, c is velocity of light in vacuum, λ is the wavelength, $Re[n_{\mathrm{eff}}]$ is the real part of effective refractive index (n_{eff}). Figure 4 shows the dispersion curve at variable diameter and fixed pitch. The dispersion value at 1.55 μm is observed 300 ps/km nm in Fig. 4a after varying the diameter d = 1.9 μm and d_1 = 1.3 μm. After changing the design parameters d = 2 μm and d_1 = 1.4 μm with fixed air hole pitch (Λ) = 2 μm, the dispersion is −1000 ps/km nm at 1.55 μm in Fig. 4b. Figure4c shows the dispersion value 250 ps/km nm at 1.55 μm. In last, Fig. 4d depicts the dispersion value 300 ps/km nm at 1.55 μm. By comparing the dispersion graphs, the best result has been observed in Fig. 4b.

Confinement loss is the ability to confine the light within the core region. As the core and cladding material are generally the same, and hence, they have same refractive index. The cladding consists of air hole rings over the entire length of fiber. The light guidance in the core is due to finite number of holes in bulk silica. Here, the cladding does not insulate the core from the surrounding material, as the holes do not merge with their neighbors. So physically we can imagine, the light leakage from the core region into the outer air hole region is unavoidable and confinement loss is occurring due to the extent of the cladding. In a solid core PCF

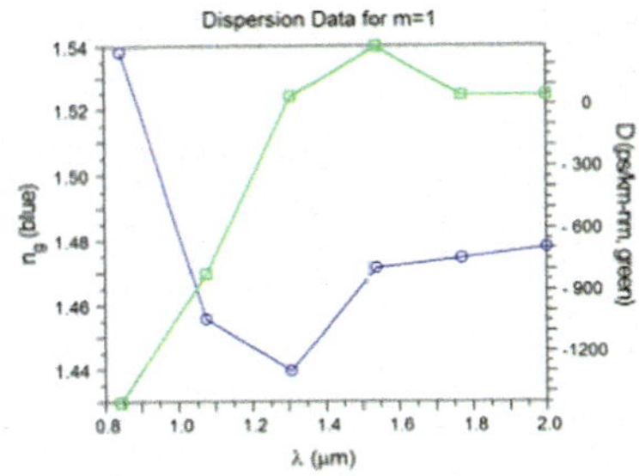

(a) d=1.9μm and d₁=1.3μm, d/Λ=0.95μm and d₁/Λ=0.65μm

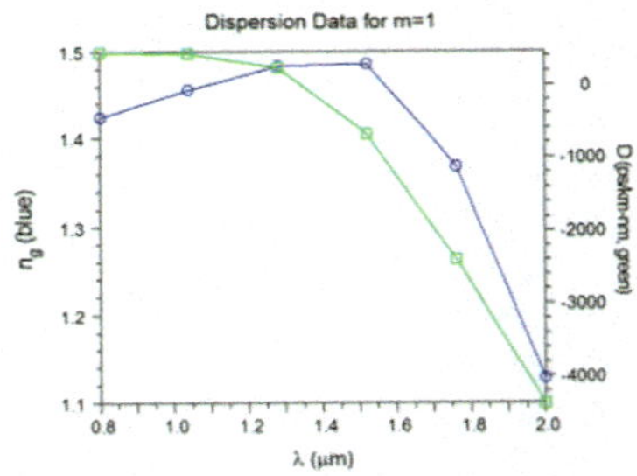

(b) d=2μm and d₁=1.4μm, d/Λ=1μm and d₁/Λ=0.7μm

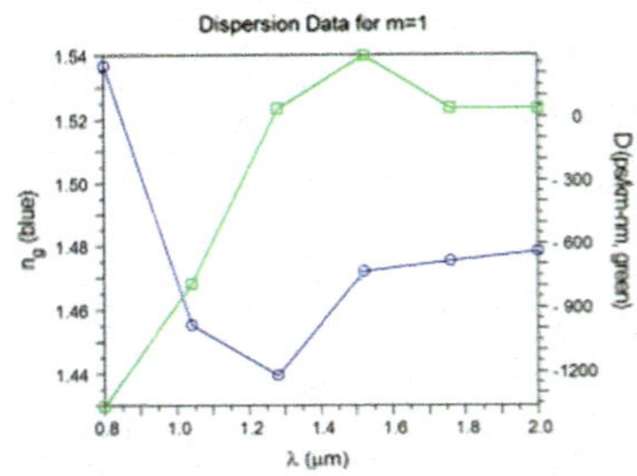

(c) d=2.1 μm and d₁=1.5μm, d/Λ=1.05μm and d₁/Λ=0.75μm

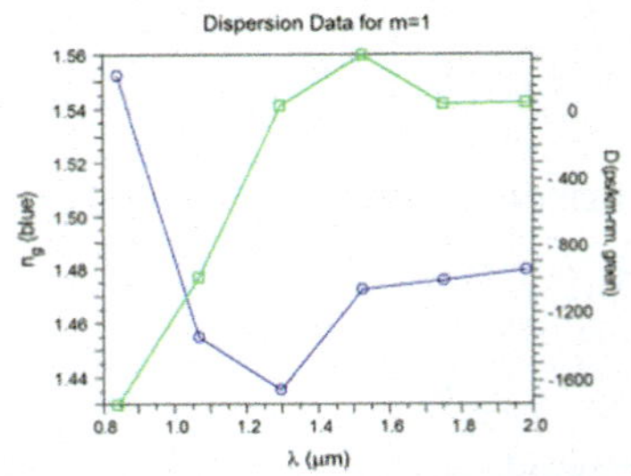

(d) d=2.2μm and d₁=1.6μm, d/Λ=1.1μm and d₁/Λ=0.8μm

Fig. 4 Dispersion slope with fixed air hole pitch (Λ) = 2 μm

for small values of d/Λ, the resulting loss can be large unless a sufficiently large number of periods are used. The confinement loss L_c, with unit of dB/m is obtained from the imaginary part of n_{eff} as follows:

$$L_c = \frac{(20 \times 10^6)}{\ln(10)} k_0 \mathrm{Im}[n_{eff}] \quad (5)$$

With the unit dB/m, where $Im[n_{eff}]$ is the imaginary part of the refractive index and $k_0 = 2\pi/\lambda$ is the wave number in free space [12].

Figure 5 shows the confinement loss of proposed structure at different air fill fraction. Figure 5a depicts that the confinement loss value is in the range of 13×10^{-7} dB/m at 1.55 μm, assuming $d = 1.9$ μm and $d_1 = 1.3$ μm, $d/\Lambda = 0.95$ μm and $d_1/\Lambda = 0.65$ μm. Figure 5b calculated at $d = 2$ μm and $d_1 = 1.4$ μm, $d/\Lambda = 1$ μm and $d_1/\Lambda = 0.7$ μm is -5×10^{-7} dB/m on the scale of 1.55 μm. When changing the diameter $d = 2.1$ μm, $d_1 = 1.5$ μm, $d/\Lambda = 1.05$ μm and $d_1/\Lambda = 0.75$ μm, the confinement loss is -9.5×10^{-7} dB/m at bandwidth 1.55 μm. Figure 5d shows the confinement loss approximate 0 at 1.55 μm when change in diameter $d = 2.2$ μm, $d_1 = 1.6$ μm, $d/\Lambda = 1.1$ μm and $d_1/\Lambda = 0.8$ μm. The confinement loss range is 10^{-7} dB/m after observing it at different air fill fraction.

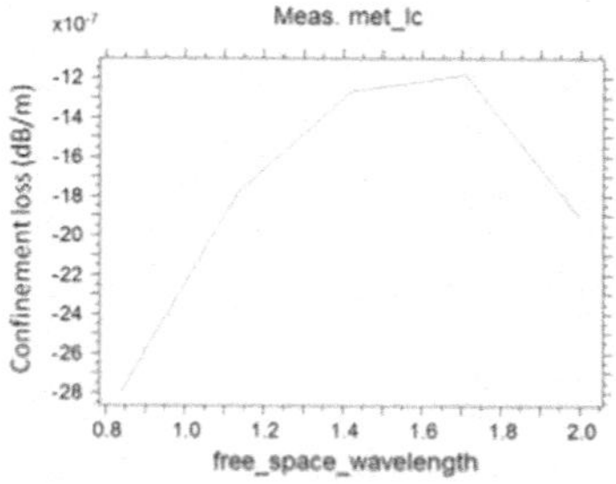

(a) d=1.9μm and d_l=1.3μm, d/Λ=0.95μm and d_l/Λ = 0.65μm

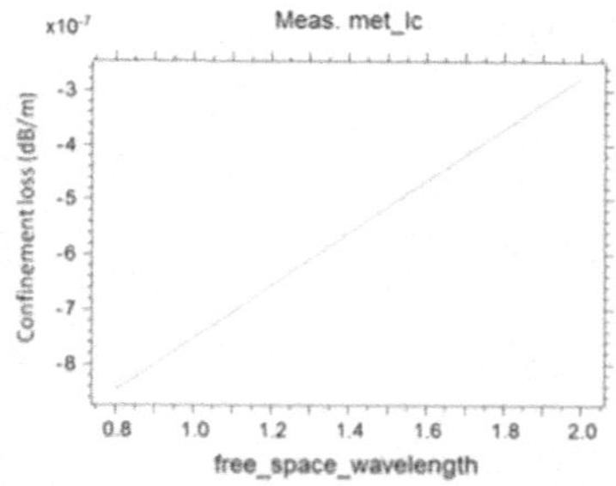

(b) d=2 μm and d_l=1.4μm, d/Λ=1μm and d_l/Λ=0.7μm

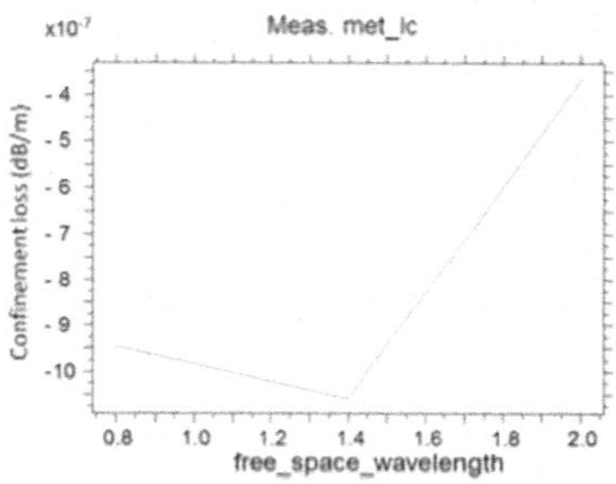

(c) d=2.1μm and d_l=1.5μm, d/Λ=1.05μm and d_l/Λ=0.75μm

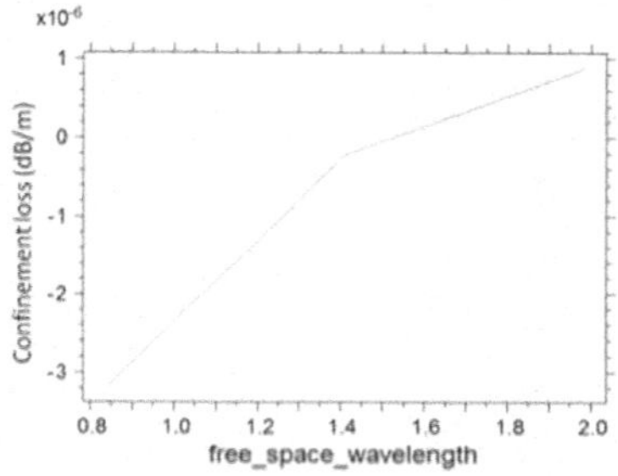

(d) d=2.2μm and d_l=1.6μm, d/Λ=1.1μm and d_l/Λ=0.8μm

Fig. 5 Confinement loss with fixed air hole pitch Λ = 2 μm

4 Conclusion

A novel structure has been proposed to evolve negative dispersion and very low confinement loss by using silicon material. The design offers a wide range of wavelengths from 0.8 to 2.0 μm. It has been shown that the confinement loss has been sharply dropped down to a very low value at the proffered communication wavelength for this design. It is examined that if we keep different air hole diameter '*d*' and by keeping the hole-to-hole spacing 'Λ' constant, we can get nearly zero and negative dispersion over a wide range of wavelengths. Thus, large negative dispersion of PCFs makes them useful prospects for use as dispersion compensating fibers in optical communication. Now, alternate technologies are required that can speed up the data transmission capacity. And since the introduction of micro-structured PCF, a revolution has begun in the field of optical communication.

References

1. D.C. Tee, M.H. Abu Bakar, N. Tamchek, F.R. Mahamd Adikan, Photonic crystal fiber in photonic crystal fiber for residual dispersion compensation over E + S + C + L + U wavelength bands. IEEE Photon. J. **5**(3) (2013)
2. J.C. Knight, T.A. Birks, P.B.J. Russell, D.M. Atkin, All silica single mode optical fiber with photonic crystal cladding. Opt. Lett. **21**, 1547–1549 (1996)
3. T.A. Birks, J.C. Knight, P.St.J. Russell, Endlessly single mode photonic crystal fiber. Opt. Lett. **22**, 961–963 (1997)
4. J.C. Knight, J. Broeng, T.A. Birks, PStJ Russell, Photonic bandgap guidance in optical fiber. Science **282**, 1476–1478 (1998)
5. R.F. Cregan, B.J. Mangan, J.C. Knight, T.A. Birks, PStJ Russell, P.J. Roberts, D.C. Allan, Single mode photonic bandgap guidance of light in air. Science **285**, 1537–1539 (1999)
6. M. Koshiba, K. Saitoh, Structural dependence of effective area and mode field diameter for holey fibers. Opt. Express **11**, 1746–1756 (2003)
7. J.-M. Hsu, W.H. Zheng, C.-L. Lee, J.S. Horng, Theoretical investigation of a dispersion compensating photonic crystal fiber with ultra-high dispersion coefficient and extremely low confinement loss. Photonics Nanostruct. Fundam. Appl. **16**, 1–8 (2015)
8. A. Medjouria, L.M. Simohamed, O. Zianeb, A. Boudrioua, Z. Becer, Design of a circular photonic crystal fiber with flattened chromatic dispersion using a defected core and selectively reduced air holes: application to supercontinuum generation at 1.55 μm. Photonics Nanostruct. Fundam. Appl. **16**, 43–50 (2015)
9. A. Medjouria, L.M. Simohamedc, O. Zianeb, A. Boudriou, Analysis of a new circular photonic crystal fiber with large mode area. Optik **126**, 5718–5724 (2015)
10. Mahmood Seifouria, Moslem Dekamina, Saeed Olyae, A new circular chalcogenide/silica hybrid microstructured optical fiber with high negative dispersion for the purpose of dispersion compensation. Optik **126**, 3093–3098 (2015)
11. A. Medjouria, L.M. Simohamed, O. Zian, A. Boudriou, Investigation of high birefringence and chromatic dispersion management in photonic crystal fiber with square air holes. Optik **126**, 2269–2227 (2015)
12. S.M. Nejad, N. Ehteshami, Novel design to compensate dispersion for index guiding photonic crystal fiber with defected core. IEEE ICMEE-2010 **2**, 417–421 (2010)

A Secure Three-Factor Remote User Authentication Scheme Using Elliptic Curve Cryptosystem

Rifaqat Ali and Arup Kumar Pal

Abstract Recently, three factors such as biometric, smart card, and password based authentication schemes have drawn considerable attention in the field of information security. In this paper, the authors have presented an elliptic curve cryptosystem based authentication scheme using biometric, smart card, and password and also analyzed the formal and informal security of the authentication scheme. In this scheme, the parameters of elliptic curve are derived from the biometric features like iris, fingerprints, etc., which is suitable to withstand the forgery. The formal and informal security analysis are done based on the BAN logic and suggested propositions, respectively. The security analysis ensures that the presented scheme can withstand various kinds of malicious attacks. In addition, the scheme is also comparable with other related schemes in the context of communication cost, computation cost, and smart card storage. The scheme is suitable to ensure high degree of security with reduced comparatively overhead.

Keywords Authentication · BAN logic · Biometric · Key agreement
Elliptic curve cryptography (ECC) · Smart card

1 Introduction

In recent, the e-commerce and m-commerce based applications are become widely popular among users due to the rapid advancement of Internet technology, computer devices, smart phones, etc. Password based authentication is one of the essential security mechanisms during secure communication with these

R. Ali (✉) · A. K. Pal
Department of Computer Science and Engineering,
Indian Institute of Technology (Indian School of Mines),
Dhanbad 826004, Jharkhand, India
e-mail: rifaqatali27@gmail.com

A. K. Pal
e-mail: arupkrpal@gmail.com

V. Nath (ed.), *Proceedings of the International Conference on Microelectronics, Computing & Communication Systems*, Lecture Notes in Electrical Engineering 453,
https://doi.org/10.1007/978-981-10-5565-2_2

e-commerce and m-commerce applications. In 1981, Lamport [1] presented the first remote user authentication scheme for insecure network. In his scheme, server maintains a password table to authenticate the legitimate user. Since then, in order to improve the system security, computation, and communication efficiency, a large number of smart card and password based authentication have been presented in the literature [2–5]. However, the security flaws in password authentication based protocols have been exposed seriously due to the management of password in improper way. One of the common issues in password based applications is to select suitable password. The selection of long and random password is highly secured but such type of password is not practically convenient to remember for a use. Sometimes, it may happen that the user may share his password with the other people, in that scenario; there is no way to identify who is the legal user. In order to resolve the single password authentication problems, several biometric-based remote user authentication have been presented by several researcher [6–10]. Generally, biometric based remote user authentication is extremely more secure and reliable than the traditional authentication scheme. The advantageous of using biometric keys over the traditional password are like biometric keys cannot be lost or forgotten and even it is not possible to copy, share, and guess the biometric key. The biometric system is basically a pattern recognition system which operates by obtaining biometric data from an individual extracting a feature set from the obtained data and comparing these features with the template set in the database.

In order to design a secure and efficient authentication protocols, many researchers have considered several cryptographic techniques such as *ECC*, *RSA*, non-invertible hash function, and several other mathematical operations such as *XOR* and concatenate. *ECC* provides same level of security with smaller key size than *RSA* (1024-bits *RSA* key is equal to the 160-bits *ECC* key). In 2012, Li [11] presented a two-factor remote user authentication scheme based on *ECC* and claimed that presented protocol is secure against various kinds of security attacks and provides mutual authentication and user anonymity with low computation cost. However, Zhang et al. [12] point out that Li's protocol cannot provide mutual authentication and propose an improved scheme based on *ECC*. They claimed that their scheme provide all security attributes with lower computation cost. In recent years, many *ECC* based mutual authentication and key agreement scheme have been presented in the literature [13–15]. In order to improve the security of remote authentication scheme *ECC* combines with biometric. In 2014, Arshad and Nikooghadam [16] presented *ECC* based three-factor remote authentication and key agreement scheme, which is improvement of Tan et al. scheme [17]. They claimed that their protocol resists various kinds of security flaws with better complexity. Last few years, many biometric and *ECC* based authentication scheme have been presented in the literature for distinct application systems [18, 19]. In this paper, we have also presented of an *ECC* based three-factor mutual authentication scheme where this scheme is verified through formal and informal security analysis.

Rest of the paper is described as follows: In Sect. 2, the *ECC* based three-factor authentication scheme is presented. The security validation using BAN logic demonstrates in Sect. 3. Moreover, Sect. 4 shows informal security analysis of the presented scheme. The performance comparison is presented in Sect. 5. Finally, the conclusion is drawn in Sect. 6.

2 ECC-Based Three-Factor Authentication Scheme

This section presents the three factor authentication scheme based on elliptic curve cryptosystem. The authentication scheme consists of four phases namely registration phase, login phase, authentication and key agreement phase, and password change phase.

2.1 *Registration Phase*

In this phase, a user U_i sends a request to the authentication server for registration or re-registration purpose. Initially, the user freely chooses his/her identity ID_i, password PW_i, and also imprints his/her personal biometrics F_i at the sensor. Then user calculates $\mathrm{RPW}_i = h(\mathrm{PW}_i \| r_i)$ and submits $\{\mathrm{ID}_i, \mathrm{RPW}_i, F_i\}$ to the server through secure channel. Here r_i is considered as a random number generated by the user. The server performs the following operations after receiving the message from the user:

i. Firstly, the server finds out the coordinate points (*x*, *y*, and angle) from biometric feature F_i. From this value, the coefficient of the elliptic curve coordinates value *A*, *B*, and *G* points [10] are derived.
ii. The server computes $T_i = H(F_i)$, where *H* is biohashing function.
iii. Next, the server computes $C_i = X \cdot G$, $D_i = h(\mathrm{ID}_i) \| \mathrm{RPW}_i) \cdot G$, $E_i = C_i + D_i$, $S_i = \mathrm{C_i} + h(\mathrm{RPW}_i) \cdot G$, where *X* is master key of the server.
iv. Finally, the server issues a smart card which contains $\{G, E_i, H(\cdot), h(\cdot), S_i, T_i, E_k(\cdot)/D_k(\cdot)\}$ and sends smart card to the user via a secure channel.
v. After receiving the smart card, the user enters r_i into his/her smart card and finally the smart card contains $\{G, E_i, H(\cdot), h(\cdot), S_i, T_i, E_k(\cdot)/D_k(\cdot), r_i\}$.

2.2 *Login Phase*

The login phase is invoked when the user wants to login to the remote server. The following steps are performed:

i. The user inserts his/her smart card into smart card reader and inputs the personal biometric F_i on the specific device to verify the user's biometric.
ii. Verifying $T_i = H(F_i)$.
iii. If the above condition does not hold, it means that the user U_i does not pass the correct biometric verification and the phase is terminated. If it is holds, the user passes the correct biometric verification and inputs his/her identity ID_i and password PW_i to perform the following operation.
iv. After receiving the user's identity ID_i and password PW_i, the smart card computes the following: $\text{RPW}_i^* = h(\text{PW}_i \| r_i)$, $D_i^* = h(\text{ID}_i \| \text{RPW}_i^*) \cdot G$, $C_i^* = E_i - D_i^*$, $S_i^* = C_i + h(\text{RPW}_i^*) \cdot G$ and checks the condition whether the computed S_i^* matches with stored S_i. If it is same, then it implies that the user entered correct $\{\text{ID}_i, \text{PW}_i\}$ and allows him\her to move for the next steps. Otherwise, it will terminate the session.
v. The smart card generates a random nonce N_c and computes the following: $\text{AID}_i = \text{ID}_i \oplus h(N_c)$, $M_1 = h(N_c) \cdot G \cdot \text{ID}_i$ and $M_2 = M_1 \cdot \text{RPW}_i$, $M_3 = E_{Cx}(N_c, M_1)$ and $M_4 = h(Cx \| M_2 \| N_c)$, where Cx is the x coordinate of $C_i = X \cdot G = (Cx, Cy)$.
vi. Then the user sends (AID_i, M_3, M_4) to the server over the public channel.

2.3 Authentication and Key Agreement Phase

This phase achieves mutual authentication and session key agreement between the user and the server after performing all the steps which are presented below.

i. After receiving the login request message $\{\text{AID}_i, M_3, M_4\}$, the server first decrypts M_3 using the key C_x and retrieves $\{N_c, M_1\}$. Then the server computes $\text{ID}_i^* = \text{AID}_i \oplus h(N_c)$, $M_1^* = h(N_c)G$. ID_i^*, $M_2^* = M_1^* \cdot \text{RPW}_i$, $M_4^* = h(Cx \| M_2^* \| N_c)$ and checks whether $\text{ID}_i^* = \text{ID}_i$ and $M_4^* = M_4$ or not. If this condition is hold then moves into the next step. Otherwise, terminates the session.
ii. The server generates a random nonce N_S and computes the following: $N_{sc} = N_S \oplus N_c$ and $H_i = h(N_S \| \text{RPW}_i \| M_1)$. Then server sends $\{N_{sc}, H_i\}$ to the user.
iii. After receiving the message $\{N_{sc}, H_i\}$ from the server, the user first computes $N_S = N_{sc} \oplus N_c$ using its own previously generated random nonce N_c. Then user checks $H_i^* = h(N_S \| \text{RPW}_i \| M_1)$ using its own RPW_i, M_1 and previously generated random nonce N_c.
iv. The user computes $\text{SK} = h(N_S \| N_c \| \text{RPW}_i \| M_2)$, $Z_i = \text{SK} \cdot G + C_i$ and sends Z_i to the server for session key verification.
v. The server computes $\text{SK} = h(N_S \| N_c \| \text{RPW}_i \| M_2)$, $Z_i^* = \text{SK} \cdot G + C_i$ and compares $Z_i^* = Z_i$. If the comparison is matched it means that the session key verification holds correctly.

2.4 Password Change Phase

In this phase, whenever the user U_i wants to change his/her password, then he/she inserts his/her smart card into smart card reader and submits identity ID_i, password PW_i, and biometric information F_i and subsequently the smart card reader performs the following steps:

i. Verifying $T_i = H(F_i)$.
ii. If the above condition does not hold, it means that the user U_i does not pass the correct biometric verification and the phase is terminated. If it holds, the user passes the correct biometric verification and performs the next steps.
iii. The user enters his/her identity ID_i and password PW_i, then the smart card computes the following: $\mathrm{RPW}_i^* = h(\mathrm{PW}_i \| r_i)$, $D_i^* = h(\mathrm{ID_i} \| \mathrm{RPW}_i^*) \cdot G$, $C_i^* = E_i - D_i^*$, $S_i^* = C_i^* + h(\mathrm{RPW}_i^*) \cdot G$ and checks the condition whether the computed S_i^* is matched with stored S_i. If the comparison holds, it implies that the user has entered the correct $\{\mathrm{ID}_i, \mathrm{PW}_i\}$ and allows him/her to move into the next steps. Otherwise, it will terminate the session.
iv. The user inputs a new password $\mathrm{PW}_i^{\mathrm{new}}$, then the smart card computes: $\mathrm{RPW}_i^{\mathrm{new}} = h(\mathrm{PW}_i^{\mathrm{new}} \| r_i)$, $D_i^{\mathrm{new}} = h(\mathrm{ID_i} \| \mathrm{RPW}_i^{\mathrm{new}}) \cdot G$, $E_i^{\mathrm{new}} = E_i - h(\mathrm{ID}_i \| \mathrm{RPW}_i) \cdot G + h(\mathrm{ID}_i \| \mathrm{RPW}_i^{\mathrm{new}}) \cdot G$, $S_i^{\mathrm{new}} = E_i - h(\mathrm{ID_i} \| \mathrm{RPW}_i) \cdot G + h(\mathrm{RPW}_i^{\mathrm{new}}) \cdot G$. Finally, the smart card replaces S_i with S_i^{new} and E_i with E_i^{new} into memory of the smart card and keeps rest of the smart card parameters unchanged.

3 Security Analysis Based on BAN Logic

This section verifies the validity of the presented scheme through Burrows-Abadi-Needham (*BAN*) logic [20]. The notation and logical postulates used in BAN logic is illustrated in Table 1. The BAN logic is a set of rules which can be used to verify whether information exchanged scheme is trustworthy and secured against various kinds of malicious attacks. To implement the *BAN* logic, the following steps are performed:

Step 1: In the BAN logic, the goals of our scheme can be presented as follows:

$$\text{Goal 1: } U_i \mid\equiv \left(U_i \overset{\mathrm{SK}}{\leftrightarrow} S\right)$$

$$\text{Goal 2: } U_i \mid\equiv S \mid\equiv \left(U_i \overset{\mathrm{SK}}{\leftrightarrow} S\right)$$

$$\text{Goal 3: } S \mid\equiv \left(U_i \overset{\mathrm{SK}}{\leftrightarrow} S\right)$$

$$\text{Goal 4: } S \mid\equiv U_i \mid\equiv \left(U_i \overset{\mathrm{SK}}{\leftrightarrow} S\right)$$

Table 1 Notation of BAN logic

$\frac{P\mid\equiv P \overset{K}{\leftrightarrow} Q,\ P\triangleleft\{X\}_K}{P\mid\equiv Q\mid\sim X}$	**Message Meaning Rule**: If A believes that K is shared by P and Q and sees X encrypted with K, then P believes that Q once said X
$\frac{P\mid\equiv Q\Rightarrow X,\ P\mid\equiv Q\mid\equiv X}{P\mid\equiv X}$	**Jurisdiction Rule**: If P believes that Q has jurisdiction over X and P believes Q on the truth of X, then P believes on X
$\frac{P\mid\equiv \#X}{P\mid\equiv \#(X,Y)}$	**Freshness Conjuncatenation Rule**: If P believes that freshness of X, then P believes that freshness of *(X, Y)*
$\frac{P\mid\equiv \#X,\ P\mid\equiv Q\mid\sim X}{P\mid\equiv Q\mid\equiv X}$	**Nonce Verification Rule**: If P believes that freshness of X and P believes that Q once said X, then P believes that Q believes X
$\frac{P\mid\equiv \#X,\ P\mid\equiv Q\mid\equiv X}{P\mid\equiv P \overset{K}{\leftrightarrow} Q}$	**Session Key Rule**: If P believes that X is fresh and P believes that Q believes X, which is the necessary parameters of session key. Then P believes that share the session key with Q

Step 2: We transform our presented scheme to the idealized form as follows:

$$\text{Message 1.}\quad U \rightarrow S : \text{AID}_i,\ M_3\text{: } \{N_c,\ M_1\}_{Cx},\ M_4$$

$$\text{Message 2.}\quad S \rightarrow U : N_{sc},\ H_i\text{: } \langle N_S,\ M_1\rangle_{\text{RPW}_i}$$

Step 3: We make the following assumption to analyze the presented scheme.

$$A_1\text{: } S| \equiv \#(N_c,\ N_s)$$

$$A_2\text{: } U_i| \equiv \#(N_c,\ N_s)$$

$$A_3\text{: } S| \equiv S \overset{\text{Cx}}{\leftrightarrow} U_i$$

$$A_4\text{: } U_i| \equiv U_i \overset{\text{RPW}_i}{\leftrightarrow} S$$

$$A_5\text{: } S| \equiv U_i \Rightarrow N_c$$

$$A_6\text{: } U_i| \equiv S \Rightarrow N_s$$

Step 4: The main proofs are described as follows:

According to Message 1, we obtain

$$S_1\text{: } S \triangleleft \left(\text{AID}_i, M_3\text{: } \{N_c, M_1\}_{Cx}.M_4\right)$$

According to the assumption A_3, S_1 and the message meaning rule to obtain

$$S_2: S| \equiv U_i| \sim \{N_c, M_1\}$$

According to assumption A_1 and freshness conjuncatenation rule to obtain

$$S_3: S| \equiv \#\{N_c, M_1\}$$

According to S_2, S_3 and nonce verification rule to obtain

$$S_4: S| \equiv U_i| \equiv \{N_c, M_1\},$$

where N_c is the necessary parameter of the session key of the presented scheme.
According to the assumption A_5, S_4 and jurisdiction rule to obtain

$$S_5: S| \equiv \{N_c, M_1\}$$

According to the assumption A_1, S_4 and session key rule to obtain

$$S_6: S| \equiv \left(U_i \overset{SK}{\leftrightarrow} S\right) \quad \textbf{Goal 3 achieved.}$$

According to the assumption A_1, S_6 and nonce verification rule to obtain

$$S_7: S| \equiv U_i| \equiv \left(U_i \overset{SK}{\leftrightarrow} S\right) \quad \textbf{Goal 4 achieved.}$$

According to Message 2, we could obtain

$$S_8: U \triangleleft (N_{sc}, H_i: \langle N_S, M_1 \rangle_{RPW_i})$$

According to assumption A_4, S_8 and message meaning rule to obtain

$$S_9: U_i| \equiv S| \sim \{N_S, M_1\}$$

According to assumption A_2 and freshness conjuncatenation rule to obtain

$$S_{10}: U_i| \equiv \#\{N_S, M_1\}$$

According to S_9, S_{10} and nonce verification rule to obtain

$$S_{11}: U_i| \equiv S| \equiv \{N_S, M_1\},$$

where N_s is the necessary parameter of the session key of the presented scheme.

According to assumption A_6, S_{11} and jurisdiction rule to obtain

$$S_{12}: U_i| \equiv \{N_S, M_1\},$$

According to assumption A_2, S_{11} and session key rule to obtain

$$S_{13}: U_i| \equiv \left(U_i \overset{\text{SK}}{\leftrightarrow} S\right) \quad \textbf{Goal 1 achieved}.$$

According to the assumption A_2, S_{13} and nonce verification rule to obtain

$$S_{15}: U_i| \equiv S| \equiv \left(U_i \overset{\text{SK}}{\leftrightarrow} S\right) \quad \textbf{Goal 2 achieved}.$$

In this paper, the presented scheme has achieved the preferred goals as shown by the BAN logic. So this formal proof of the presented scheme is suitable to provide mutual authentication and session key agreement between participant entities securely.

4 Informal Security Analysis

In this section, we have further analyzed the presented scheme in the context of informal security. The presented security analysis demonstrates the effectiveness of such kind of authentication scheme in terms of security.

Proposition 1 An outsider attacker cannot extract user's password PW_i, identity ID_i, and server's secret key X from the smart card parameters $\{E_i, S_i, T_i, r_i\}$, login request message $\{\text{AID}_i, M_3, M_4\}$, and reply message $\{N_{sc}, H_i\}$ between the user and server.

Proof An outsider attacker obtains the user's smart card and extract secret information $\{E_i, S_i, T_i, r_i\}$ from the smart card by some means, and he also intercepts the login request message $\{\text{AID}_i, M_3, M_4\}$ and reply message $\{N_{\text{sc}}, H_i\}$ between the user and server. But an attacker cannot extract the identity ID_i, password PW_i, and secret key X as follows:

1. From $S_i = C_i + h(\text{RPW}_i) \cdot G = X \cdot G + h(\text{PW}_i \| r_i) \cdot G$, given r_i, an attacker has to guess server's secret key X and user's password PW_i at the same time to solve the mentioned equation. It is not computationally feasible in polynomial time to solve two unknown parameters password PW_i and secret key X from one equation.
2. $E_i = C_i + D_i = X \cdot G + h(\text{ID}_i \| \text{RPW}_i) \cdot G = X \cdot G + h(\text{ID}_i \| h(\text{PW}_i \| r_i)) \cdot G$. In this case, the attacker has no knowledge of server's secret key X, user's identity ID_i and password PW_i. Therefore, an attacker guesses three unknown values at the same time to solve the above equation which is not feasible in polynomial time.

3. $AID_i = ID_i \oplus h(N_c)$, an attacker cannot extract user's identity ID_i because it is computationally hard due to non invertible hash function.
4. $M_3 = E_{Cx}(N_c, M_1)$, where Cx is the x coordinate of $C_i = X \cdot G = (Cx, Cy)$. It is hard to compute secret key X due to elliptic curve discrete logarithm problem (ECDLP).
5. $M_4 = h(Cx\|M_2\|N_c) = h(X \cdot G\|h(N_c) \cdot G \cdot ID_i \cdot h(PW_i\|r_i)\|N_c)$. An attacker cannot extract the identity ID_i, password $PW_{i,}$ and secret key X due to elliptic curve discrete logarithm problem (ECDLP) and non-invertible hash function.
6. $H_i = h(N_S\|RPW_i\|M_1) = h(N_S\|h(PW_i\|r_i)\|h(N_c) \cdot G \cdot ID_i$, An attacker has to guess three unknown parameters N_S, N_c and ID_i at the same time to extract the user's password from the equation $H_i = h(N_S\|RPW_i\|M_1)$ which is computationally infeasible in polynomial time.

Proposition 2 An insider attacker cannot extracts server's secret key X from his/her own smart card parameters $\{E_i, S_i, T_i, r_i\}$, login request message $\{AID_i, M_3, M_4\}$, and reply message $\{N_{sc}, H_i\}$ between the user and server.

Proof In this attack model, a legal but malicious user tries to extract the private key X of the sever by using his own identity , password $_i$, smart card parameters $\{E_i, S_i, T_i, r_i\}$, login request message $\{AID_i, M_3, M_4\}$, and reply message $\{N_{sc}, H_i\}$. In the following, we show that the malicious user cannot get the secret key X.

1. From $S_i = C_i + h(RPW_i) \cdot G = X \cdot G + h(PW_i\|r_i) \cdot G$, given r_i. It is hard to compute secret key X due to the elliptic curve discrete logarithm problem (ECDLP).
2. Similarly, the attacker cannot extract secret key X from $\{E_i, M_3, M_4\}$ parameters due to the same reason.

4.1 User Un-Traceability Attack

It is our assumption that an adversary has trapped two login request message $\{AID_i, M_3, M_4\}$ and $\{AID'_i, M'_3, M'_4\}$ during the execution of protocol and try to trace the both message are belonging to same user or not, where $AID_i = ID_i \oplus h(N_c)$, $M_3 = E_{Cx}(N_c, M_1)$ and $M_4 = h(Cx\|M_2\|N_c)$. It may be noted that the each parameters $\{AID_i, M_3, M_4\}$ are dependent on the random nonce N_c. Since the random nonce are distinct in each authentication session and valid for that session only. So in the presented scheme, the login request message $\{AID_i, M_3, M_4\}$ and $\{AID'_i, M'_3, M'_4\}$ are dissimilar in each authentication session. Therefore, an adversary cannot trace the user after intercepting the login message.

4.2 Privileged Insider Attack

Today, most of the authentication protocols are not secure due to the privileged insider attack. So it is very important to keep user's confidential information secret from the server. If a malicious administrator obtains the user's password by some means then he/she may use that password for accessing the other application servers where the user must registered himself/herself to every application server using the same identity ID_i and password PW_i. During the registration phase of the presented scheme, we provide $RPW_i = h(PW_i \| r_i)$ instead of plaintext password PW_i to the server, where r_i is a random number. So the insider attacker cannot extract PW_i from RPW_i due to the non-invertible hash function.

4.3 User-Server Impersonation Attack

In this attack model, we assume that the attacker intercepts the login request message $\{AID_i, M_3, M_4\}$ and reply message $\{N_{sc}, H_i\}$ and tries to impersonate as a legal user or server. However, *the Proposition* 1 shows that an attacker cannot extract user's password PW_i, identity ID_i, and server's secret key X. Thus, an adversary cannot computes valid login request $\{AID_i, M_3, M_4\}$ and reply message $\{N_{sc}, H_i\}$ without knowing user's password PW_i, identity ID_i, and server's secret key X. Therefore, an attacker fails to impersonate as legitimacy entity of the presented scheme.

4.4 Password and Identity Guessing Attack

In the presented scheme, we have assumed that each user uses very low entropy identity ID_i and password PW_i which is easily guessable in polynomial time. However, *the Proposition* 1 shows that an attacker cannot extract the user's identity ID_i and password PW_i from the smart card parameters $\{E_i, S_i, T_i, r_i\}$ and also from the communicated messages $\{AID_i, M_3, M_4, N_{sc}, H_i\}$ between the user and the server. Therefore, the presented scheme is secure against password and identity guessing attack.

4.5 Replay Attack

In the replay attack, the attackers intercepted the previous login message and later on transmit the same message to the server and try to impersonate as the legitimate entity. Suppose an attacker sends the previous intercepted message

$\{\text{AID}_i, M_3, M_4\}$ to the server, after receiving the message the server matches the received message with stored message. If it matches, then server rejects the attacker's login request. The presented protocol is secure against replay attack due to the random nonce N_c and N_S which are generated by user and server, respectively. Random nonce confirms that each login message is distinct in each session and valid for that session only.

4.6 Stolen Smart Card Attack

To access the remote server, an attacker computes the login message $\{\text{AID}_i, M_3, M_4\}$ by using the extracting parameters $\{\text{E}_\text{i}, S_i, \text{T}_\text{i}, r_i\}$ from the stolen smart card. However, an attacker cannot compute valid login message $\{\text{AID}_i, M_3, M_4\}$ as follows:

1. $\text{AID}_i = \text{ID}_i \oplus h(N_c)$, an attacker requires identity ID_i for computing AID_i. *The Proposition* 1 shows that the attacker cannot extract identity ID_i from the smart card parameters $\{E_i, S_i, T_i, r_i\}$, login request message $\{\text{AID}_i, M_3, M_4\}$, and reply message $\{N_{sc}, H_i\}$. So, an attacker cannot compute AID_i without the knowledge of identity ID_i.
2. $M_3 = E_{Cx}(N_c, M_1)$, where Cx is the x coordinate of $C_i = X \cdot G = (Cx, Cy)$ and $M_1 = h(N_c) \cdot G \cdot \text{ID}_i$. An attacker requires identity ID_i and secret key X for calculating the message M_3. *The Proposition* 1 shows that the attacker cannot extract ID_i and X from the smart card parameters $\{E_i, S_i, T_i, r_i\}$,login request message $\{\text{AID}_i, M_3, M_4\}$, and reply message $\{N_{sc}, H_i\}$. So an attacker cannot compute M_3 without the knowledge of identity ID_i and secret key X.
3. Similarly, an attacker cannot compute $M_4 = h(Cx \| M_2 \| N_c)$ without the knowledge of ID_i and X.

Therefore, we may conclude that the presented scheme is secure against stolen smart card attack.

4.7 Efficient Login and Password Change Phase

During the login and password change procedure of the presented scheme, the smart card reader detects the error very quickly if an attacker inputs the wrong information such as biometric template F_i, identity ID_i, and password PW_i to the card reader. As the result, an attacker cannot generate a fake login message which reduces extra computation and communication overhead as well as network congestion. Thus, the presented scheme provides efficient login and password change phase.

4.8 Session Key Recovery Attack

In the presented scheme, the security of the session key $\text{SK} = h(N_S \| N_c \| \text{RPW}_i \| M_2)$ based on the non-invertible hash function. Moreover, the session key depends on the password PW_i, identity ID_i, random nonce N_S and N_c, which is generated by server and user, respectively. However, *the Proposition* 1 shows that the attacker cannot extract user's password PW_i and identity ID_i from the smart card parameters $\{E_i, S_i, T_i, r_i\}$ and communicated messages $\{\text{AID}_i, M_3, M_4, N_{sc}, H_i\}$ between the user and the server. Without the knowledge of password PW_i and identity ID_i, an attacker cannot compute the session key SK. Thus, the presented scheme is secure against session key recovery attack.

4.9 Perfect Forward Secrecy Attack

The perfect forward secrecy means, if the system's confidential information is compromised, then the secrecy of previous established session key should not be affected. In the presented scheme, we assume that user's password PW_i and identity ID_i are compromised by the attacker. Yet, the adversary cannot compute the previous session key $\text{SK} = h(N_S, N_c, \text{RPW}_i, M_2)$ without the knowledge of random nonce N_c and N_S, which are generated by user and server, respectively. Therefore, the presented scheme is secure against perfect forward secrecy attack.

4.10 Session Key Verification and Agreement

In this protocol, the user and server agreed upon a common session key $\text{SK} = h(N_S \| N_c \| \text{RPW}_i \| M_2)$ and verifies it using the following condition $Z_i = \text{SK} \cdot G + C_i$. To compute Z_i, an attacker has to know $\text{SK} = h(N_S \| N_c \| \text{RPW}_i \| M_2)$, $C_i = X \cdot G$ and computes $\text{SK} = h(N_S \| N_c \| \text{RPW}_i \| M_2)$, user's identity ID_i and password PW_i are needed and to compute $C_i = X \cdot G$ server's secret key X is needed. *The Proposition* 1 shows that an adversary cannot extract user's password PW_i, identity ID_i, and server's secret key X from the smart card parameters $\{E_i, S_i, T_i, r_i\}$ and communicated messages $\{\text{AID}_i, M_3, M_4, N_{sc}, H_i\}$ between the user and the server. Since the only authorized entities can compute SK and Z_i. This demonstrates that the user and server can correctly verify the established session key.

5 Performance Analysis

In this section, we have compared the security and performance of the presented scheme with other existing relevant schemes [12–16]. In Table 2, we have summarized the communication cost, computation cost, and memory storage cost of the presented scheme and other relevant schemes. To analyze the computational complexity, we define the notation T_H is the one-way secure hash function, T_S is the symmetric key encryption/decryption operation, and T_{PM} is the point multiplication operation on elliptic curve cryptosystem. The presented protocol requires $4T_H + 3T_{PM}$, $12T_H + 8T_{PM} + 2T_S$, and $7T_H + 4T_{PM}$ operation for registration phase, login and authentication phase, and password change phase, respectively. In Table 3 and Fig. 1, we have provided communication cost and smart card storage cost comparison of the presented scheme with other relevant scheme [12–16]. In order to measure the communication cost, we have assumed that the length of identity ID_i, password PW_i, random nonce, elliptic curve point, and hash function h (·) are all 160 bits length. Moreover, symmetric key encryption/decryption takes 512 bits. It is noticeable from Table 2 that the presented scheme achieves comparatively better communication cost than other schemes [13, 16]. In Table 3, we compared the presented scheme with existing related schemes in the context of different security functionalities. It is noticeable that the presented protocol is secure against relevant security attacks and achieves several security attributes than other schemes.

6 Conclusion

In this paper, we have presented a secure biometric based remote user authentication scheme using elliptic curve cryptosystem. The security of the presented scheme validated through both formal and informal way. The formal security analysis using BAN logic, which confirms that, the presented scheme achieves mutual authentication and session key agreement securely. The informal security analysis ensures that the presented scheme can resist various kinds of malicious attacks. The performance comparison demonstrates that the presented scheme is more secure and efficient than other relevant schemes. Moreover, the presented scheme can update the password on the user's demand without contacting the server.

Table 2 Performance comparison: memory space, communication cost, and computational cost

TCC	$19T_H + 4\ T_{PM} + 11T_S$	$11T_H + 8T_{PM} + 8T_S$	$18T_H + 6T_{PM}$	$25T_H + 4T_{PM} + 4T_S$	$30T_H + 12T_{PM}$	$23T_H + 15T_{PM} + 2T_S$
CC	160 * 3 + 512 * 2 = 1504	160 * 5 + 512 = 1312	160 * 7 = 1120	160 * 4 + 512 * 2 = 1664	160 * 7 = 1120	160 * 5 + 512 = 1312

Note: *SCMR* smart card memory requirement; *CCRP* computation cost of registration phase; *CCLA* computation cost of login and authentication phase; *CCPH* computation cost of password change phase; *TCC* total computation cost; *CC* communication cost

Table 3 Security feature comparisons among the presented scheme and other relevant scheme

Security requirement	Zhang et al. [12]	Zhang et al. [13]	Xu et al. [14]	Wang [15]	Arshad and Nikooghadam [16]	Presented scheme
A_1	Yes	No	Yes	No	Yes	Yes
A_2	Yes	No	Yes	No	Yes	Yes
A_3	Yes	No	Yes	Yes	Yes	Yes
A_4	Yes	No	Yes	Yes	Yes	Yes
A_5	No	No	Yes	Yes	Yes	Yes
A_6	No	Yes	Yes	Yes	Yes	Yes
A_7	No	Yes	Yes	Yes	Yes	Yes
A_8	Yes	Yes	No	Yes	No	Yes
A_9	No	Yes	No	Yes	No	Yes

Note: A_1 resist password guessing attack, A_2 resist identity guessing attack, A_3 resist impersonation attack, A_4 resist privileged insider attack, A_5 resist replay attack, A_6 resist user un-traceability attack, A_7 resist forward secrecy attack, A_8 session key verification, A_9 efficient login and password change phase

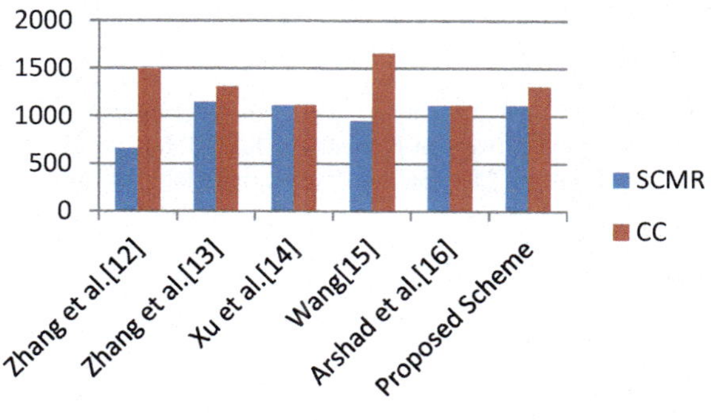

Fig. 1 Communication and storage overhead (bits) of different authentication scheme. Note: *SCMR* smart card memory requirement, *CC* communication cost

References

1. L. Lamport, Password authentication with insecure communication. Commun. ACM **24**(11), 770–772 (1981)
2. C.-C. Lee, L.-H. Li, M.-S. Hwang, A remote user authentication scheme using hash functions. ACM SIGOPS Oper. Syst. Rev. **36**(4), 23–29 (2002)
3. M. Peyravian, C. Jeffries, Secure remote user access over insecure networks. Comput. Commun. **29**(5), 660–667 (2006)
4. X.-M. Wang et al., Cryptanalysis and improvement on two efficient remote user authentication scheme using smart cards. Comput. Stand. Interfaces **29**(5), 507–512 (2007)
5. S. Kumari, M.K. Khan, X. Li, An improved remote user authentication scheme with key agreement. Comput. Electr. Eng. **40**(6), 1997–2012 (2014)
6. C.T. Li, M.S. Hwang, An efficient biometrics-based remote user authentication scheme using smart cards. J. Netw. Comput. Appl. **33**(1), 1–5 (2010)
7. C.H. Lin, Y.Y. Lai, A flexible biometrics remote user authentication scheme. Comput. Stand. Interfaces **27**(1), 19–23 (2004)

8. B.T. Nathan, R. Meenakumari, S. Usha, *Formation of Elliptic Curve Using Finger Print for Network Security*. In Process Automation, Control and Computing (PACC), 2011 International Conference on IEEE, pp. 1–5
9. X. Li, J.W. Niu, J. Ma, W.D. Wang, C.L. Liu, Cryptanalysis and improvement of a biometrics-based remote user authentication scheme using smart cards. J. Netw. Comput. Appl. **34**(1), 73–79 (2011)
10. U. Subramaniam, K. Subbaraya, A biometric based secure session key agreement using modified elliptic curve cryptography. Int. Arab J. Inf. Technol. (IAJIT) **12**(2) (2015)
11. C.-T. Li, A new password authentication and user anonymity scheme based on elliptic curve cryptography and smart card. IET Inf. Secur. **7**(1), 3–10 (2013)
12. L. Zhang et al., Two-factor remote authentication protocol with user anonymity based on elliptic curve cryptography. Wireless Pers. Commun. **81**(1), 53–75 (2015)
13. Y. Zhang et al., An efficient password authentication scheme using smart card based on elliptic curve cryptography. Inf. Technol. Control **43**(4), 390–401 (2014)
14. X. Xu, P. Zhu, Q. Wen, Z. Jin, H. Zhang, L. He, A secure and efficient authentication and key agreement scheme based on ECC for telecare medicine information systems. J. Med. Syst. **38** (6) (2014)
15. L. Wang, Analysis and enhancement of a password authentication and update scheme based on elliptic curve cryptography. J. Appl. Math. (2014)
16. H. Arshad, M. Nikooghadam, Three-factor anonymous authentication and key agreement scheme for telecare medicine information systems. J. Med. Syst. **38**(12) (2014)
17. Z. Tan, A user anonymity preserving three-factor authentication scheme for telecare medicine information systems. J. Med. Syst. **38**(3), 1–9 (2014)
18. Y. Lu et al., An enhanced biometric-based authentication scheme for telecare medicine information systems using elliptic curve cryptosystem. J. Med. Syst. **39**(3), 1–8 (2015)
19. H.L. Yeh et al., Robust elliptic curve cryptography-based three factor user authentication providing privacy of biometric data. IET Inf. Secur. **7**(3), 247–252 (2013)
20. M. Burrows, M. Abadi, R. Needham, A logic of authentication. ACM Trans. Comput. Syst. **8** (1), 1836 (1990)

Implementation of Fingerprint-Based Biometric System and Its Integration with HRMS Application at RDCIS, SAIL

S. Selvi, Manas Rath, N. N. J. Hemrom, A. Bhattacharya and A. K. Biswal

Abstract RDCIS is a R&D unit of Steel Authority of India Limited (SAIL). RDCIS has implemented Human Resource Management System for management of online leave applications, tour applications, dependant declaration, medical validation of employees, etc., for quickly building the workflows and processes of HR functions. At RDCIS, there are around 481 employees and 223 Contract Workers working under different Contractors. RDCIS main office is located at Ranchi, RDCIS plant centres are located at Bhilai, Bokaro, Rourkela, Burnpur, Durgapur and Bhadravati. SAIL has its own network connecting all the steel plants and city offices. RDCIS has own VLAN network and connected to RDCIS plant centres through SAIL-Net. Before the implementation of biometric system, the employee attendance was captured in the attendance register. Attendance from plant centres and city offices was sent through e-mail system every month to RDCIS, Ranchi. The practice for capturing attendance was manual, and every morning and evening, employee signs his in-time and out-time in the attendance register kept in the Head of Group (HoG) room. The employee also has to update the register for other information like leave details and tour details. At the end of the month, the secretariat of the group compiles all the employees' attendance data and crossverifies with the leave book. There is a particular format given by the personal department for filling up the employee attendance details. The group fills the attendance of the employees as per the format and sends the attendance sheet to the finance department. The finance department enters the attendance details in the payroll system for the preparation of salary. The disadvantage of manual attendance system is many. Some are the attendance register gets lost, mismatch between the leave book and the attendance register, wrong entry of details in the attendance register, etc. Also sometimes, the attendance is manipulated and sent to payroll system. To address all these problems, a project was taken to implement biometric

S. Selvi (✉) · M. Rath · N. N. J. Hemrom · A. Bhattacharya · A. K. Biswal
RDCIS, Steel Authority of India Limited, Ranchi 834002, India
e-mail: selvi@sail-rdcis.com

V. Nath (ed.), *Proceedings of the International Conference on Microelectronics, Computing & Communication Systems*, Lecture Notes in Electrical Engineering 453,
https://doi.org/10.1007/978-981-10-5565-2_3

attendance system along with Human Resource Management System. This will help in continual monitoring of employee attendance and to meet current and future requirements of personnel department. In this regard, RDCIS has implemented biometric attendance system in all locations of RDCIS to monitor the employee's attendance centrally from Ranchi. The system has been integrated with payroll system for preparation of employee's salary. This paper discusses in detail the challenges in implementing biometric system at RDCIS. The Web-based biometric attendance system will help RDCIS for better time management of human resources. The software has been developed with three-tier approach. The software tools used are Oracle Database, HTML and JSP. The software has been deployed with Tomcat Apache Server on Windows Operating System.

Keywords Human Resource Management System (HRMS) · Management Information System (MIS) · Computer & Information Technology (C&IT) Research and Development Centre for Iron & Steel (RDCIS) · Steel Authority of India Limited (SAIL) · Human Resource (HR) · Personnel & Administration (P&A) · Java Server Pages (JSP) · General Manager (GM) · Head of Group (HoG)

1 Introduction

RDCIS has about 481 employees, which include both Executives and Non-Executives and 223 Contract Workers working under different Contractors. Further RDCIS has five plant centres at Bhilai, Bokaro, Burnpur, Durgapur and Rourkela and two city offices at Delhi and Kolkata. At RDCIS, online leave application, tour application, dependant declaration and leave encashment have been successfully implemented through HRMS application software. The HRMS software is available to all plant centres and city offices across Sail-Net and Internet. Also, with the implementation of HRMS, the department is in a position to provide faster employee services and online access to various information of the employees with proper security mechanism. Further, considering the reducing manpower in the department as well as erroneous reporting of data, due to its manual nature, RDCIS has planned to go for comprehensive, integrated and online system for generation of attendance for payroll system.

The corporate guideline of SAIL is to implement Biometric Attendance Monitoring System at all plants/units. In this regard, RDCIS has implemented biometric attendance system at RDCIS and at all RDCIS plant centres. This biometric system is integrated with HRMS and payroll system. It is an open- and soft-coded system, which can be horizontally transferred to other steel plants having similar working and operational practices with minor modifications.

2 Need for Biometric System

There are many reasons that organizations have to go for Biometric Attendance Monitoring System. Biometric-based solutions are able to provide for confidential financial transactions and personal data privacy. Here are a few benefits of the system as follows:

- Maintain time record.
- Enhancement of security.
- For confidential financial transactions and personal data privacy.
- Attendance tracking.
- Convenient and considerably more accurate than manual method.
- Time saving.
- Transparency.
- Employee satisfaction.

3 Approach Adopted

Proposed biometric system has following system components:

System Field Devices
Biometric Readers—Fingerprint Biometric Time Recording Terminals with Time Display.

Enrolment Kit—Captures & registers the Fingerprint.

System Software

- Application Software with Rules Engine—Process and analyse the transaction data based on company rules. It maintains database of employees and enables the administrator to configure the terminals.
- Data base MS-SQL—this is central database for all terminals and stores transaction data and employee master data.
- Fingerprint Enrolment Software—Captures fingerprints and stores in central server.
- HRMS Integration Module—This module HRMS and biometric application software enable the application to access employee master, company policy, leave policy, shift scheduling configured in HRMS. It transfers attendance transaction data to HRMS from the application software in specified format in specific intervals of time (Fig. 1).

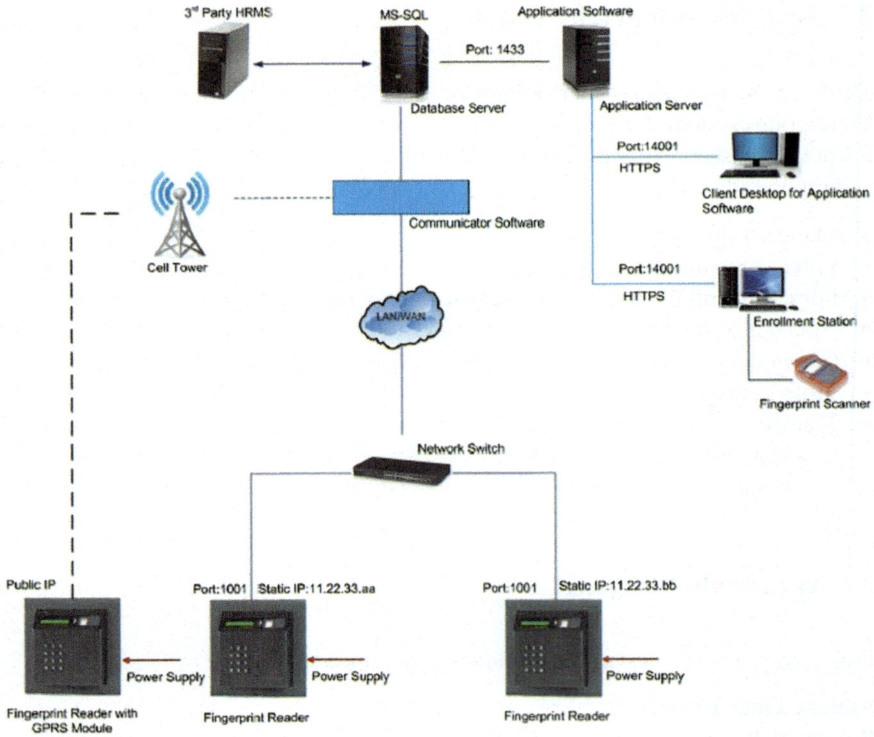

Fig. 1 System architecture

4 Biometric System Features

- Dedicated to time and attendance operations.
- Rugged design, vandal resistant polycarbonate casing suitable for interior or exterior environment.
- No external wiring connections, user interface for configuration of the panel.
- Connection by internal terminal block with locking screw.
- Facility to configure controller and biometric module through on board Web server.
- Data retention in case of power failure using flash memory.
- Inbuilt battery backup for 2–8 h (optional).
- Data transfer option—LAN/WAN/USB/GPRS.

5 System Process Flow

The different activities are performed at biometric server level, field level, administrator level, HRMS/Payroll level. The central server performs the following activities:

1. Store all transaction databases.
2. Store all terminals database.
3. Process the transaction data and convert into specified format for HRMS.
4. Send the process data to HRMS.

6 Enrolment and Verification

(1) Verification mode

In the verification mode, the system validates a person's identity by comparing the captured biometric data with her own biometric template(s)-stored system database. In such a system, an individual who desires to be recognized claims an identity, usually via a PIN , a user name, a smart card, etc., and the system conducts a one-to-one comparison to determine whether the claim is true or not. Identity verification is typically used for positive recognition, where the aim is to prevent multiple people from using the same identity (Fig. 2).

(2) Identification mode

In the identification mode, the system recognizes an individual by searching the templates of all the users in the database for a match. Therefore, the system conducts a one-to-many comparison to establish an individual's identity (or fails if the subject is not enrolled in the system database) without the subject having to claim an identity.

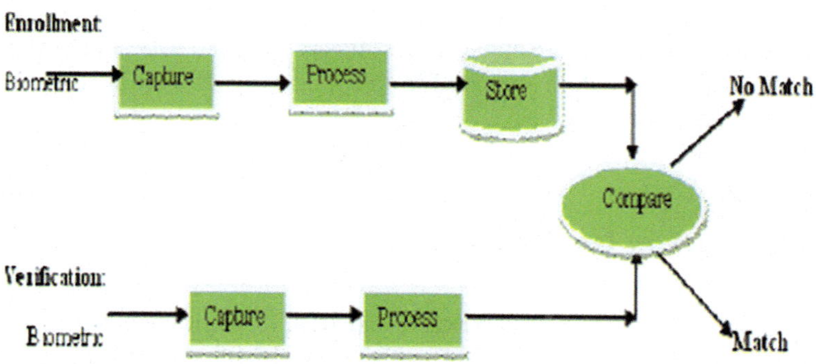

Fig. 2 Biometric enrolment and verification process

7 Fingerprint Enrolment Kit with Terminal Management Software

Duplicate search facility is there during enrolment in order to avoid multiple identifying referring to the same individual.

- Enrolment of 1/2 finger per user.
- Facility of finger selection for enrolment.
- Facility of location/department/designation/division/access group-wise allocation of finger templates to their respective readers.
- Facility of restoring same group of finger templates when old finger reader is replaced with new one.
- Robust enrolment to the consolidation of biometric template over three images.
- Ability to tune the false acceptance rate according to the application security requirements.
- Live controls related to the image quality and the finger positioning during enrolment.

8 HRMS Integration

The application software has three-tier clients—server architecture. The integration with Oracle HRMS enables both-way data transfer. The integration is possible in many ways—through SFTP Up loader, Through Web Service or through Table View. 1. Through SFTP Up loader:—The application will export the data in a required format into a file and in a specific folder through secured file transfer protocol (SFTP) to avoid any modification/manipulation. This folder will be shared with the HRMS/Payroll application which will be 24 $\times$ 7 running application and will constantly monitor this folder for new incoming file. Once the file is found, it will open the file and start importing the data into database. This application can be configured to keep monitoring every 30 min which also can be configured. The format of data required in HRMS shall be specified initially before integration. The base file in HRMS from where application software will take data, and the target file in HRMS where data shall be exported from application software shall be specified as per the requirement (Fig. 3).

9 Challenges in Implementation

RDCIS main office is located at Ranchi, and RDCIS plant centres are located at Bhilai, Bokaro, Rourkela, Burnpur, Durgapur and Bhadravati. SAIL has its own network connecting all the steel plants and city offices. RDCIS has own VLAN

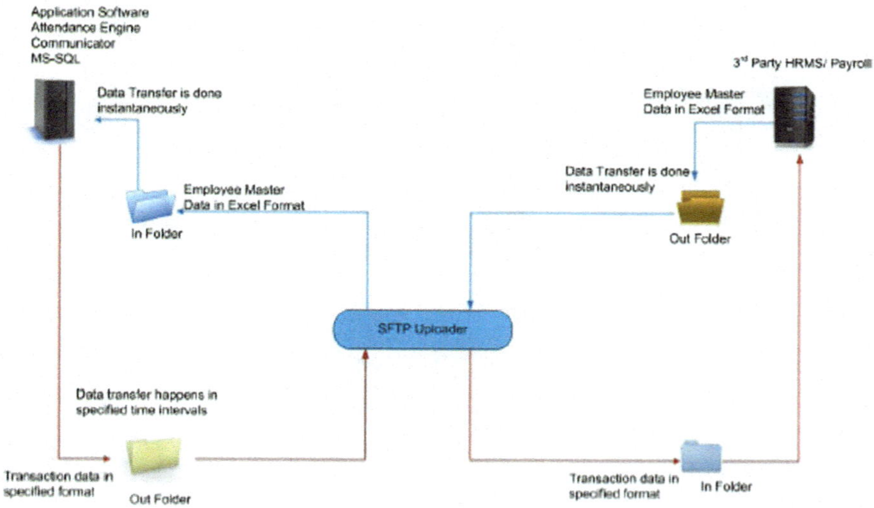

Fig. 3 Integration scheme

network and is connected to RDCIS plant centres through SAIL-Net. The biometric devices are installed in all the plant centres, in different locations of RDCIS building and at SAIL satellite township. Biometric server is located in C&IT, RDCIS and Ranchi. All the biometric devices in RDCIS building are connected through VLAN. All biometric devices at plant centres are connected to Ranchi server through SAIL-Net. All devices at SAIL satellite township, Ranchi, are connected through GPRS Scheme to biometric server. The punching time at all locations are different and decided based on plant timings for steel plants. The attendance rule is framed by P&A dept. for late punching. The punched details in the biometric device are downloaded to the central server every morning and evening. There are also employees working in shift in RDCIS as well as at SAIL satellite township. There are three shifts maintained: Shift A, Shift B and Shift C. Shift employees are also taken care in the biometric attendance system (Fig. 4).

The biometric punch details are then transferred to HRMS database. The biometric data is then processed as per the timing rule defined in the system. The punching details are monitored by HoG, HoA and ED. There is different type of MIS generated in the system for attendance (Fig. 5).

The employees coming late can apply for Outside Duty through OD module of HRMS. The Leave Module in HRMS gives leave details. The Tour Module provides information of employees on tour. Taking together leave, tour, OD, punch details, the attendance is prepared. Employees can view their attendance in HRMS. Once the attendance is finalized, the HRMS system transfers the data to payroll system. The attendance is then used for salary preparation (Fig. 6).

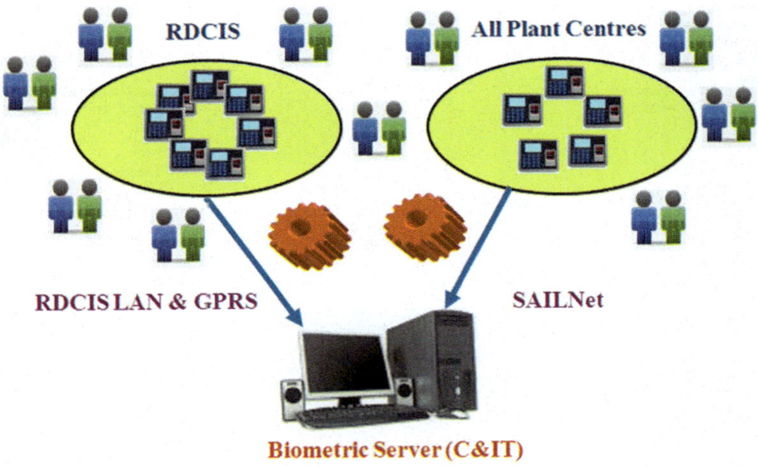

Fig. 4 Biometric server data downloading scheme

Fig. 5 Scheduling of data transfer from biometric server to HRMS server

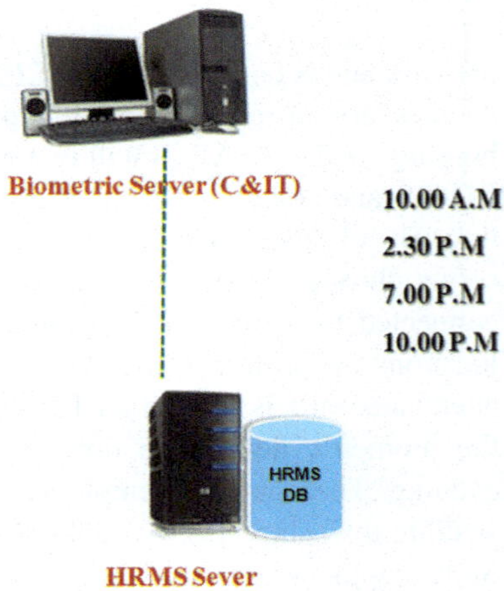

10 Impact in Organization

- Online access control & attendance monitoring of employees and Contract Workers.
- Availability of various employee services.
- Online availability of information.
- Speed of processing and integrity of data.
- Transparency.
- Information for decision-making (MIS).

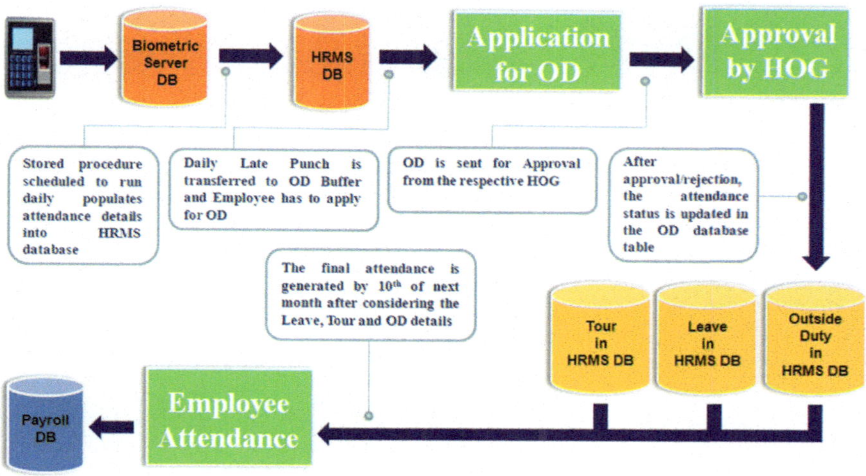

Fig. 6 Methodology to transfer data from biometric server to payroll database

11 Conclusion

RDCIS has implemented biometric attendance system as part of HR system to improve the current practice and to automate the business functions of P&A department. Biometric technologies present a number of advantages to the organization, ranging from stronger user authentication, greater convenience for users, to improved security and operational efficiencies. Biometric is an emerging technology with many opportunities for growth. Biometric attendance system will help organization in eliminating data entry errors and calculations, increase security and productivity, save cost and time, increase employee satisfaction.

References

1. R. Barker, Tasks and Deliverables. Addison Wesley
2. C. Longman, R. Barker, Function and Process Modeling, Addison Wesley
3. Java Server Pages, O'Reilly Media
4. Oracle Manuals on Designer R2, Oracle 11g Database
5. Oracle Manuals on Oracle 10g Developer Suite. http://www.w3schools.com/
6. http://tomcat.apache.org/
7. Procedures and practices followed at P&A department

Fabrication and Investigation of Low-Voltage Programmable Flash Memory Gate Stack

Rasika Dhavse, Kumar Prashant, Chetan Dabhi, Anand Darji and R. M. Patrikar

Abstract In this paper, fabrication, characterization, and analysis of a FGMOS gate stack employing ultra-thin tunnel oxide of 3 nm thickness are discussed. Apart from basic C-V and G-V profiles, high-frequency hysteresis curve has been investigated and device-level parameters are extracted. Use of ultra-thin tunnel oxide has facilitated direct tunneling mechanism at program/erase voltages of 10 V for 200 ms and −8 V for 40 ms, respectively. Excellent memory window of 1.2 V has been obtained. Frequency-dependent capacitance and reliability-related profiles are also studied. The device is useful for power-efficient non-real-time applications like data logging, biometric security, backup servers.

Keywords Flash memory cell · Floating gate MOSFET · Low voltage programmable · Memory window · Silicon nanocrystals · Thin oxide Quantum dots

1 Introduction

Scaling floating gate (FG) devices vertically down to the sub-nano-technological dimensions imposes typical problems related to their data integrity and sustainability[1–10]. The vertical scaling limitation in flash technology is rooted from extreme requirements put on the tunnel oxide layer. It should be thin enough to allow charge alterability during program/erase (P/E) operations and thick enough to retain stored charge during read disturb and idle conditions [5] throughout their lifetime of minimum 10 years [11, 12]. ITRS Table 2013 data shown in Table 1 suggests that after 65 nm technology node, tunnel oxide has stagnated at 5–6 nm for 2D planar NAND flash and 8–9 nm for NOR flash [11]. Alternate memory

R. Dhavse (✉) · K. Prashant · C. Dabhi · A. Darji
Electronics Engineering Department, SVNIT, Surat, India
e-mail: rsk@eced.svnit.ac.in

R. M. Patrikar
Electronics Engineering Department, VNIT, Nagpur, India

V. Nath (ed.), *Proceedings of the International Conference on Microelectronics, Computing & Communication Systems*, Lecture Notes in Electrical Engineering 453,
https://doi.org/10.1007/978-981-10-5565-2_4

Table 1 Oxide scaling stagnation in flash memory [11]

Year of production	2014	2016	2018	2020
2D NAND flash				
Minimum array 1/2 pitch (nm)	17	14	12	12
Tunnel oxide thickness (nm)	6–7	5–6	5–6	5–6
IPD thickness (nm)	10	9	9	9
P/E voltage (V)	15–17	15–17	15–17	15–17
Endurance (P/E Cycles)	1.0E+04	1.0E+04	5.0E+03	5.0E+03
NOR flash				
Gate length (nm)	110	100	100	90
Tunnel oxide thickness (nm)	8–9	8–9	8–9	8–9
IPD thickness (nm)	13–15	13–15	13–15	11–13
P/E voltage (V)	9	9	9	9
Endurance (P/E Cycles)	1.0E+05	1.0E+05	1.0E+05	1.0E+05

structures to replace conventional flash technology are being proposed and investigated [12] but their future depends on whether they are promising in true sense and how fast they can emerge.

Since past few years, ITRS community has fixed tunnel oxide thickness to around 8–9 nm (NOR flash) [11]. This led to stagnation of P/E voltage. It is found that P/E voltage has been scaled down by very small extent as compared to the supply voltage in CMOS technology [13]. 15 V is projected P/E voltage for a NAND flash until 2020, whereas supply voltage is just 1 V [6, 11, 14–16]. A flash memory uses charge pump to obtain higher voltages than the supply voltage. As such, they either have to deliver substantially high current or same current at a faster rate to memory bit lines without much increase in the area on silicon [17]. The key parameter in their design, especially with reduced supply voltage specifications, is voltage conversion efficiency [13, 18]. Also, there are trade-offs like switched voltage values (with respect to time), rise time minimization, charge consumption, silicon area occupation minimization, complex clocking and control section and number of stages [13, 17–22]. Thus, in current low-power supply scenario, most logical solution to address this issue is reduction in the requirement of P/E voltage itself, which implies that flash memory with thinner tunnel oxide must be used.

This paper begins with a briefing about C-V and G-V profiles of FGMOS gate stack. Later, fabrication of FGMOS gate stack using ultra-thin tunnel oxide and corresponding structural and electrical characteristics are illustrated. Complete C-V, hysteresis, and reliability analyses are presented. Various parameters are extracted from obtained curves.

2 Parameters of Analyses

The focus of this research work is on fabrication of a flash gate stack with ultra-thin tunnel oxide. A thin tunnel oxide provides a trapezoidal barrier which can be overcome by the mechanism of direct tunneling [23–26]. Direct tunneling can be easily carried out at reduced P/E voltages resulting into power-efficient memory devices. Standard CMOS process is used for fabrication.

A. Basic Profiling

C-V measurement technique is the most adapted one and aids in extraction of many parameters [1, 23–27]. As the MOS capacitor is fabricated directly on the substrate, the top metal layer contact and the backside of the substrate are used as two electrical contacts needed for a C-V test. The C-V analyzer applies a high-frequency (1 MHz or 100 kHz) drive signal to the backside of the substrate, via the chuck of a prober [26]. This high-frequency AC drive signal is superimposed on a relatively slow DC bias sweep. During this DC sweep, the capacitor goes through accumulation, depletion, and inversion modes of operation. Such a high-frequency C-V (HFCV) profile for n-type substrate MOS is as shown in Fig. 1 [24].

Various parameters of the gate stack can be calculated from this characteristics using following formulae [23–26]:

$$T_{\mathrm{ox}} = \frac{A * \varepsilon_{\mathrm{ox}}}{(1E - 19)C_{\mathrm{CG}}}, \tag{1}$$

where T_{ox} is the oxide thickness (nm), A is the gate area (cm^2), $\varepsilon_{\mathrm{ox}}$ is the permittivity of the oxide material (F/cm), and C_{ox} is the oxide capacitance (pF). Flat-band capacitance (C_{FB}) can be calculated by using Eqs. (2) and (3) which gives the Debye length parameter.

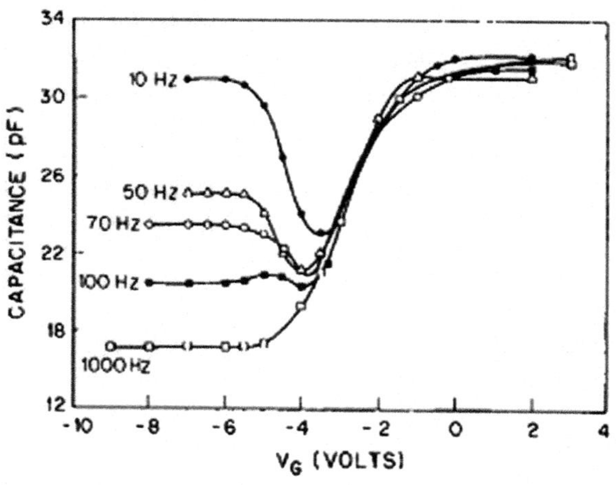

Fig. 1 HF-CV for n-type substrate MOS capacitor [24]

$$C_{\mathrm{FB}} = \frac{C_{\mathrm{CG}} \varepsilon s A (1E-4)(\lambda)}{(1E-12) C_{\mathrm{CG}} + \varepsilon s A/(1E-4)(\lambda)} \tag{2}$$

$$\lambda = \sqrt{\frac{Vt \varepsilon_0 \varepsilon_s}{q N_a}} \tag{3}$$

where $V_t = kT/q$ is the volt equivalent of temperature and N_a is the bulk doping concentration for a p-type acceptor material. Extrapolating this flat-band capacitance on the C-V curve gives value of flat-band voltage V_{FB}. V_T can be calculated from a C-V curve as follows:

$$V_T = [\pm \frac{A}{C_{\mathrm{CG}}} \sqrt{\left(\frac{4\varepsilon_s q}{N_{\mathrm{Bulk}}}\right) |\phi_B|} + 2|\phi_B| + V_{\mathrm{FB}}], \tag{4}$$

where A is the gate area (cm^2), C_{CG} is the oxide capacitance (pF), ε s is the permittivity of the substrate material (F/cm), q is the electron charge (1.60219×10^{-19} C), N_a is the bulk doping concentration for a p-type acceptor material in cm^{-3}, ϕ_B is the bulk potential (V), and V_{FB} is the flat-band potential (V).

$$C_{\mathrm{Debye}} = \frac{\sqrt{2}\varepsilon_s}{L_D} \tag{5}$$

$$W_{\max} = \frac{\varepsilon_0 \varepsilon_s A}{C_{\mathrm{dmin}}} \tag{6}$$

$$C_{\mathrm{FB}} = \frac{C_{\mathrm{CG}} C_{\mathrm{Debye}}}{C_{\mathrm{CG}} + C_{\mathrm{Debye}}} \tag{7}$$

Device with considered tunnel oxide thickness is ultimately supposed to serve in 45 nm regime, and hence, interface investigation becomes vital. When backside of capacitor is used as one of the contacts, the series resistance due to oxides, floating gate, and the substrate (though negligibly small) affect the C-V and G-V measurements. So for interface trap conductance calculations of the system, this series resistance can be compensated by using following formulae [24, 28]

$$C = \frac{(G_M^2 + \omega^2 C_M^2) C_M}{a^2 + \left(\frac{G_M}{\omega C_M}\right)^2} \;\&\; G = \frac{(G_M^2 + \omega^2 C_M^2) a}{a^2 + \left(\frac{G_M}{\omega C_M}\right)^2} \tag{8}$$

where C, C_M and G, G_M are the compensated and measured capacitances and conductances, respectively. Here, ω is 2π times the measurement frequency,

$a = G_M - \left(G_M^2 + \omega^2 C_M^2\right) R_{SERIES}$, and all data sets here are corrected with a series resistance of

$$R_{SERIES} = \frac{\left(\frac{G_M}{\omega C_M}\right)^2}{\left[1 + \left(\frac{G_M}{\omega C_M}\right)^2\right] G_M} \quad (9)$$

Then, the interface trap conductance parameter is then given by

$$\frac{G_{it}}{\omega} = \frac{\omega C_{CG} G}{G^2 + \omega^2 (C_{CG} - C)^2} \quad (10)$$

Here, C_{CG} is (εA/EOT) with A as the device area and EOT as total thickness of tunnel and control oxide.

B. Memory Behavior

Memory behavior of the device can be easily tested by carrying out P/E mechanisms and comparing HFCVs. Shift in the threshold or flat-band voltage then dictates the memory window. Threshold voltage shift is governed by the charges stored in the device, and Gauss's law can be used to estimate corresponding charge storage. Nominal read voltage lies between and is well separated from the two threshold voltage values. This is shown in Fig. 2.

C. Reliability Study

Repeated subjection to high voltages during data storage operations makes the tunnel oxide more prone to damages which may lead to the formation of a continuous leakage path causing malicious functioning of a memory cell. This is shown in Fig. 3. The defects created thereof give rise to stress-induced leakage current (SILC) which decreases data retention of flash memories by affecting the threshold

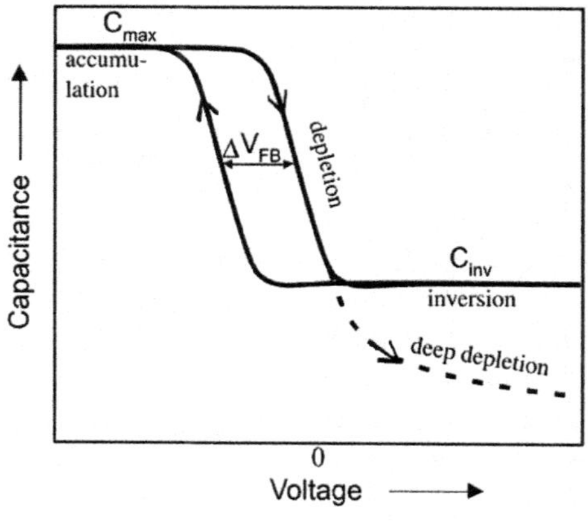

Fig. 2 HFCV hysteresis showing V_T/flat-band shift [29]

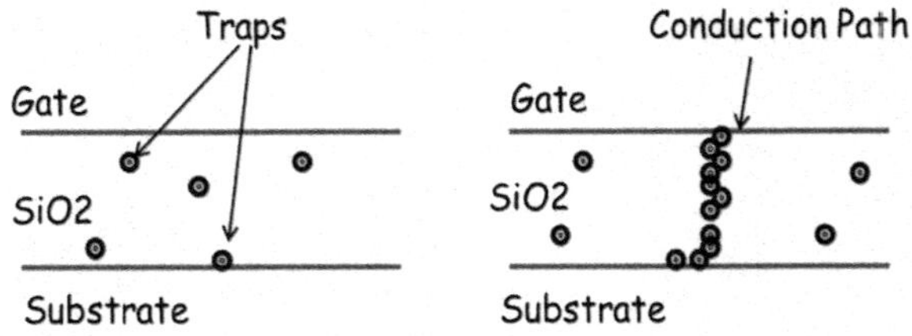

Fig. 3 Data integrity problem observed in thin dielectrics due to stress-induced trap formation

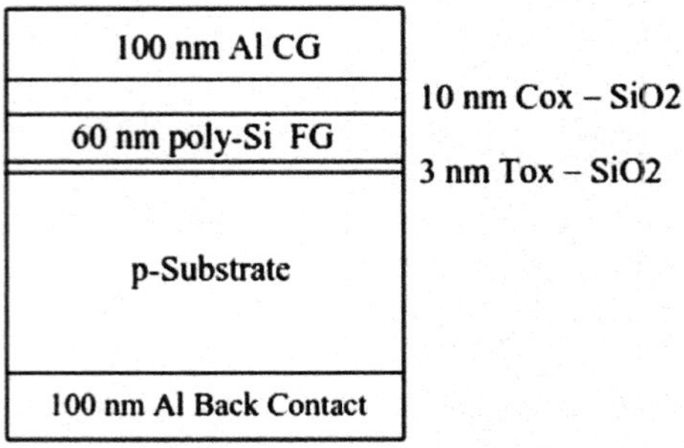

Fig. 4 Cross section of FGMOS gate stack

voltage shift of the devices [30]. If V_T shift is beyond the nominal read voltage, then it leads to memory closure. Retention performance breaks down while using very thin oxides due to quantum tunneling mechanism, especially below 3 nm [1, 2, 8–10, 23–25, 31]. Increased leakage is a common failure mode of electronic devices.

3 Experimental Work

Fabrication of the FGMOS gate stack, which is basically a capacitor (FGMOSCAP), was done in a class 1000 clean room. To ensure best epitaxial and subsequent layers, a single-side polished (SSP) p-type silicon wafer of (100) orientation of 1–5 Ω-cm was chosen. This reflects to a Boron doping concentration of 2.4×10^{15} cm^{-3} [32]. This lies in the most accepted substrate doping concentration range of 10^{14}–10^{17} cm^{-3} [15, 32, 33]. Fabricated gate stack is shown in Fig. 4. Device dimensions were 45 × 45 μm. After standard RCA and buffered HF cleaning, the wafer was subjected to a three-zone thermal oxidation furnace. Temperature of all the three zones was set to 800 °C. Dry oxygen was then passed at a rate of 40 ml/min/div, and flow was set for 20 divisions. After approximately 5 s, a 3 nm epitaxial tunnel oxide was grown. For depositing n-type poly layer, low-pressure chemical vapor deposition (LPCVD) was carried out in AMAT Polygen Chamber. The dynamics involved in the polysilicon deposition include breaking up of silane and phosphene and formation of n-type silicon that starts crystallizing on the wafer surface under controlled temperature and pressure.

Gas combination PH_3:N_2(B):N_2(Pr):SiH_4 was used in a proportion of 85:9000:5000:125 sccm for 40 s at 750 °C to give phosphorus doping concentration of the order of 10^{17} cm^{-3}. To obtain 60-nm-thick FG, extra polysilicon was etched out by using TMAH and DI water (1:9). 10 nm control oxide was deposited by inductively coupled plasma enhanced CVD (ICPCVD) with Silane: N_2 gas flow in proportion of 16:40 sccm. Deposition was carried out for 10 s at 40 mbar pressure and 30 W of RF power. Thermal evaporator was used to perform physical vapor deposition of control gate (CG). Aluminum chunks were used as raw material in a N_2 cooled evaporating furnace, with a base pressure of 5.2×10^{-2} bar. 30 min were required to deposit around 100-nm top gate layer.

Back of the wafer was etched with BHF to remove native oxide and it was followed by aluminum deposition to form the back contact.

4 Results and Analyses

A. Device Characterization

As-processed films were subjected to wafer level small signal metrology using Keithley 4200 probe station at room temperature. High-frequency C-V characteristics (at 1 MHz) of fabricated FGMOSCAP structure as shown in Fig. 5a indicate that the device has a negative flat band voltage, which goes with the theory of calculation of flat band voltage from work function difference between CG and substrate and presence of silicon oxide interface charges [23]. Table 2 shows various findings from these curves for the device under consideration, at 1 MHz using Eqs. (1)–(7). It can be observed that maximum gate capacitance is directly proportional to the device diameter. Calculated value of tunnel oxide thickness is 2.73 nm, which fairly agrees with the approximated value. Figure 5b shows a representative data set, where HFCV (@ 1 MHz) and G-V measurements of the device with 13 nm EOT are plotted. The peak in the depletion mode of conductance curve is manifestation of lossy inversion layer. This is due to capture of holes by the traps at some location x in the gate stack [24, 28]. Parameters calculated by using Eqs. (8)–(10) are tabulated in Table 3.

B. Frequency-Dependent C-V Characteristics

In inversion mode, small signal capacitance depends on whether the measurements are made at high or low frequency. The terms ‘high’ and ‘low’ are defined with respect to the generation-recombination rate of the minority carriers in the inversion layer [23–25, 34]. If gate voltage is varied rapidly, the charge in the inversion layer cannot change in response and does not contribute to small signal AC capacitance. Hence, it remains at the minimum corresponding to maximum depletion width. However, if gate bias is changed slowly, minority carriers can be generated in bulk, drift across the depletion region to the inversion layer or vice versa, and recombine. In this case, the inversion charge increases exponentially with applied gate voltage.

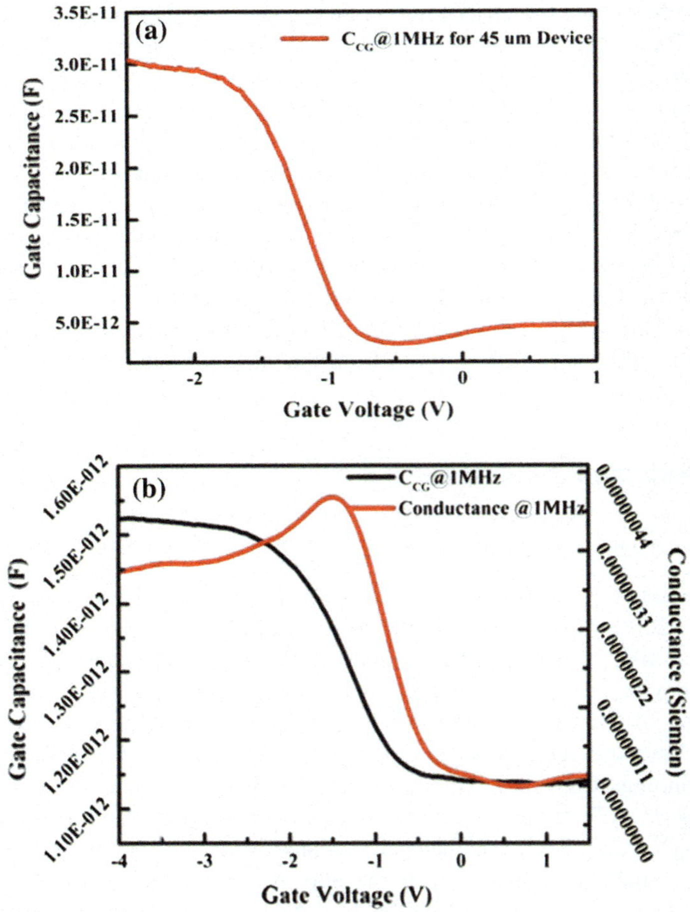

Fig. 5 **a** High-frequency C-V characteristics and **b** Capacitance and conductance graphs of a conventional FGMOSCAP structure @ 1 MHz

Table 2 Parameters extracted from C-V profile

C_{CGmax} (F)	C_{CGmin} (F)	C_d (F)	N_a (per ccm)	L_{Debye} (m)	C_{FB} (F)	V_{FB} (V)
30.0E−12	2.9568E−12	3.2751E−12	2.54E15	8.14E−6	2.534E−11	−2.02

Frequency-dependent behavior exhibiting this analysis of fabricated FGMOS capacitors is shown in Fig. 6. Although its variation is small, the capacitance value in accumulation regime is not exactly same for all samples due to non-uniformity in the oxide thickness throughout wafer surface. Other factors responsible for the capacitance variation are variable measurement offsets and their frequency

Table 3 Interface trap parameter calculations @ 1 MHz

$G_M = 4.02\text{E}{-7}$ (S)						
$C_M = 1.43762\text{E}{-12}$ (F)						
R_{SERIES} (Ω)	a	C (F)	G (S)	C_{CG} (F)	A (m^2)	G_{it}/ω
4917.20	4.00946E −7	5.9184E −20	1.65061E −14	5.4728E −10	2.025E −9	2.08327E +11

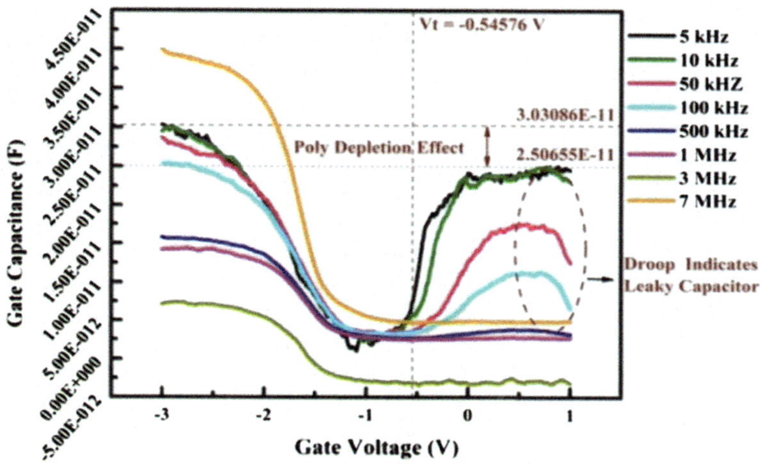

Fig. 6 Frequency-dependent behavior of FGMOS capacitor

dependence. Also, only a few samples gave considerable capacitance swing indicating that other devices are leaky. It is observed from Fig. 6 that with decrease in frequency, the inversion capacitance goes on approaching the accumulation value. Further, at medium frequencies, it can be observed that for $V_{GS} > 0.5$ V, there is droop in C-V characteristics instead of a flat. This indicates that the particular samples were leaky.

This may be attributed to the defects in the control oxide layer, which was deposited (not grown) and the rough control oxide and floating gate interface which was formed by etching process. The inversion capacitance which is series combination of C_{ox} and T_{ox} is much less than that of the accumulation due to poly depletion effect of the floating gate.

C. Memory Window Analysis

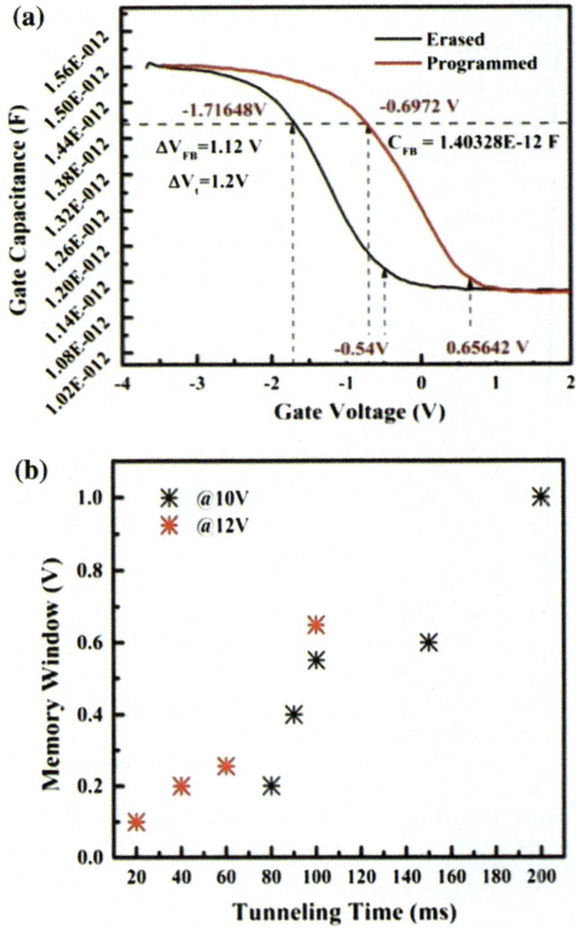

Fig. 7 **a** Hysteresis curve for 1.2 V memory window @ 1 MHz. **b** Dependence of memory window on tunneling pulse

Hysteresis curve of programmed and erased device showing memory window is shown in Fig. 7a.

The device was programmed by applying a tunneling pulse of 10 V at the control gate. For erasing, similar pulse of −8 V was applied. The DC gate bias was swept from −3.5 to 2 V during HFCV measurement, and the device was charged for 200 ms and discharged for 40 ms to get memory window of 1.2 V. The charge stored across the floating gate can be accounted for increase in the capacitance in programmed state. Here, charge stored is calculated as $C_{CG}*\Delta V_T = 1.5E-12*1.2 = 1.8E-12$ C, which in turn implies that 1.125E07 electrons are stored on the FG. Thin oxide enables direct tunneling mechanism that leads to reduction of tunneling voltages (10/−8 V against 15–17 V of ITRS standard). This, in turn, supports the scaling of tunnel oxide below the earlier stagnated values [10, 35]. This

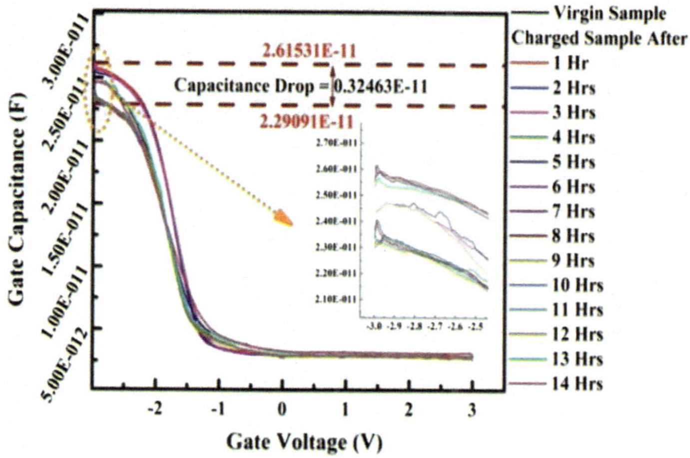

Fig. 8 Data retention of device

achievement is the most desirable quality if the device shows appropriate data retention.

D. Reliability Analysis

Device reliability was tested by executing data retention study. The observations are plotted in Fig. 8. Leakage of charge can be indirectly seen by gate capacitance drop over a period of time. It is found that gate capacitance is dropped by 3.256 pF after 14 h from charging. Extrapolation of this curve leads to data retention of approximately 85.5 h, i.e., 10^5 s. Figure 9a clearly indicates effect of stressing. Variation in the leakage current at lower voltages is collectively attributed to variable measurement offsets. Stressing degrades the quality and increases leakage current by a factor of 10 after 600 P/E cycles. With increasing number of P/E cycles, memory window of the device is found to be shrunk. Correspondingly, gate capacitance also gets affected. Samples were repeatedly programmed with 10 V, 60 ms pulse, erased with −8 V, 40 ms pulse, and subjected to HFCV measurements. Resultant C-V curves from Fig. 10b indicate that after subsequent P/E cycles, memory closure is bound to happen. However, it is observed that the capacitance value increases with number of program erase cycles. This is because with increasing P/E cycles, stress sites and charges trapped therein increase causing reduction in effective oxide thickness [23–25, 28]. Tunneling current density is found to be increased by a factor of 11(approximately).

E. Result Comparison

Ultra-thin oxide (here, 3 nm as compared to 6–7 nm recommended by ITRS) enables direct tunneling mechanism that further leads to use of lesser tunneling voltages (here, 10/−8 V as compared to 15–17 V recommended by ITRS). Figure 10 indicates the breakdown characteristics of fabricated sample of a

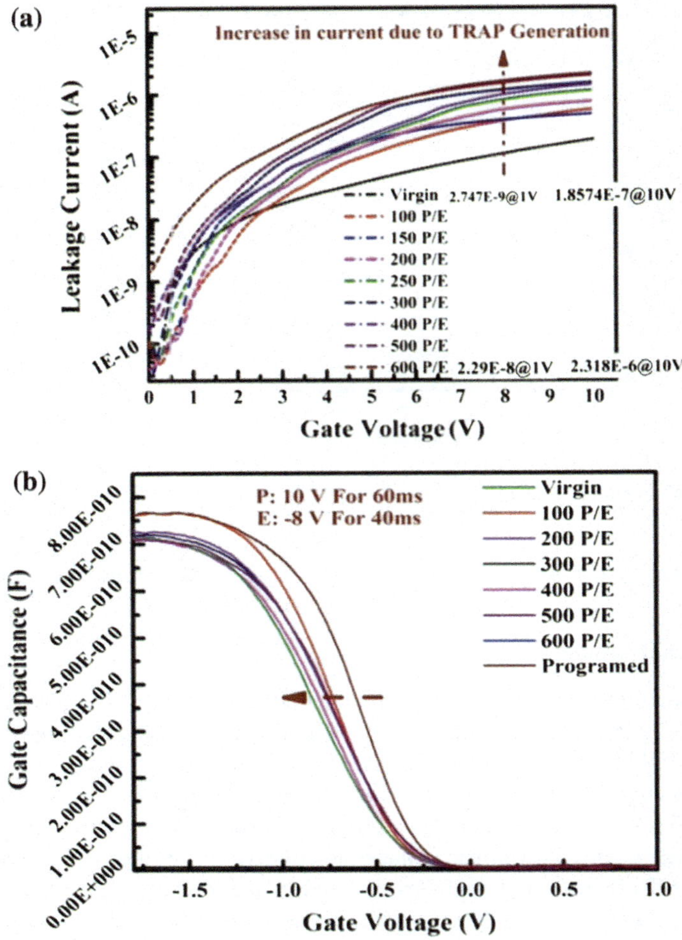

Fig. 9 **a** Reliability study of the device in terms of SILC **b** Effect of stressing on memory window @ 1 kHz

conventional FGMOSCAP. It is observed that the highest breakdown voltage for one of the samples is −12.8 V. In this device, the thicknesses of tunnel oxide and control oxide was 3 and 10 nm, respectively.

The dielectric breakdown strength of the oxide is $E = \frac{V}{d}$, where E = electric field (stress experienced by the oxide), V = applied potential difference between the capacitor electrodes, and d = distance between the electrodes. Thus, the electric field in this particular device is found to be 12.8 V/(10 + 3 nm) = 9.85 MV/cm. This is highly acceptable performance as the reported dielectric breakdown of SiO_2 lies around 8–13 MV/cm [36]. Figure 10 also shows the variation of gate current before and after oxide breakdown. The ratio of currents flowing before and after

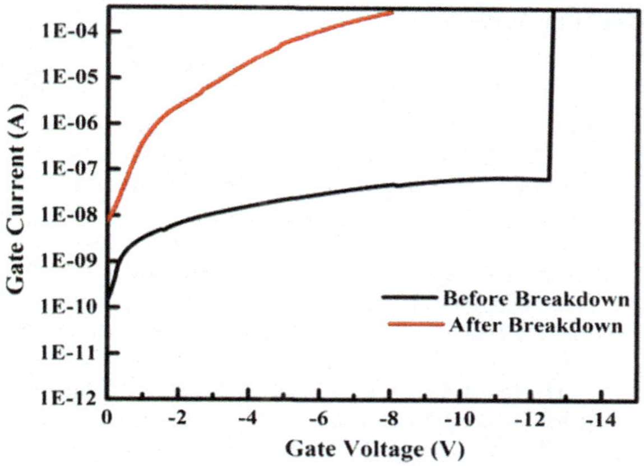

Fig. 10 I-V characteristics of conventional FGMOSCAP showing gate current before and after the breakdown

(nA and mA) clearly indicates the breakdown mechanism. The device exhibits significant amount of stress induced as well as time-dependent leakage. This is due to large interface trap density parameter which tends to gradually increase. This is a well-known issue in thin-oxide flash memories and can be addressed by employing advanced technologies like quantum dot floating gates or ONO-based memory devices.

5 Conclusion

A conventional FGMOS capacitor with ultra-thin tunnel oxide is fabricated, characterized, and analyzed. Device's vertical dimensions are suited for the next-generation FG technology devices. Results indicate that the fabricated device offers excellent memory behavior (comparable with industry expectations) at low P/E voltages which is a considerable achievement of this work. It is attributed to P/E mechanisms by direct tunneling owing to use of the ultra-thin tunnel oxide layer (3 nm against 7 nm ITRS standard). This will further minimize the on-chip charge pump overheads. Though stressing has degraded the device marginally, low-voltage P/E mechanism has dominated data retention feature. If this gate stack is used in a real flash memory device, it would be useful in low-power low-speed VLSI applications like data logging, biometric security, back-up servers.

Acknowledgements This device was fabricated and characterized at IIT Bombay Nano-Fabrication Centre (IITB-NF), Mumbai, India.

References

1. K. Sung-Mo, L. Yusuf, S.-M. Yoo, *CMOS Digital Integrated Circuits: Analysis and Design*, 3rd edn. (TMH, New Delhi, India, 2003)
2. M. Jan, *Rabaey, Anantha Chandrakasan, Borivoje Nikolić, Digital Integrated Circuits: A Design Perspective*, 2nd edn. (Pearson Prentice-Hall, Delhi, India, 2008)
3. Datalight Inc (2008) Choosing NAND or NOR Flash Memory: Tradeoffs and Strategies. Available: http://www.datalight.com. 8 July 2008
4. Toshiba America Electronic Components, Inc. (2011). NAND vs. NOR Flash Memory Technology Overview. Available: www.mitchellcreativegroup.com. 21 May 2011
5. Jan De Blauwe, Nanocrystal nonvolatile memory devices. IEEE Trans. Nanotechnol. **1**(1), 72–77 (2002)
6. T-H. Hou, J. Lee, J.T. Shaw, E.C. Kan, Flash Memory Scaling: From Material Selection to Performance Improvement (2007). Available: www.researchgate.net. Dec 2007
7. Y. He, Z. Zhang, L. Wang, W. Li, J. He, An interesting phenomenon in the C-V Measurements of Nanocrystal Based MOS capacitor. Proc. IEEE Workshop Electron. Devices Semicond. Technol. 129–132 (2007)
8. International Technology Roadmap for Semiconductors (ITRS) (2001), Available: http://www.itrs.net [Online]
9. International Technology Roadmap for Semiconductors (ITRS) (2003), Available: http://www.itrs.net [Online]
10. International Technology Roadmap for Semiconductors (ITRS) (2005), Available: http://www.itrs.net [Online]
11. International Technology Roadmap for Semiconductors (ITRS) (2013), Available: http://www.itrs.net [Online]
12. Y. Fujisajki, Overview of emerging semiconductor non-volatile memories, IEICE Electron. Express. **9**(10), 908–925 (2012)
13. O-Y. Wong, R. Wong, W-S. Tarn, C-W. Kok, An overview of charge pumping circuits for flash memory applications, Proc. IEEE Int. Conf. ASIC (ASICON), 116–119 (2011)
14. International Technology Roadmap for Semiconductors (ITRS) (2010). Available: http://www.itrs.net [Online]
15. International Technology Roadmap for Semiconductors (ITRS) (2011). Available: http://www.itrs.net [Online]
16. International Technology Roadmap for Semiconductors (ITRS) (2012). Available: http://www.itrs.net [Online]
17. O. Khouri, S. Gregori, R. Micheloni, D. Soltesz, G. Torelli, *Low Output Resistance Charge Pump for Flash Memory Programming* (Design and Testing, Proc. IEEE Int. Workshop on Memory Technology, 2001), pp. 99–104
18. G. Palumbo, D. Pappalardo, Charge pump circuits: an overview on design strategies and topologies. IEEE Circ. Syst. Mag. **10**(1), 31–45 (2010)
19. S.A. Bhalerao, A.V. Chaudhary, R.M. Patrikar, A CMOS low voltage charge pump. Proc. IEEE 6th Int. Conf. Embed. Syst. 20th Int. Conf. VLSI Des. 941–946 (2007)
20. B.R. Gregoire, A compact switched-capacitor regulated charge pump power supply. IEEE J. Solid-State Circ. **41**(8), 1944–1953 (2006)
21. S.W. Choi, D.J. kim, J. Chung, B.S. Han, J. Park, Efficiency optimization of charge pump circuit in NAND FLASH memory. IEICE Electron. Express. **8**(16), 1343–1347 (2011)
22. I-Y. Chug, J. Shin, New charge pump circuits for high output and large current drivability. IEICE Electron. Express. **6**(12), 800–805 (2009)
23. S.M. Sze, *Semiconductor Devices: Physics & Technology*, 2nd edn. (Wiley, Inc., 2002)
24. H. Nicollean, J.R. Brews, *MOS (Metal Oxide Semiconductor) Physics and Technology* (Wiley-Inderscience Publication, 1982)
25. B.G. Streetman, S. Banerjee, *Solid State Electronic Devices*, 6th edn. (Prentice Hall Publications, 2006)

26. Keithley 4200-SCS User Manual, (2000)
27. K. Yang, C. Hu, MOS capacitance measurements for high-leakage thin dielectrics. IEEE Trans. Electron. Devices. **46**(7), 1500–1501 (1999)
28. S.H. Bae, R.J. Hillard, C.S. Oldsen, M.C. Benjamin, S. Thirupapuliyur, N. Ho, P.A. Kraus, Interface trap characterization of alternate gate dielectrics with elastic gate MOS metrology. AIP Conf. Proc. **788**, 191–194 (2005)
29. Joel L. Plawsky. Electrical properties and diffusion (2004), Renslaesser Institute of Polytechnic. Available: http://homepages.rpi.edu [Online]. (Mar 2004)
30. Y. W. Park, J. Lee, Device considerations of planar NAND flash memory for extending towards sub-20 nm regime. Proc. IEEE Int. Mem. Workshop. 1–4 (2013)
31. C-Y. Lu, T-C. Lu, R. Liu, Non-volatile memory technology- today and tomorrow. Proc. IEEE Int. Symp. Phys. Fail. Anal. Integr. Circ. **1**, 18–23 (2006)
32. W.R. Thurber, R.L. Mattis, Y.M. Liu, J.J. Filliben, *The Relationship Between Resistivity and Dopant Density for Phosphorus-and Boron-Doped Silicon*. (National Bureau of Standards Special Publication, 1981, pp. 400–464)
33. Synopsys Sentaurus TCAD User Manuals, Version E-2012, Dec 2012
34. B. Rong, Capacitance-voltage characterization for MOS capacitor on p-type high resistivity silicon substrate. Proc. IEEE Int. Conf. Solid State Integr. Circ. Technol. **1**, 198–201 (2004)
35. H. Hanafi, S. Tiwari, Imran Khan, Fast and long-retention time nano-crystal memory. IEEE Trans. Electron Devices. **43**(9), 1553–1558 (1996)
36. D. Demaria, E. Cartier, Mechanism for stress-induced leakage currents in thin silicon dioxide films. JAP. **78**(6), 3883–3894 (1995)

An Effective Method for Maintenance Scheduling of Vehicles Using Neural Network

Sushma Kamlu and Vijaya Laxmi

Abstract The maintenance scheduling of vehicles of a transportation system has its own significance as far as effective operation of a transportation system is concerned. Presently, inspection planning is used to plan for maintenance activity of vehicles in a transportation system. It helps the operator to organize maintenance activity and increase the ability to identify a proactive failure situation. In order to avoid the dilemma like premature aging and failure of vehicles in transportation system responsible for spontaneous and costly maintenance charges, at regular intervals, it is imperative to carry out preventive maintenance (PM). This paper presents the application of neural network technique for automatic maintenance scheduling of vehicles. This paper presents an economic method for solving maintenance scheduling of medium type vehicles by exploiting the neural network technique.

Keywords Maintenance scheduling · Preventive maintenance · Neural network

1 Introduction

Preventive maintenance (PM) of vehicle transportation system has a significant impact on transportation concert. Inspection planning is a method of transportation, which is used to access the safety and operational circumstances of vehicles. Usually, an organization/industry with transportation system keeps the record of the maintenance plan and fleet history for all the vehicles manually in order to avoid failure of services, which has its own drawback. The maintenance plan may also vary from recommended maintenance plan provided by the manufacturer depend-

S. Kamlu (✉) · V. Laxmi
Department of Electrical & Electronics Engineering,
Birla Institute of Technology, Mesra, Ranchi, India
e-mail: sskadwane@gmail.com

V. Laxmi
e-mail: vlaxmi@bitmesra.ac.in

V. Nath (ed.), *Proceedings of the International Conference on Microelectronics, Computing & Communication Systems*, Lecture Notes in Electrical Engineering 453,
https://doi.org/10.1007/978-981-10-5565-2_5

ing upon numerous factors and circumstances. Basically, scheduled maintenance, repair maintenance, and on-condition maintenance are the three main types of maintenance activities. Failure of vehicles can be classified into in-service failure and incidental failure. An in-service failure is detected at a time, when the replacement will cause interruption of services, while an incidental failure is identified on expiration of the equipment or component. To overcome it, after survey, data may be used to implement new techniques like fuzzy logic and neural network for automatic scheduling of maintenance of vehicle. The research work has been till now done only for fleet management of vehicles [1]. Therefore, in this work a technique has been developed for maintenance scheduling of the vehicles, taking medium load of vehicles into consideration.

The neural network has a large number of interconnected nodes, which exhibit the capability to learn and simplify from training patterns or data. Like fuzzy logic, neural network can execute processing which resembles the human brain. Nowadays, applications of the neural network are preferred worldwide, for text to speech conversion, vision processing network, and nonlinear self-tuning adaptive control. For constrained optimization problems, Hopfield-type recurrent networks have been used based on "penalty methods." The neural network has two advantages; one is its learning capabilities and second is distributed representation. Among three learning classes, the unsupervised learning procedure assembles internal models to capture regularities in their input vectors without having any prior information. These are appropriate to get clusters of data indicative of the existence of fuzzy rules. In distributed representation, a pattern of activity is distributed over numerous computing elements, where each computing element represents many different values. Some applications of scheduling like job shop scheduling [2] have been solved, with the help of a competitive learning rule presents an effective and sound solution. On the basis of state input information, i.e., training data of neurons of the operation of Hopfield neural networks [3] select neuron output. Nowadays, various techniques have been developed to resolve the problem of scheduling such as linear programming, classical local search heuristic algorithm, and multiprocessor task scheduling [4].

In this paper, all uncertainties related to the parameters, such as past running hours, operating condition of the vehicle, and fuel consumption rate, have been considered, and a neural network technique has been developed for automatic scheduling of maintenance of vehicles. The maintenance of a vehicle may get affected by some or all of these factors: (i) speed of travel, (ii) topography, (iii) weight of load. In this paper, geographical information system (GIS)-based knowledge has been considered as inputs, which are terrain/texture of the road, mileage of the vehicle, and applied load on the vehicle in order to develop an economic system for automatic maintenance of vehicles. The primary goals of maintenance scheduling are to preserve vehicles in safe operating mode, ensuring each vehicle to operate at crest efficiency, maximizing vehicle life, and minimizing vehicle service failures.

The paper has been structured as follows. The next section deals with the overview of neural network model. Section 3 consists of the problem formulation, and Sect. 4 deals with the solution of maintenance scheduling using neural network. Section 5 deals with the case study with results, and the last section consists of conclusion of the proposed work.

2 Overview on Neural Network Model

Biology provides an important principle which helps to improve the research and development in the field of artificial neural networks (ANNs) [5]. A number of neurons connected in certain interconnection prototype are involved in an ANN model as shown in Fig. 1.

Basically, an ANN model is an iterative algorithm for the search phase. ANN learning refers to the method of modifying the weights of connections between the nodes of a specified network. Biological learning mechanisms were described in Hebb's principle, and weight modification rule for artificial neural network can be given by, $\nabla w_{i,j} = kx_i x_j$, where k is a small constant, which corresponds to the strength of the connection from jth node to ith node and x_i and x_j are the commencement level of respective nodes.

In the back-propagation algorithm with feed-forward neural network architecture, input nodes generate the dimensionality of input patterns [6], and number of nodes in the output layer depends upon the problem under consideration. Each hidden node and output node applies a sigmoid function to its net input. The main reason to use S-shaped sigmoid function is that it is continuous, monotonically increasing invertible, differentiable function which asymptotically approaches its saturation values as net $\rightarrow \pm\infty$.

The back-propagation learning rule is an iterative gradient algorithm designed to minimize the mean squared error between the actual output of a multilayer neural network and the desired output by modifying network weights [7, 8]. It uses supervised learning in which network is trained using data for which inputs as well as desired outputs are known. Once the network weights are frozen, it can be used to compute output values for new input values. Input–output pairs constitute the training set $\{(x_p, d_p): p = 1, \ldots, P\}$. Training algorithm works irrespective of the

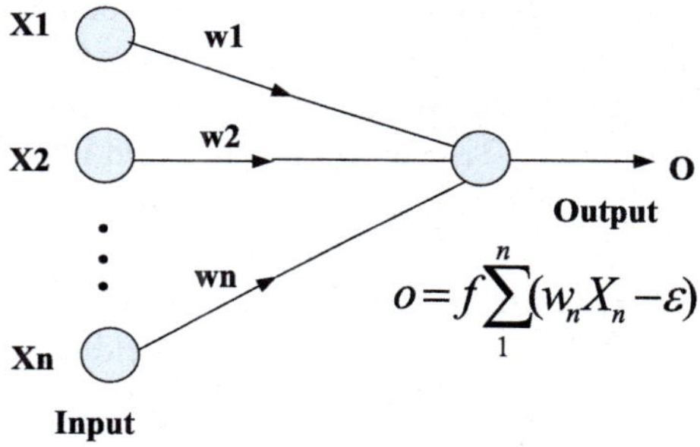

$$o = f\sum_{1}^{n}(w_n X_n - \varepsilon)$$

Fig. 1 Feed-forward multilayer perceptron

weight values that preceded training, which may at the start have been assigned randomly. An error may be obtained from the difference $\ell_{p\cdot j} = |o_{p\cdot j} - d_{p\cdot j}|$ between the *j*th components of the actual and desired output.

3 Problem Formulation for Maintenance Scheduling of Vehicles

In this paper, factors considered for development of neural network model for the maintenance scheduling of vehicles are discussed in detail. As discussed in Sect. 1, terrain of the road, mileage of the vehicle, and the weight of the load on a vehicle are the major factors which affect the working condition of the equipments or components of a vehicle like vehicle engine, fuel filter, oil filter breather, and transmission gears. Basically, the operating condition of the vehicle is based on the running condition such as strength of the ground, texture of surface, slope of the ground, rate of consumption of fuel, type of load. Hence, these factors affect the maintenance plan of a vehicle. The maintenance schedule of vehicles running on a flat terrain may be different than that of vehicles running on rough terrain. The vehicle may have good mileage on flat terrain, but it may not be able to have the same mileage on an increase or decrease in the gradient. Hence, the maintenance activities may differ from the recommended maintenance plan by the manufacturer of the vehicle depending upon the running condition of the vehicle.

Therefore, a new technique has been developed to have an effective maintenance plan by using real-time GIS-based data. The GIS-based vehicle tracking system gives the information about the geographical location, average speed, etc. The slope of the road is calculated from that information as discussed later in this section. The factors which are considered in this work to affect the maintenance plan are explained in this section and later implemented with feed-forward back-propagation neural network model. The factors affecting the maintenance of vehicle as explained above are as follows:

A. *Terrain*

Classification of the terrain is anticipated for classification of surficial resources by weather conditions, biological accretion, manual and volcanic bustle. It includes residual resources weathered from rock; transported materials composed of mineral, gravity, or any combination of these agents, landforms and geology course of action [9, 10]. Information about the terrain assists in handling of vehicles in safe mode effectively. Terrain can be classified into different classes based on three important factors, which are as follows: (i) ground strength, which is the information about the bearing capacity of the soil and specifies the potential level of environmental damage which affects on the efficiency of vehicles, (ii) surface roughness, which determines the distribution of infield obstacles and affects the vehicle stability and travel speeds, and (iii) slope, which is one of the major factors affecting travel speeds and vehicle stability.

Five slope classes are recognized by the national terrain classification system (NTCS) [11]. They are as follows: class 1 for 0–20% gradient, class 2 for 13–20% gradient, class 3 for 21–35% gradient, class 4 for 36–50% of gradient, and class 5 for gradient 51% and above.

B. *Load*

Classification of vehicles depends upon several entities and is usually based on the gross vehicle weight rating (GVWR) [10]. As per the American Automotive Manufacturers Association (AAMA), the GVWR is the maximum weight a vehicle is allowed to bring, including the weight of the vehicle, driver, payload, and fuel. The load is classified as a light load, medium load, and heavy load as given in Fig. 2.

In this work, medium type of load has been considered. The medium type load is further classified as class IV, class V, and class VI as depicted in Fig. 2. Generally, medium-duty vehicles have gasoline or diesel engine, single rear axle and it is for commercial use. School buses, star buses, and trucks are some types of medium-load vehicles.

The exact weight of a loaded or unloaded truck is directly obtained from the weighting system. For a bus, applied load has been calculated by adding the actual weight of unloaded bus and the number of tickets sold out multiplied by average weight of the person considered as 65 kg. It is given by Eq. (1) below.

$$\text{Load} = \text{number of tickets sold per trip} \times 65 + \text{weights of the vehicle} \quad (1)$$

C. *Mileage*

The fuel economy in an automobile is described by mileage of a vehicle. In other words, it is the distance travelled in miles per liter or kilometer per liter (km/L). Mileage of a vehicle is classified as shown in Fig. 3.

Fig. 2 Classification of load

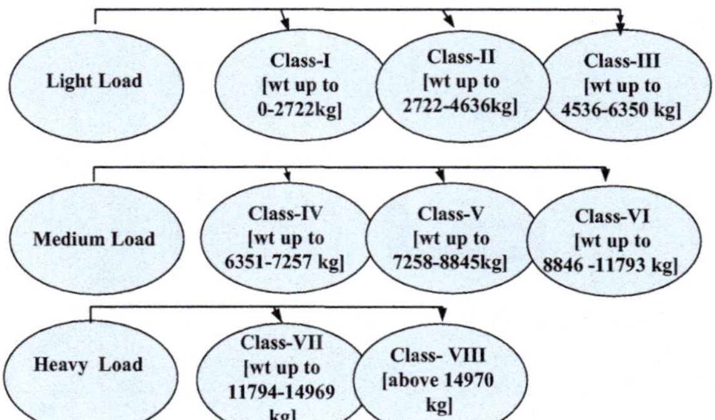

There are several factors which may affect the fuel economy of a vehicle, which are as follows:

- Fuel economy largely reduces in town due to frequent use of the accelerator and brakes.
- Excessive idling decreases kilometer per liter (km/L) or miles per gallon (MPG).
- Fuel economy reduces due to cold weather and frequent short trips.
- The distance as well as the average speed between source and destination also affects fuel economy.
- Time of travel, since the traffic much differs at noon than the traffic in the morning and night.
- The number of traffic signals at a crossing, which may increase the traveling time of vehicles.
- Type of location, whether dense or transparent.

For maintenance of vehicles, different factors like aerodynamic, engine strength collectively affect the mileage, that's why this factor has not considered separately. Figure 4 gives the details of a GIS-based vehicle tracking system, which consists of global positioning system (GPS) receivers and global system for mobile communication (GSM) modem. The GPS is a space-based satellite navigation system, which provides location and timely information in all climatic environments. It can be useful everywhere on or close to the Earth where there is an unobstructed line of sight to four or more GPS satellites. The GPS receiver of the tracking unit collects

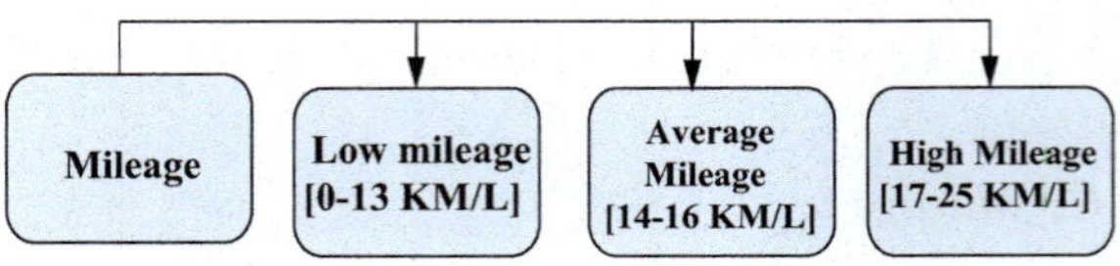

Fig. 3 Classification of mileage

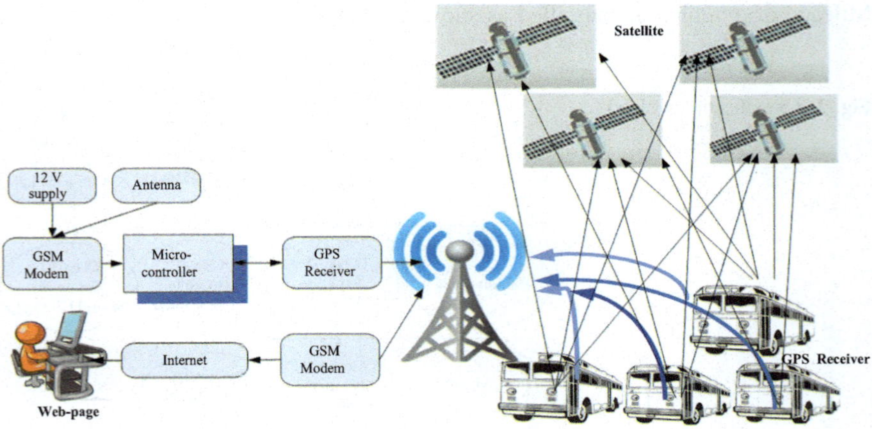

Fig. 4 Working of GIS-based vehicle tracking system

the latitude, longitude, and speed information about the vehicle and sends it to the microcontroller, and then the GSM module communicate with the microcontroller to access and send this data to the server via a previously established general packet radio service (GPRS) connection.

Daily report has been generated from GIS-based vehicle tracking system to collect the details about vehicle location, date and time, direction, speed, and distance traveled as per Table 1.

Table 2 shows to utilize the data obtained from GIS-based vehicle tracking system like vehicle location in degrees, minute, and seconds to calculate the percentage gradient, slope, and corresponding speed of the vehicle by converting it into decimal degrees. These longitude and latitude are represented as X and Y coordinates and utilized for calculation of percentage gradient and slope angle along with corresponding speed of the vehicle.

Table 1 Daily report generated from GIS-based vehicle tracking system

Sl. No.	Vehicle location	Date & time	Direction	Speed	Distance (km)
1	23°25′25.4″N 85°25′58.1″E	19-07-2013 14:12:25	257.27	25.94	290.03
2	23°25′34.1″N 85°25′53.8″E	19-07-2013 14:10:44	129.58	8.26	289.64
3	23°25′46.7″N 85°25′37.2″E	19-07-2013 14:09:03	94.59	35.19	288.88
4	23°25′46.4″N 85°25′31.1″E	19-07-2013 14:08:43	78.71	18.97	288.69
5	23°25′46.7″N 85°25′26.8″E	19-07-2013 14:08:24	91.31	24.25	288.56
6	23°25′46.5″N 85°25′18.1″E	19-07-2013 14:07:47	89.7	35.08	288.31
7	23°25′46.9″N 85°25′08.8″E	19-07-2013 14:07:19	97.93	35.3	288.03
8	23°25′48.0″N 85°25′03.4″E	19-07-2013 14:06:51	102.19	16.81	287.86
9	23°25′49.5″N 85°24′57.2″E	19-07-2013 14:06:11	101.54	14.24	287.68
10	23°26′02.6″N 85°24′58.0″E	19-07-2013 14:04:05	191.99	8.08	287.20
11	23°26′02.8″N 85°24′57.6″E	19-07-2013 13:36:55	16.98	9.58	286.93
12	23°25′58.7″N 85°24′56.9″E	19-07-2013 13:36:27	11.47	23.53	286.79
13	23°25′51.0″N 85°24′55.4″E	19-07-2013 13:35:41	37.03	13.83	286.51
14	23°25′49.2″N 85°24′58.0″E	19-07-2013 13:35:08	293.16	12.19	286.38
15	23°25′48.0″N 85°25′03.0″E	19-07-2013 13:34:40	281.4	18.86	286.22
16	23°25′47.0″N 85°25′07.3″E	19-07-2013 13:34:13	279.23	35.03	286.10
17	23°25′46.3″N 85°25′17.4″E	19-07-2013 13:33:45	268.89	41.09	285.79
18	23°25′46.5″N 85°25′25.0″E	19-07-2013 13:33:18	281.08	15.36	285.56
19	23°25′45.3″N 85°25′40.4″E	19-07-2013 13:31:19	321.46	27.58	285.00
20	23°25′34.6″N 85°25′53.4″E	19-07-2013 13:30:01	322.46	17.88	284.50
21	23°25′26.6″N 85°26′01.7″E	19-07-2013 13:29:22	316.69	38	284.15
22	23°24′59.9″N 85°26′16.1″E	19-07-2013 13:27:40	71.33	32.38	282.88
23	23°24′58.4″N 85°26′11.0″E	19-07-2013 13:27:22	72.9	31.65	282.73
24	23°24′57.1″N 85°26′06.7″E	19-07-2013 13:27:03	90.96	15.23	282.58
25	23°25′00.5″N 85°26′18.2″E	19-07-2013 13:18:39	254.69	27.94	282.20

Table 2 Calculation of percentage gradient, slope, and corresponding speed of GIS data

Sl. no.	Vehicle geographical location	X-coordinate	Change in X (ΔX)	Y-coordinate	Change in Y (ΔY)	%Gradient ΔY/ΔX	Slope in degree	Speed in km/hr
1	23.42371,85.4328	23.42371	0.00242	85.4328	0.0012	50	26.375	25.94
2	23.42613,85.4316	23.42613	0.0035	85.4316	0.0046	131.428	52.733	8.26
3	23.42963,85.427	23.42963	7.00E-05	85.427	0.0017	2428.57	87.64	35.19
4	23.42956,85.4253	23.42956	7.00E-05	85.4253	0.0012	1714.285	86.66	18.97
5	23.42963,85.4241	23.42963	0.00006	85.4241	0.0024	4000	88.567	24.25
6	23.42957,85.4217	23.42957	0.00013	85.4217	0.0026	2000	87.137	35.08
7	23.4297,85.4191	23.4297	0.0003	85.4191	0.0015	500	78.69	35.3
8	23.43,85.4176	23.43	4.10E-04	85.4176	0.0017	414.63	76.44	16.81
9	23.43041,85.4159	23.43041	0.0013	85.4159	3.00E-04	4.0541	12.99	14.24
10	23.43406,85.4161	23.43488	7.70E-04	85.4156	4.00E-04	5.4054	23.07	8.08
11	23.43411,85.416	23.43411	0.0011	85.416	2.00E-04	∞	90	9.58
12	23.43297,85.4158	23.43297	0.0021	85.4158	4.00E-04	19.0476	10.78	23.53
13	23.43084,85.4154	23.43084	5.00E-04	85.4154	7.00E-04	140	54.46	13.83
14	23.43034,85.4161	23.43034	3.50E-04	85.4161	0.0014	400	75.96	12.19
15	23.42999,85.4175	23.42999	2.70E-04	85.4175	0.0012	444.4	77.319	18.86
16	23.42972,85.4187	23.42972	1.90E-04	85.4187	0.0028	1473.68	86.118	35.03
17	23.42953,85.4215	23.42953	6.00E-05	85.4215	0.0021	3500	88.36	41.09
18	23.42959,85.4236	23.42959	3.50E-04	85.4236	0.0043	1228.5	85.346	15.36
19	23.42924,85.4279	23.42924	0.003	85.4279	0.0036	120	50.194	27.58
20	23.42627,85.4315	23.42627	0.0022	85.4315	0.0023	104.5	46.273	17.88
21	23.42407,85.4338	23.42407	0.0074	85.4338	0.004	54.0541	28.39	38
22	23.41664,85.4378	23.41664	4.20E-04	85.4378	0.0014	333.33	73.3	32.38
23	23.41622,85.4364	23.41622	3.60E-04	85.4364	0.0012	333.33	73.3	31.65
24	23.41586,85.4352	23.41586	3.6E-004	85.4352	0.0012	333.33	73.3	15.23
25	23.41681,85.4384	23.4168	9.4E-004	85.4384	0.0032	340.43	73.6	27.94

4 Neural Network-Based Solution for Maintenance Scheduling of Vehicle

This section deals with the implementation of neural network for application in maintenance scheduling of vehicles for a transportation system. A feed-forward, back-propagation multilayer perceptron has been designed, whose architecture is shown in Fig. 5. The behavior of the back-propagation algorithm depends on the values of learning rate and momentum. Variable sets a limit on the number of times the back-propagation algorithm iterates through the training data set. Variable learning rate controls how fast the weights and bias values change. Variable momentum adds an additional boost to the rate of change of the weights and bias values. After the neural network has been trained, the network checks for accuracy of the output by calculating the percentage of correct prediction. For each epoch, every data item in the training set is processed as per the following steps: (i) "shuffle," which rearranges the sequence array into a random order by selecting a training item and extracting inputs and targets, (ii) "compute outputs," in which inputs are fed, and (iii) "update weights," which is used to modify the weights and bias by using the target values so that the outputs more closely match the target values. This particular training approach, where back-propagation updates occur for every training item is based on the difference between the computed outputs and the target outputs [7, 12].

An alternative method is to read all training data, accumulate an overall difference between all computed outputs and all target outputs, and then perform a single back-propagation update. After a neural network has been trained, the next step is to estimate how well the model will perform on new data.

A neural network with three input nodes (one for each input), ten hidden nodes, and four output nodes (one for each possible output class) has been considered for maintenance scheduling of vehicles. The input variables of the neural network are terrain, load, and mileage of the vehicle, and output is the kilometer run after which maintenance is needed. The output has been classified into four maintenance stages S1, S2, S3, and S4 representing the required maintenance after 5500, 6500, 7500, and 8500 km run, respectively. The neural network uses the hyperbolic tangent function for hidden node activation and for output node linear functions are used for activation.

Fig. 5 Architecture for feed-forward, back-propagation

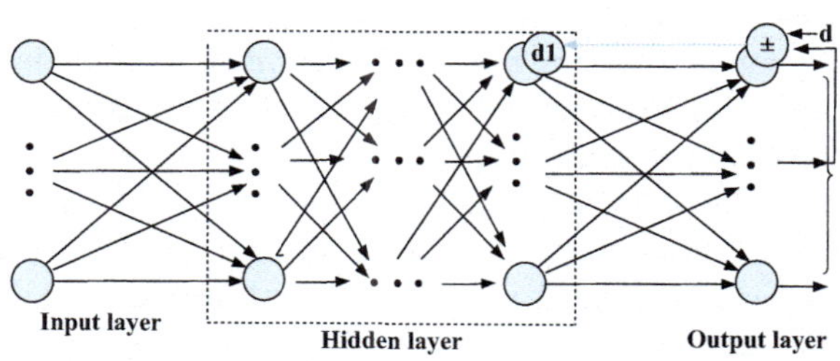

The neural network's weights and bias values are initialized to small (between 0.001 and 0.0001) random values. Then, the back-propagation algorithm is used to search for weights and bias values which generate neural network outputs that closely match the output values in the training data. In this approach, 450 input data are used as training data and 150 data are used as test data. Training with back-propagation is an iterative process. Due to large input data used for training, the training process stops after 15–25 iterations, or when the mean squared error term drops below 0.001.

In this work, 450 training data are used in neural network design. Some sample training data are shown in Table 3. As mentioned in Table 3, preconditions are defined, which are terrain, mileage, and load, with respect to these consequences are represented as running a kilometer after which maintenance is required. For example, if terrain in terms of % gradient is 20, mileage is 12.5 km/L, and total load on the vehicle as per GVWR is 8894 kg, the corresponding consequence is 7000 km, which gives the information about the required maintenance of vehicle after running 7000 km.

5 Case Study and Results

In this section, the case study with the results has been discussed in detail. As mentioned in Sect. 1, the generalized maintenance activities are scheduled maintenance, repair maintenance, and on-conditional maintenance. During scheduled maintenance, work to be carried out after every 8500–9000 km is as follows: air intake system, air filter change, steering system, brake system, charge air cooler, whereas after 18,000 km maintenance, work is to be done for engine oil and filter change, antifreeze concentration, transmission system, body mounting, wheel alignment, etc. Over 36,000 km run, vehicle has to go through maintenance work for the drive belt tension, tensioned bearing, fan hub, valve clearance adjustment, etc.

A transportation system has been chosen for case study having 6 star buses. It indicates that vehicle B4 had gone for the first maintenance for the year 2011 after 24,863 km, whereas, it went to the second maintenance activity after 2414 km. The detail of case study has been presented in Table 4. The graph for first two maintenance activities for six vehicles has been shown in Fig. 6. It shows that the ages of the four vehicles are 10–12 years and two vehicles are new 4 and 5 years old, respectively. Vehicles of 10–12 years old have low mileage up to 10–12 km/L, which runs for 52 km daily, whereas 4-to-5-year-old vehicles have mileage at 14–16 km/L, and they run for 110 km daily considering six working days per week, i.e., 10-to-12-year-old vehicles and 4-to-5-year-old vehicles run for 312 and 660 km per week, respectively.

Table 4 indicates the age and details of past maintenance activities of the vehicles in terms of kilometers in the years 2011 and 2012. For example, vehicle B1, whose age is 5 years, had its first maintenance done in 38,612 km, and it went for second maintenance at 44,653 km in the year 2011, i.e., vehicle gone for maintenance after 6041 km in the year 2011. Then, it went for next maintenance at 47,327 km in the year 2012, i.e., after 2674 km maintenance activity was done. Similarly, vehicle B4 had its first maintenance during the year 2011 after 24,863 km; it had second maintenance activity after 2413 km. It means that usually a transportation system is not performing similar maintenance activities for all its vehicles. These are the major errors which may adversely affect the life and operational circumstances of the vehicles. Therefore, neural network-based VMS for automatic scheduling has been developed to assess the safety and operational condition of the vehicles and to increase the ability of proactive failure situation identification. The results obtained by the neural network model with 450 training data and 150 test data are shown in Figs. 6 and 8. After simulation of neural network model in MATLAB, a performance plot has been obtained which indicates the region of convergence. It shows that the training was satisfactory because the validation and test curves are very much similar. Due to large input data used for training, the training process stops after 15–25 iterations, or when a mean squared error term drops below 0.001. Figure 7 shows that the performance criteria were achieved at iteration 13. The training continued for 5 more iteration before the training stopped. According to Table 3, three inputs and corresponding output have been used as the training data in the neural network model, which gives the outputs as plots as shown in the Figs. 8a, b, c, and d indicating the result of training data, validation data, testing data, and for all data.

The dashed line in each axis corresponds to the target which is equal to the difference between the desired output and the actual output. The solid line corresponds to the best fit linear regression line between outputs and targets. The R indicates the relationship between the outputs and the targets. If R is equal to 1, it indicates that there is an exact linear relationship between targets and outputs. If value of R is near to zero, it indicates a nonlinear relationship between targets and outputs. The corresponding R for training, validation, testing, and for all data are 0.8878, 0.85902, 0.85857, and 0.87992, respectively.

The details of transportation system vehicles taken as case study pertaining to maintenance scheduling are given in Table 5. The previous two maintenances are represented by Nm_1 and Nm_2; the statuses of first two maintenances have been used to determine the subsequent maintenance, Nm_3, through the neural network, where nm signifies the maintenance count. $S_i[nm]$ is the type of maintenance as mentioned in Sect. 4. Q_i [NM] is the period of weeks, after which vehicle has to go for maintenance. For example, according to Table 5, vehicle 2 whose age is 12 years, in the year 2011–2012 had two maintenances. First maintenance held after a 7500 km run with maintenance of type S3 after 7 working weeks in 2011, and second maintenance held after 5085 km run of type S1 after 5 working weeks. Nm_3 is the result obtained by neural network, which gives the plan for next maintenance.

According to the result obtained by neural network vehicle 2 has to go for maintenance after 5 working weeks of type S1. Column for Nm_3 in Table 5 gives the results obtained by neural network, which gives the information regarding the maintenance state and duration of the working time period of vehicle in weeks. For example, the previous two maintenances of the first vehicle were of S4 and S1 types

Table 3 Data set for neural network

Sl. no.	Preconditions			Consequence	Sl. no.	Preconditions			Consequence
	Terrain	Mileage	Load	Required maintenance		Terrain	Mileage	Load	Required maintenance
1.	20	12.5	8894	7000	26.	34	12	8700	6500
2.	10	10	6351	6500	27.	10	15.5	11,700	6500
3.	10	8	6500	6600	28.	13	13	11,790	6430
4.	12	11	6800	6890	29.	17	15	10,000	7200
5.	17	9	6500	7320	30.	19	14	10,800	7320
6.	18	10	7200	7470	31.	9	14	11,210	7470
7.	19	11	6800	7500	32.	8	15	11,790	7500
8.	14	12	6600	8501	33.	20	21	6000	9000
9.	13	13	6987	8890	34.	17	24	6890	9320
10.	18	14	6178	8534	35.	15	25	6360	9430
11.	20	15	6350	8890	36.	17	22	6400	9500
12.	18	13.8	6670	9000	37.	16	20	6800	10,000
13.	19	12.9	6352	9320	38.	20	18	6390	9800
14.	9	18	6579	8534	39.	10	19	6980	9768
15.	17	15	6789	8560	40.	18	20	6351	8890
16.	16	17	6439	8510	41.	10	22	7257	8560
17.	18	20	6900	9000	42.	12	23	8845	8510
18.	20	22	7180	9320	43.	14	24	8001	8800
19.	13	25	7200	9430	44.	16	25	8700	9430
20.	11	11	7257	6600	45.	18	28	8830	9500
21.	16	7	8845	6890	46.	19	20	7389	10,000
22.	14	8	8001	7000	47.	17	18	7360	9800
23.	18	10	8700	6800	48.	15	17	7999	9768
24.	12	9	7999	7320	49.	9	25	8894	7510
25.	13	10	8200	7470	50.	13	22	11,200	8000

Table 4 Details of case study

Vehicle (buses)	Age (in years)	Gap between first and second maintenance in the year 2011 (in kilometers)	Gap between second and third maintenance in the year 2012 (in kilometers)
B1 (JH01 W8924)	5	38612–44653 = 6041	44653–47327 = 2674
B2 (JH01B2579)	12	297764–304448 = 6684	304448–309533 = 5085
B3 (JH 01E2110)	10	285633–299645 = 14012	299645–309418 = 9773
B4 (JH01A 7585)	12	347064–371927 = 24863	371927–374340 = 2413
B5 (JH01C2281)	11	52742–78560 = 25818	78560–86655 = 8095
B6 (JH 01C7231)	4	47327–55253 = 7926	55253–72624 = 17371

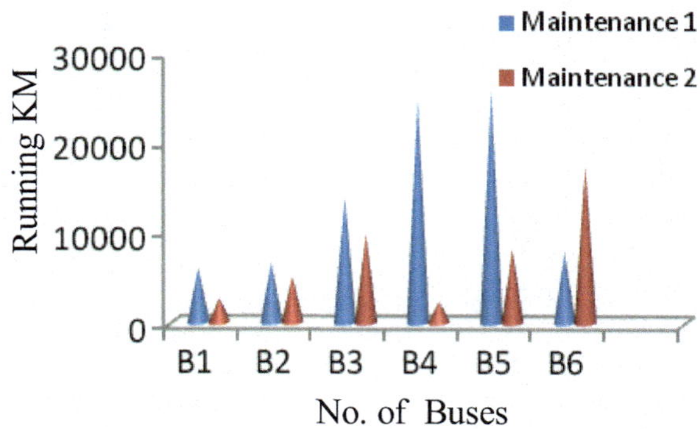

Fig. 6 Graph for past maintenance scheduling of vehicle

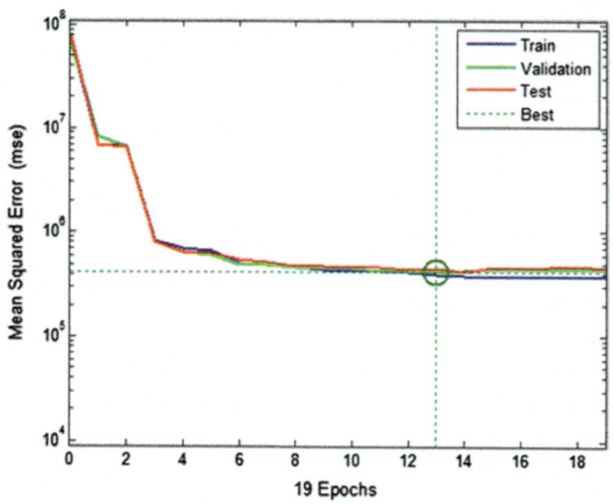

Fig. 7 Performance plot of neural network-based model

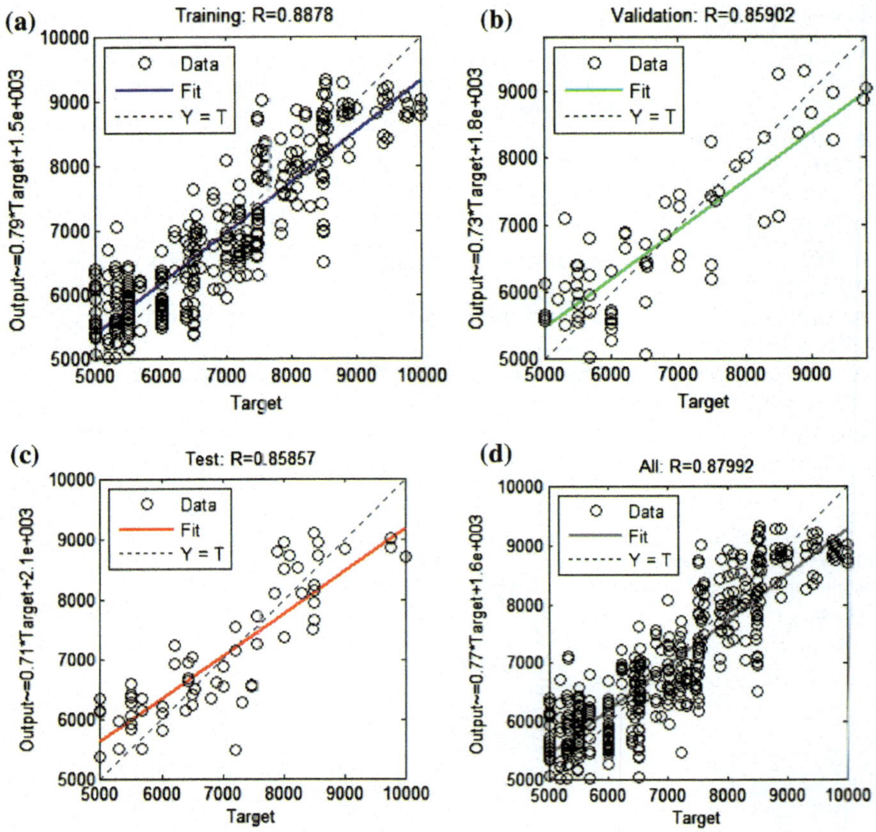

Fig. 8 **a** Target versus output for training data **b** Target versus output for validation **c** Target versus output for test data **d** Target versus output for all

and were done after 8 and 5 weeks from the date of its previous maintenance, respectively. But, from the neural network model, it should have been done of S3 and S1 types after 12 and 10 working weeks, respectively. This way extra maintenance cost was paid for S4 type maintenance instead of S3 type maintenance for the first vehicle.

Similarly, for rest all the vehicles also extra maintenance cost was paid. Hence, the proposed method helps the transportation system to save funds as per the type of maintenance is concerned defined as categories S1, S2, S3, and S4, i.e., regulatory failure, engineering failure, catastrophic failure, and opportunistic maintenance. The maintenance cost over 8500–9000 km, 18,000 km and 36,000 km run as per Indian currency is around Rs. 12,000, 35,000, 25,000, respectively, excluding on conditional and repair maintenance. As the gap between the two maintenances increases with respect to kilometer run, maintenance cost also increases.

Table 5 Details of vehicles of transportation system

Vehicles [i]	Age of vehicle (years)	Different classes of maintenance based on running kms.				Nm_1				Nm_2				Nm_3	
						Status of starting period for previous maintenance		Neural network-based maintenance		Status of starting period for previous maintenance		Neural network-based maintenance		Next proposed maintenance using fuzzy model	
		S1	S2	S3	S4	S_i[nm]	Q_i[nm]	S_n[nm]	Q_n[nm]	S_i[nm]	Q_i[nm]	S_n[nm]	Q_n[nm]	S_i[nm]	Q_i[nm]
1	5	5500	6500	7500	8500	S4	8	S3	12	S1	5	S1	10	S3	10
2	12	5500	6500	7500	8500	S3	21	S1	18	S1	16	S1	19	S2	15
3	10	5500	6500	7500	8500	S4	45	S1	16	S4	8	S1	17	S2	16
4	12	5500	6500	7500	8500	S4	80	S1	17	S1	5	S1	16	S1	15
5	11	5500	6500	7500	8500	S4	82	S2	22	S3	26	S1	16	S2	16
6	4	5500	6500	7500	8500	S4	11	S1	8	S4	26	S1	8	S3	11

Hence, fuzzy approach-based VMS for automatic maintenance scheduling in terms of the type of maintenance not only improves vehicles working condition, but it is also economically beneficial of a transportation system of vehicles.

6 Conclusion

The vehicle maintenance scheduling (VMS) of the transportation system is based on several factors, such as past running hours, operating condition of vehicle, and fuel consumption rate. Since, each parameter has intrinsic nonlinearity; the neural network approach seems to be efficient to bring about the best possible VMS. The proposed neural network model is a generic one to incorporate additional parameters to improvise the VMS by utilizing the knowledge-based input in that region. The case study with result corroborates the potential use of the approach for application in the transportation system.

References

1. S. Salhi, Incorporating vehicle routing into the vehicle fleet composition problem. Eur. J. Oper. Res. **66**(3), 313–330 (1993)
2. R.-M. Chen, Y.-M. Huang, Competitive neural network to solve scheduling problems. Neurocomputing **37**, 177–196 (2001)
3. J.G. Park, J.M. Park, D.S. Kim, C.H. Lee, S.W. Suh, M.S. Han, Dynamic neural network with heuristic. IEEE Int. Conf. Neural Networks **7**, 4650–4654 (1994)
4. R.C. Correa, A. Ferreira, P. Rebreyend, Scheduling multiprocessor tasks with genetic algorithms. IEEE Trans. Parallel Distrib. Syst. **10**(8), 825–837 (1999)
5. T.M. Willems, J.E. Rooda, Neural networks for job-shop scheduling. Control Eng. Pract. **2**, 31–39 (1994)
6. T.C.T. Du, P.M. Wolfe, Implementation of fuzzy logic systems and neural networks in industry. Comput. Ind. **32**(3), 261–272 (1997)
7. J.A. Stegemann, N.R. Buenfeld, A glossary of basic neural network terminology for regression problems. Neural Comput. Appl. **8**(4), 290–296 (1999)
8. S.K. Lahiri, K.C. Ghanta, Artificial neural network model with parameter tuning assisted by genetic algorithm technique: study of critical velocity of slurry flow in pipeline. Asia-Pacific J. Chem. Eng. **5**(5), Sept/Oct 2010, 763–777 (2009)
9. O. Slaymaker, The role of remote sensing in geomorphology and terrain analysis in the Canadian cordillera. Int. J. Appl. Earth Obs. Geoinf. **3**(1), 11–17 (2001)
10. N.N. Clark, J.M. Kern, C.M. Atkinson, R.D. Nine, Factors affecting heavy-duty diesel vehicle emissions. J. Air Waste Manage. Assoc. **52**(1), 84–94 (2002). ISSN 1047-3289
11. Forest Engineering, Chapter 7, "terrain classification" with link pp. 42–43. http://www.sappi.com/regions/sa/SappiSouthernAfrica/Sappi%20Forests/Tree%20Farming%20Guidelines/Part%203_Forest%20Engineering_Chapter%207_Terrain%20Classification.pdf
12. K.F. Reinschmidt, Neural networks: next step for simulation and control. Power Eng. (United States). **95**(11) (1991). OSTI ID: 5281949
13. O. Aloquili, A. Elbanna, and A. Al-Azizi, Automatic vehicle location tracking system based on GIS environment. IET. Softw. **3**(4), 255–263 (2009)

Improved Clustering for Categorical Data with Genetic Algorithm

Abha Sharma and R. S. Thakur

Abstract Clustering is the most significant unsupervised learning where the aim is to partition the data set into uniform groups called clusters. Many real-world data sets often contain categorical values, but many clustering algorithms work only on numeric values which limits its use in data mining. The *k*-modes algorithm is one of the very effective for proper partitions of categorical data sets, though the algorithm stops at locally optimum solution as depended on initial cluster centres. Proposed algorithm utilizes the genetic algorithm (GA) to optimize the *k*-modes clustering algorithm. The reason is, considering noise as cluster centres gives the high cost which will not fit for the next iteration and also not gets stuck to the suboptimal solutions. The superiority of proposed algorithm is demonstrated for several real-life data sets in terms of accuracy and proves it is efficient and can reveal encouraging results especially for the large datasets.

Keywords Clustering · Categorical data · Genetic algorithm · *k*-modes algorithm

1 Introduction

The ever-growing data in almost all fields significantly contribute towards future decision-making, extracting hidden, but potentially useful information embedded in the data. In depth of the clustering problem, many clustering methods usually require the designer to provide the name and number of clusters as input. Unfortunately, the designer has no idea about the inherent structure of huge data sets. As well as clustering result is sensitive to the selection of the initial cluster centres. This sensitivity may make the algorithm converge to the local optima. So, the most challenging and difficult task is the determination of the number and name

A. Sharma (✉) · R. S. Thakur
Maulana Azad National Institute of Technology, Bhopal, India
e-mail: abha_sharma31@yahoo.com

R. S. Thakur
e-mail: ramthakur2000@yahoo.com

V. Nath (ed.), *Proceedings of the International Conference on Microelectronics, Computing & Communication Systems*, Lecture Notes in Electrical Engineering 453,
https://doi.org/10.1007/978-981-10-5565-2_6

of clusters in a data set, which is a basic input parameter for most clustering algorithms.

Clustering [1–3] is an important unsupervised classification technique which groups the data objects in database such a way that objects of similar pattern in some sense reside in one cluster and objects in different clusters are dissimilar in same sense [4, 5]. Clustering has been effectively applied on variety of engineering and scientific applications such as bio-informatics, astronomy, medical imaging, remote sensing, physics, etc. Data matrix and dissimilarity matrix are basically two types of data structure for clustering, if the data is not in this format then need to preprocess the data in above suitable format [6]. Clustering algorithm generally classified into two categories hierarchical and partitioning. Hierarchical clustering algorithm builds a hierarchy of partition at each level.

This paper emphasis on partition clustering was entire data set is partitioned in some specified number of bunches or clusters. A key issue in partitioning clustering algorithms is to manually initialize the number of clusters because it has a direct influence on the creation of final clusters. To find the similarity, it is essential to first calculate the Euclidean or any other distance measure (D) between two objects x and y, which is defined by $E = \|x - y\|$ [5, 6]. Smaller the distance, greater the similarity between the two objects and vice versa. Based on the various requirements and nature of the data, several clustering algorithms have been proposed. One of the well-known partitioning clustering algorithms is k-means [7] algorithm which is best suitable for very large numeric data sets but not appropriate for data set with categorical attributes because it is not possible to find mean of categorical values. Many partitioning clustering algorithm developed for categorical data and the traditional method to deal categorical attributes is treating it as binary numbers but cannot produce significant outcome because its injustice to calculate the huge attribute and data objects only in 0 and 1.

Cluster centre initialization problem remains the same which can severely affect the final clusters. To solve this problem, genetic algorithm (GA) shown in Fig. 1 has been used to break the limitation of one and last chance to assume the cluster centres, convert the local optimal solution into global optimal solution.

GA is one of the evolutionary algorithms based on genetics originally developed by Holland [8] which can apply to various optimization problems. Its domain-independent nature stimulates its usage in many areas such as VLSI design, pattern recognition, machine learning, etc. We propose an algorithm which is modification of GA for clustering, proves that it congregates to a global optimal solution. This paper presented an algorithm of choosing initial modes using one of the evolutionary algorithms.

The paper is organized as follows: Sect. 2 presents Basics of the work, Sect. 3 Proposed method, Sect. 4 Experimental details, Sect. 5 concludes the paper.

Genetic algorithm

Start

1 i=0

2 population initialization p(i)

3 fitness computation p(i)

4 i=i+1

5 if termination condition reached go to step 10

6 select p(i) from p(i-1)

7 Apply crossover technique on p(i)

8 Apply mutate on p(i)

9 go to step 3

10 stop (output)

Fig. 1 Fundamental steps of genetic algorithm

2 Background

The objective is to find *k* partitions that minimize the total within cluster variation (TWCV) with the help of some GA operators; string representation, population size, selection operator, and crossover as one-step *k*-modes algorithm.

A generalized mechanism is presented in this paper to recognize the worst cluster in a categorical dataset to ignore it. This work utilizes the strength of genetic algorithm (GA) and the simplicity of *k*-modes clustering algorithm to determine most appropriate cluster centres. The objective function is defined in terms of the distance measure among the data objects.

A. *Dissimilarity measure*

Let *A* and *B* are categorical objects, and *m* are categorical attributes. The distance between them can be defined by the total mismatches of the corresponding categories of the two data points [9]. Formally,

$$d_1(A, B) = \sum_{j=1}^{m} d(a_j, b_j) \tag{1}$$

where

$$\delta(a_j, b_j) = \begin{cases} 0, & (a_j = b_j) \\ 1, & (a_j \neq b_j) \end{cases} \tag{2}$$

If the data set contains frequencies of categories, then the distance calculation will be done as follows [9].

B. *Mode of a set*

Consider X is set of categorical objects described by categorical attributes, C_1; C_2; ...; C_u.

Definition 1 Mode of X is a vector $M = [m_1, m_2, \ldots m_n]$ that minimises $D(X, M) = \sum_{j=1}^{m} d_1(X_i, M)$, where M may or may not the element of X [9].

C. *Find Mode for a set*

Let $nt_{k,j}$ be the number of data objects having the kth category $t_{k,j}$ in attribute C_j and the relative frequency of $t_{k,j}$ in X is fr $C_j\Im \leftarrow t_{k,j}\ |X = \frac{nt_{k,j}}{n}$ [9].

Theorem 1 The function $D(X, M)$ is minimised if $fr\left(C_j = m_j | X\right) \geq fr(A_j = t_{k,j}|X)$ for $q_j \neq c_{k,j}$ for all j = 1, 2, …, m.

D. *The k-modes algorithm*

When (1) is used as the dissimilarity measure for categorical objects, the cost function becomes

$$P(W, M) = \sum_{l=1}^{k}\sum_{i=1}^{n}\sum_{j=1}^{m} w_{i,l}\delta(\boldsymbol{x}_{ij}, \boldsymbol{m}_{lj}) \tag{3}$$

where $w_{i,j} \in W$ and $M_l = [m_{l1}, m_{l1}, \ldots, m_{lu}] \in M$.

Similar to k-means algorithm both the k-modes algorithm by Huang [9] and Chaturvedi et al. [10] produces local optimal solution dependent on the initial modes [9, 10] and the order of objects in the data set.

In case of simple matching distance measure, the data objects can be misclassified, therefore cannot always represent the real semantic distance between data points. The reason is that the simple matching distance measure is either 0 or 1, gives less desired results. K-modes is partitioning clustering algorithm where initialisation of cluster is necessary. If the initialisation is wrong, i.e. if some noise or outlier data is assumed to be a cluster centre, then the final results will affect drastically. Hence, k-mode is unable to handle noise and outlier.

Genetic algorithms are randomized search and fitness represents the TWCV. Larger the fitness means denser the clusters produce better clustering results [11]. The proposed approach is described below.

3 Proposed GA-Based Clustering

The basic steps of GA shown in Fig. 2, which also implemented in the GA-based clustering to combine the multiple initialisation of clustering, viewing each initialisation as an independent clustering of data set to get more tight clusters. Utilization of searching capability of GA is shown in Fig. 1 for appropriately determine cluster centres.

This paper proposed GA clustering with *K*-modes algorithm for decomposing the data set into clusters over *n* clustering's obtained by random initializations of the *K*-modes algorithm where each iteration assumed to be generation. There are some other ways to multiply initialize unsupervised learning such as: (a) iterate algorithm number of times with different initializations, (b) mingle several initialisation of any algorithm [8]. This paper used the concept (b), with the environment of GA for creating multiple partitions of the categorical data. In our current implementation of the *k*-modes algorithm, we include GA for global optimal solution. Initialize all the objects as cluster centres turn by turn then if the initialization is wrong then that assumption of initialization will be ignored and not considered for the next iteration.

GA Based Categorical data Clustering Algorithm
Input:
k: Number of cluster centres
P: Size of population
D: Data set
Tmax: Number of iterations
Method:
1) Initialize each chromosome with k randomly chosen data objects as string.
2) For T=1 to Tmax (maximum iteration)
a) for chromosome c=1 to P
i) allocate the data object to the cluster with the nearest mode.
ii)Recompute modes of chromosome c using frequency based method.
iii) compute the fitness of c.
b)Produce the new generation of chromosome using genetic operators.
Output: *clusters of dataset*

Fig. 2 Proposed algorithm

Proposed work developed a theoretical framework for the (1) creation of dense cluster and (2) remove the drawback of wrong assumption of initial modes and its evaluation, based on the concept of GA. The basic steps of GA are also followed in GA clustering which shown in Fig. 2 described now in detail. The same data set has been taken for implementing the GA-based data clustering, with population size p_i where i is chromosome, chosen randomly from the data set. With each individual chromosome, clustering is done and new mode is obtained. Afterwards the crossover is applied using cost function of k-mode algorithm. The new chromosomes obtained by applying the roulette wheel selection. Now all the updated chromosomes or say offspring is an input for the next generation and its fitness value is calculated. Then again the selection, crossover is applied to get a set of chromosomes for the third generation. These steps are repeatedly applied until a termination condition reached. Some number of iteration has been taken as termination condition.

A. *Encoding*

In this work, the chromosomes are encoded as binary membership matrix with size $N \times k$ [4] where the number of objects is N and number of cluster assumed is k that satisfy Eq. (2) and each binary value is individual gene, i.e. value of gene is 1 if the object x_i belongs to that cluster otherwise 0. Let us consider the following example.

Example 1 Suppose $N = 4$ and $k = 2$ then the binary membership matrix and the string representation for a chromosome is Yellow Small Stretch Adult Purple Large Dip Child from lenses real-world data set shows the two clusters (Yellow Small Stretch Child) and (Purple Large Dip Child). Each categorical data in the chromosome is a gene.

B. *Population initialisation*

The search space is the collection of all the $N \times k$ chromosomes that satisfy Eq. (2) [4]. The initial population is chosen randomly. This process is repeated for each of the P chromosomes in the population.

C. *Fitness calculation*

To describe the fitness function, need to follow two phases, initially the clusters are formed according to the centres encoded in the concern chromosome, then the cluster centres encoded in the chromosome are replaced by the modes of the respective clusters. Therefore, assign each point $x_i = 1, 2, 3, \ldots, n_i$ with mode m_j such that

$$||x_i - m_i|| < ||x_i - m_t||, \quad t = 1, 2, 3, \ldots k$$

Further, the cluster centre encoded in the chromosomes is replaced by the mode using frequency method in Eq. (3) only for attribute j

$$\text{fr}\, C_j^* = t_{k,j}|X = \frac{nt_{k,j}}{n}$$

These C_j replaces the previous mode C_i in the string. Let us see the following example.

Example 2 Let the first cluster mode in the Example 1 are (Yellow Small Stretch Adult) and other objects in the particular cluster are (Yellow Small Dip Adult) and (Purple Large Stretch Child). Consequently, the updated cluster mode is (Yellow Small Stretch Adult) with the frequency method.

Later, the cost or within cluster variation [9] is calculated as

$$P(W, M) = \sum_{l=1}^{k} \sum_{i=1}^{n} \sum_{j=1}^{m} w_{i,l} \delta(\boldsymbol{x}_{ij}, \boldsymbol{m}_{lj})$$

The fitness function is defined as $f = 1/P(W, M)$, i.e. less the cost more fit will be the chromosome.

D. *Selection*

Spinning the roulette wheel T times is the fundamental selection method. The proposed algorithm used the roulette wheel selection method based on the above concept. Each time a chromosome is selected for the next iteration, with the number of copies proportional to its fitness. Suppose $P_j(0 \leq j \leq T)$ be the cumulative probabilities such that

$$f(x) = \begin{cases} 0 & \text{for} \quad m = 0 \\ \frac{\sum_{i=1}^{m} F(si)}{\sum_{i=1}^{t} F(si)} & \text{for} \quad m = 1, 2, 3, \ldots, t \end{cases}$$

E. *Crossover process*

Crossover is applied after the selection process in which information is exchanged between two parent chromosomes to generate two offsprings. Similar to genetic k-means algorithm [5], this paper use one-step k-modes algorithm as the crossover operator.

F. *Termination point*

All the operators such as crossover, fitness, selection for our proposed algorithm have been run for maximum number of iterations. The best mode up to the last iteration is the final optimised solution to the clustering problem. Then elitism is applied in each iteration to forward best chromosome to the final iteration for better.

4 Experiments

Experiments show the performance of the proposed approach which implemented in MATLAB R2014a on a personal computer with Intel core 2 duo 2.0 GHz and 1 GB RAM. The real-life data sets such as lenses, balloon and lung cancer data set from UCI repository [12] are used for the purpose of demonstrating the effectiveness of the algorithm where Figs. 3, 4 and 5 are the bar charts showing the authenticity of the present algorithm.

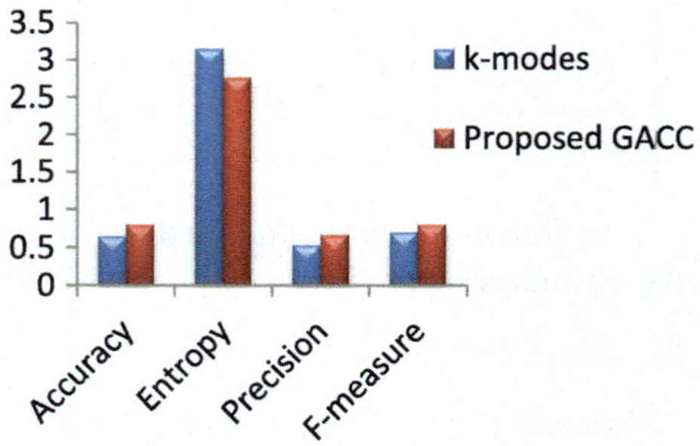

Fig. 3 Validity measures for proposed algorithm on Lung cancer data set

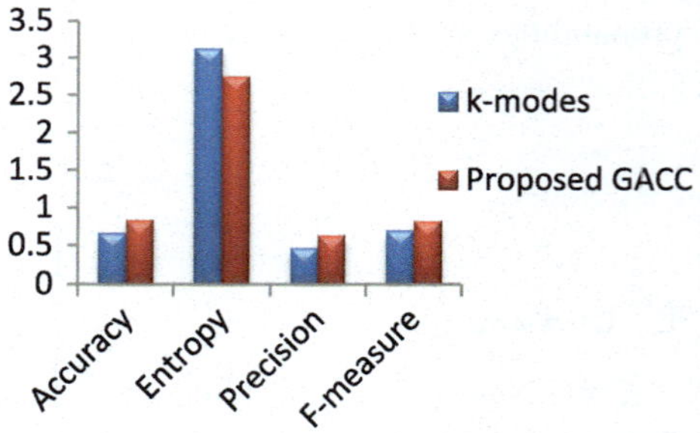

Fig. 4 Validity measures for proposed algorithm on Balloon data set

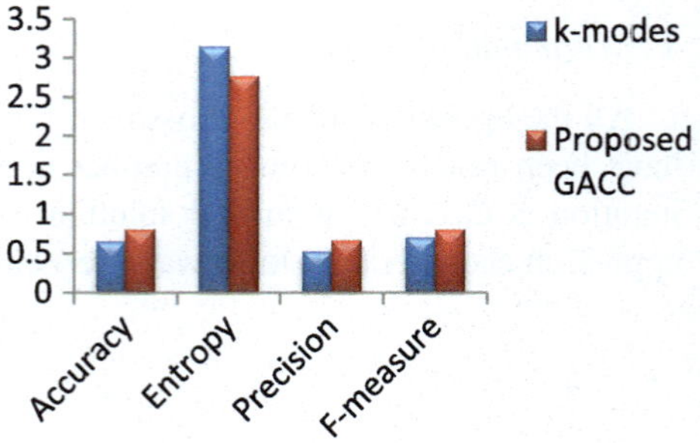

Fig. 5 Validity measures for proposed algorithm on balance data set

As the crossover operator kills all the offspring having high cost, the proposed algorithm was implemented with the following parameters: Population size (N_{pop}) = 10, maximum number of iteration (M_{gen}) assumed 100 in all the experiments. The results obtained is much hopeful to find the appropriate cluster centres having good hike in accuracy with reduced entropy compared to existing well-known *k*-modes algorithm which showed overall better clustering results.

5 Conclusion

Initial cluster centres can poorly effect the clustering results therefore provide local optimal solution. Finding out the appropriate cluster centres in the data set is attracting attention in many research areas. This paper proposed an algorithm for clustering of categorical data which uses the simplicity of *k*-modes and robustness of GA to find globally optimum partitions of the data set. This approach does not serve as the best method in terms of time and space, but experiments proves noticeably results in terms of accuracy and appropriate initialisation of cluster centres. We used the some popular data sets like lenses and balloon data sets from online UCI repository of data sets. Both types of validation measures of clustering has been performed using precision, entropy, purity and cost evaluation, all satisfying the proposed algorithm as well as the accuracy is increased by 23%.

Acknowledgements The work is supported by research grant from MPCST Bhopal, under grant no. 1080/CST/R&D/2012 dated 30-06-2012.

References

1. L. Kaufman, P. Rousseeuw, *Finding Groups in Data: An introduction to cluster analysis* (Wiley, USA, 2005)
2. D. Lam, C. Donald, Wunsch, *Clustering, Signal processing theory and machine learning*, vol. 1 (2014), pp. 1115–1149
3. S. Theodoridis, *"Clustering Basic concepts", Pattern Recognition*, 3rd edn. (2006), pp. 483–516
4. K. Krishna, M. Narasimha Murty, Genetic K-means algorithm. IEEE Trans. Syst. Mans Cybern. Part B: Cybern. **29**(3), 433–439 (1999)
5. U. Maulik, S. Bandyopadhyay, Genetic algorithm-based clustering technique. Pattern Recogn. **33**, 1455–1465 (2000)
6. J. Han, M. Kamber, *Data Mining: Concepts and Techniques* (Morgan Kaufmann Publisher, San Francisco, CA, 2001)
7. J.B. MacQueen, *Some Methods for Classification and Analysis of Multivariate Observations.* Proceedings of 5th Berkeley Symposium on Mathematical Statistics and Probability, vol. 1 (1965), pp. 281–297
8. S.S. Khan, A. Ahmad, *Computing Initial points using Density Based Multiscale Data Condensation for Clustering Categorical data.* 2nd International Conference on Applied Artificial Intelligence, ICAAI (2003)

9. Z. Huang, Extensions to the *k*-means algorithm for clustering large data sets with categorical values. Data Min. Knowl. Disc. **2**, 283–304 (1998)
10. A. Chaturvedi, P. Green, J. Carroll, *k*-modes clustering. J. Classif. **18**, 35–55 (2001)
11. Rajashree Dash, Rasmita Dash, Comparative analysis of K-mean genetic algorithm based data clustering. Int. J. Adv. Comput. Math. Sci. **3**, 257–265 (2012)
12. https://archive.ics.uci.edu/ml/datasets.html

Universally Verifiable Certificateless Signcryption Scheme for MANET

Susmita Mandal, Sujata Mohanty and Banshidhar Majhi

Abstract The mobile ad hoc network (MANET) is a collection of wireless mobile nodes that communicate with one another through a standard transmission medium such as Wi-Fi, cellular, or satellite communication. However, their basic characteristics make them vulnerable against numerous attacks accordingly raising the need of security. In this paper, we propose a certificateless signcryption scheme based on the difficulty of solving the Diffie–Hellman problem. The simulation result proves that the scheme is secure against active and passive attacks using AVISPA (Automated Validation of Internet Security Protocols and Applications) tool.

Keywords Mobile ad hoc network · Certificateless signcryption · AVISPA tool

1 Introduction

Mobile ad hoc networks (MANETs) are infrastructureless, wireless multi-hop networks which provide communication between two or more mobile computers utilizing standard network protocols. These networks are profitable in situations of battlefield communication, counterterrorism, emergency rescue, and conference meetings [1]. MANETs are insecure in nature due to its limited resource, absence of central administration, and the flexibility for nodes to join, leave, and move inside the network where some of the nodes get compromised by adversary. Thus, compared with the wired networks, MANETs are more vulnerable to security attacks [2]. Since they are deployed in an open network, messages sent from a

S. Mandal (✉) · S. Mohanty · B. Majhi
Department of Computer Science and Engineering, National Institute of Technology Rourkela, Rourkela, Odisha, India
e-mail: susmitamandal108@gmail.com

S. Mohanty
e-mail: sujata.nitrkl@gmail.com

B. Majhi
e-mail: bmajhi@nitrkl.ac.in

V. Nath (ed.), *Proceedings of the International Conference on Microelectronics, Computing & Communication Systems*, Lecture Notes in Electrical Engineering 453,
https://doi.org/10.1007/978-981-10-5565-2_7

source to a destination must be confidential by maintaining the integrity of the information using digital signature. The characteristic nature of wireless ad hoc networks makes them exceptionally vulnerable against attacks ranging from passive eavesdropping to active interference. However, modern cryptography can satisfy essential security criteria to ensure data confidentiality, integrity, authentication, and non-repudiation.

In traditional public key cryptosystem (PKC), usually, a trusted third party sometimes called certificate authority (CA) manages the certificates of all nodes, issuing a digitally signed certificate to bind the identity of a node with the public key of CA. Identity-based cryptosystem was first proposed by Shamir [3] which has solved the problem of user identity as public key but depends on a private key generator as a third party. Zhang et al. [4] proposed an ID-based certificateless solution that allows public keys to be derived from their network IDs and some other common information using bilinear pairing. However, they ignore the issue of private key escrow, and combining the secret sharing method increases the network traffic. Later, Zhang et al. [5] proposed a self-organized certificateless public key encryption scheme (CL-PKE) which lacks security on transmitted master secret key shares and therefore is prone to impersonation attack. Several identity-based cryptosystems have been proposed in [6, 7], but they suffer from the key-escrow problem. Therefore, neither public key cryptosystem nor identity-based cryptosystem fits the requirements of mobile ad hoc network being self-organized and autonomous. In 2003, Al-Riyami and Paterson [8] introduced a certificateless public key cryptography (CL-PKC) that meets the necessary requirements lacked in traditional public key and identity-based cryptosystem. However, to enhance the communication, a message is encrypted followed by signature using the encrypt-then-sign approach which increases the complexity of the system. To address this issue, Zhang [9] proposed signcryption which aims at efficiently providing security with encryption and signature. Lv et al. [10] designed a virtual PKG (VPKG)-based escrow-free certificateless public key cryptosystem which uses threshold-based cryptosystem to secretly share the secret keys. The inclusion of PKG violates the property of decentralized administration of MANET. In 2008, Barbosa and Farshim introduced the concept of certificateless signcryption (CL-SC) using bilinear pairings and proved its security with a random oracle model [11]. However, CL-PKC and CL-SC resolve the problem of key escrow using a trusted third party (KGC) to help a user to generate his/her partial private key. The user selects some secret information with the partial private key to generate a complete private key. Therefore, the private key of a user is being secured from being compromised. Though several certificateless signcryption schemes have been proposed in the literature [12–15], most of the schemes use KGC which makes it impractical for MANET as it does not support a centralized administration. To resolve this issue, several researchers introduced the concept of threshold-based (t, n) cryptosystem where initially a master key is generated and distributed among n nodes [16–18]. To form the master key, at least t out of n nodes must provide their secret key to compute the master key. This solves the problem of key escrow and centralized management but increases the network traffic to communicate with at least t nodes. Miao et al. [19] introduced a self-organized certificateless key management scheme which is free from any trusted third party such as certificate authority or key generator center.

Thus, considering all constraint and properties of MANET, we propose a certificateless signcryption scheme by eliminating the need of a key generation center (KGC) to generate user's partial private key. The scheme allows a node to generate its own public and private key using its identity. The security of the proposed scheme is proved against active and passive adversary using AVISPA tool.

The rest of the paper is organized as follows: Sect. 2 discusses basic preliminaries used in the paper. In Sect. 3, the proposed certificateless signcryption scheme for MANET is described. Section 4 presents the correctness proof of proposed scheme. In Sect. 5, the simulation result using HLPSL specification roles is specified using AVISPA tool. Finally, we conclude in Sect. 6.

2 Preliminaries

In this phase, we briefly review basic techniques used throughout the paper.

1. Discrete Logarithm Problem

Let G be a cyclic group of order q with a generator g, the discrete logarithm problem states that for every h belongs to G, there is a unique $x \in Zp$, such that it is infeasible to compute x from $g^x = h$.

2. Diffie–Hellman Problem

Let p be a prime and g be an integer, the Diffie–Hellman problem states that it is infeasible to compute g^{ab} from the known elements of g^a and g^b.

3 Proposed Work

This section demonstrates the certificateless signcryption scheme for MANET where the identity of the ad hoc network node could be any unique information about the node such as IP or MAC address. As all the nodes in MANETs are equal , each node acts as a host and a router. Therefore, every node can generate their public/private key pairs when they join the network. Table 1 lists the main notations used throughout the paper.

Our proposed scheme consists of the following four algorithms: *setup phase, key generation phase, signcryption phase, and unsigncryption phase*. Each phase is run by designated sender and receiver.

Table 1 Notations used in our scheme

Notations	Description
p, q	Two large primes
ID_x	Network identity of node x where $x \in \{S, R\}$
S, R	Sender and receiver
V_x	Secret key of node x where $x \in \{S, R\}$
P_x	Public key of node x where $x \in \{S, R\}$
$H(\cdot)$	One-way hash function
$\|\|$	Concatenation function
$\oplus$	XOR operation

A. *Setup Phase*:

Assume every node in MANET shares these two large prime numbers p and q such that $p = 2q + 1$ and a cyclic generator g.

B. *Key Generation Phase*:

When an ad hoc node joins the group, it first selects a secret key $x_s \in Z_q^*$ and computes the private key V_s as in

$$V_s = ID_S \oplus H(x_s) \tag{1}$$

Then, compute its public key $P_s = (g)^{V_S} \bmod p$ and specifies P_s as the identity. Therefore, the key pair of a sender node is represented as (V_S, P_S).

C. *Signcryption Phase*:

This algorithm takes a plaintext M, publicly known parameters $\{p, q, g\}$, identity of sender node (ID_s) with its public and private keys (V_S, P_S) alongside, public key and identity of receiver node (ID_r, P_r). A time stamp is added to provide a timing on signed document such that it guarantees integrity and authentication of a document at a particular time.

Step 1. Sender node S chooses two random nonces N_a, w and computes the signcrypted message adding a time stamp T_S. Firstly, S computes a function K_1:

$$K_1 = H(g^{N_a}) \tag{2}$$

Step 2. It is assumed that the public key of receiver node is known to the sender, and therefore, a shared session key Y_{sr} is computed using Diffie–Hellman hard assumption.

$$Y_{\mathrm{sr}} = (P_{\mathrm{r}})^{V_s} \bmod p \tag{3}$$

Step 3. Upon creating K_1 and $Y_{\mathrm{sr},}$ a cryptographic function K_2 is computed.

$$K_2 = H\left(g^{N_a}\right) * Y_{\mathrm{sr}} \tag{4}$$

The plaintext M is encrypted using symmetric key algorithm taking K_2 as the key.

$$C = E_{K_2}(M) \tag{5}$$

Step 4. Using the random nonce w and time stamp T_{s}, the node S creates an equation to attach the time stamp with the ciphertext in a way to attain integrity.

$$r = C * g^{(w + T_{\mathrm{s}})} \tag{6}$$

A partial signature is initially generated with time stamp using the function r.

$$S_1 = (r * V_{\mathrm{s}}) \tag{7}$$

Step 5. A function Q and Z is computed to shape the final signature as follows:

$$Q = (w - S_1) \tag{8}$$

$$Z = N_a - (V_a * r) \tag{9}$$

The signcryption σ is send using set of computed parameters to the receiver node.

$$\sigma = \{Z, Q, C, r, T_{\mathrm{s}}\} \tag{10}$$

Step 6. The sender node computes a function f to prove the validity of signcrypted text in case of dispute. Then, it is send to receiver.

$$f = H(C||K_1||K_2||r)$$

D. *Unsigncryption Phase*:

This algorithm takes a ciphertext C, publicly known parameters (p, q, g), sender node's identity ID_s, public key P_s, and receiver node's public and private key pair (V_r, P_r) as input to retrieve the plaintext M. If the ciphertext is valid, it is accepted else an error message is send to the sender node S by suspending the session.

Step 1. The receiver node R verifies the integrity of received message by validating the attached time stamp.
Initially, the receiver using the received signcrypted parameters computes r'.

$$r' = C * (P_s)^r * (g)^{Q+T_s} \tag{11}$$

Step 2. Upon computing the value of r', the receiver node compares the value of received r with computed r'. If it matches, the ciphertext is accepted else rejected and the session is closed. To obtain the plaintext M, the following functions are computed.

$$K_1' = H\left((g)^Z * (P_s)^{r'}\right) \tag{12}$$

Step 3. The shared session key Y_{rs} is computed using sender's public key P_s.

$$Y_{rs} = (P_s)^{V_r} \bmod p \tag{13}$$

Step 4. Using the session key, the symmetric key K_2' is computed.

$$K_2' = H\left((g)^Z * (P_s)^{r'}\right) * Y_{rs} \tag{14}$$

Step 5. Finally, the symmetric key decryption algorithm is applied on the ciphertext.

$$M = D_{K_2}(C) \tag{15}$$

The proposed scheme can be universally verified in case the sender node denies sending the signcrypted text using a trusted verifier. This method can be carried out between the receiver node R and the verifier where the receiver node needs to provide the signcryption value $\sigma = \{Z, K_2', C, r', P_s\}$ and the received verifiable function f to the third-party verifier.

$$f' = H\left(C \middle\| H\left((g)^Z * (P_s)^{r'}\right) \middle\| K_2' \| r'\right)$$

The computed value f' is then compared with f as $f' = f$; if both are equal, the message is accepted else an error message is sent to receiver.

4 Correctness Analysis

Firstly, the receiver node computes r' and compares with received r as $r' = r$ for integrity; if both are equal, the message is accepted else an error is sent.

$$r' = C * (P_s)^r * (g)^{Q+T_s}$$
$$r' = C * g^{(V_s*r)} * g^{(w-S_1+T_s)}$$
$$r' = C * g^{(V_s*r)} * g^{(w-V_s*r+T_s)}$$
$$r' = C * g^{(V_s*r+w-V_s*r+T_s)}$$
$$r' = C * g^{(w+T_s)}$$

Secondly, $K_1{}'$ is computed to verify the symmetric key used for encryption.

$$K_1' = H\left((g)^Z * (P_s)^{r'}\right)$$
$$K_1' = H\left(g^{(N_a-V_a*r)} * (g)^{V_a*r'}\right)$$
$$K_1' = H\left(g^{(N_a-V_a*r+V_a*r')}\right)$$
$$K_1' = H\left(g^{(N_a)}\right)$$

Thirdly, K_2' is computed to decrypt the ciphertext.

$$K_2' = H\left((g)^Z*(P_s)^{r'}\right) * Y_{rs}$$
$$K_2' = H\left(g^{N_a}\right) * Y_{rs}$$

Therefore, the above three proofs state that integrity and confidentiality of the signcrypted text are achieved.

5 Simulation Using AVISPA Tool

The formal verification of the proposed certificateless signcryption scheme for MANET is depicted using AVISPA software in this section. We implemented our scheme in Figs. 1 and 2 using AVISPA tool [20, 21] with a role-based language called HLPSL (High-Level Protocol Specification Language).

AVISPA is a push-button tool for the automated validation of the Internet security-sensitive protocols and applications. It uses a special language called High-Level Protocol Specification Language and integrates the different back ends

```
role sender(A,B:agent,
            Yba,Yab:symmetric_key,
            Add,Sub,Mul,H:hash_func,
            Pa,Pb:public_key,
            Snd,Rcv:channel(dy))

played_by A
def=
local State:nat,
    G,IDa,IDb,Xa,Va,Xb,Vb,R,K1,K2,Ts,C,W,
M,Q,S1,E,Z,E1,R1,Na:text

const sender_va, receiver_vb,  subs1, subs2:protocol_id

init State := 0
transition
1.State=0/\Rcv (start)=|>
  State':=1 /\Xa':=new()
        /\Va':=xor(IDa,H(Xa))
        /\Pa':=exp(G,Va)
        /\Na':=new()
        /\K1':=H(exp(G,Na))
        /\Yab':=exp(Pb,Va)
        /\K2':=Mul(K1,Yab)
        /\C':=({M}_K2)
        /\S1':=Mul(R,Xa)
        /\W':=new()
        /\Ts':=new()
        /\E':=Add(W,Ts)
        /\R':=Mul(exp(G,E),C)
        /\Q':=Sub(W,S1)
        /\Z':=Sub(Na,Mul(Xa,R))
        /\Snd({Z.Q.C.R.Ts}_Yab)
        /\secret({Va},subs1,{A,B})
end role
```

Fig. 1 Role specification of sender node

```
role receiver(A,B:agent,
              Yba,Yab:symmetric_key,
              Add,Sub,Mul,H:hash_func,
              Pa,Pb:public_key,
              Snd,Rcv:channel(dy))

played_by B
def=
local State:nat,
    G,IDa,IDb,Xa,Va,Xb,Vb,R,K1,K2,Ts,C,W,
        M,Q,S1,E,Z,E1,R1,Na:text

    const sender_va, receiver_vb, subs1,subs2:protocol_id
init State := 0
transition

1.State=0/\Rcv({Z.Q.C.R.Ts}_Yab)=|>
State':=1/\Xb':=new()
          /\Vb':=xor(IDb,H(Xb))
          /\Pb':=exp(G,Vb)
          /\Yba':=exp(Pa,Vb)
          /\E1':=Add(Q,Ts)
           /\R1':=Mul(C,Mul(exp(Pa,R1')),exp(G,E1'))
          /\K1':=H(Mul(exp(G,Z)),exp(Pa,R1'))
          /\K2':=Mul(K1',Yba)
           /\ secret({Vb},subs2,{B,A})
end role
```

Fig. 2 Role specification of receiver node

that implement a variety of state-of-the-art automatic analysis techniques. This language is based on roles where the AVISPA tool mechanically translates the HLPSL into a lower-level specification using HLPSL2IF translator. Afterward, it generates an intermediate format (IF). The present version integrates four back ends, namely On-the-y Model-Checker (OFMC), CL-based Attack Searcher (CL-AtSe), SAT-based Model-Checker (SATMC), and Tree-Automata-based Protocol Analyzer (TA4SP). The architecture of AVISPA is demonstrated in Fig. 3.

Analysis of Simulation Result:

The simulation results for the formal security verification analysis of our scheme using OFMC and CL-AtSe is depicted in Figs. 4 and 5. The summary of the results under OFMC and CL-AtSe proves that the protocol is safe to withstand popular active attacks, such as the masquerading, replay, man-in-the-middle attacks, and passive attacks.

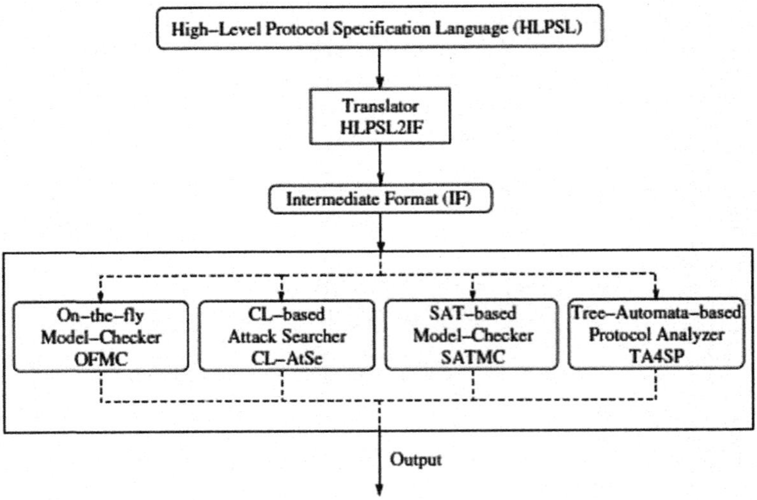

Fig. 3 Architecture of AVISPA

```
% OFMC
% Version of 2006/02/13
SUMMARY
 SAFE
DETAILS
 BOUNDED_NUMBER_OF_SESSIONS
PROTOCOL
 C:\SPAN\testsuite\results\modifiedManet.if
GOAL
 as_specified
BACKEND
 OFMC
COMMENTS
STATISTICS
 parseTime: 0.00s
 searchTime: 0.03s
 visitedNodes: 14 nodes
 depth: 4 plies
```

Fig. 4 OFMC simulation result

```
%CL-AtSe
SUMMARY
 SAFE

DETAILS
 BOUNDED_NUMBER_OF_SESSIONS
 TYPED_MODEL

PROTOCOL
 C:\SPAN\testsuite\results\modifiedManet.if

GOAL

 As Specified

BACKEND
 CL-AtSe

STATISTICS

 Analysed   : 0 states
 Reachable  : 0 states
 Translation: 0.00 seconds


 Computation: 0.00 seconds
```

Fig. 5 CL-AtSe simulation result

Performance Analysis:

The performance of our proposed scheme is represented in a tabular form, where the value of hash is negligible.

T_{exp}: exponentiation computation,
T_{mul}: multiplication function,
T_h: hash function (Table 2).

Table 2 Performance analysis of our scheme

Categories	Proposed scheme
Cost of signcryption	$3T_{exp} + 3T_{mul} + 2T_h$
Cost of unsigncryption	$3T_{exp} + 2T_{mul} + 2T_h$
Cost of key generation	$1T_{exp} + 1T_h$
Number of public parameters	3

Table 3 Comparison with existing schemes

Properties	[4]	[5]	[10]	[19]	Our scheme
Network model	ID-PKG	ID-PKG	Virtual PKG	CL-PKC	CL-SC
Confidentiality	Yes	No	Yes	Yes	Yes
Authentication	Yes	Yes	Yes	Yes	Yes
Impersonation attack	No	Yes	No	No	No
Non-repudiation	No	No	No	Yes	Yes
Centralized administration	Yes	Yes	Yes	No	No
Key escrow	Yes	No	No	No	No
Network traffic	High	High	Low	Low	Low

Comparison With Existing Schemes:

The proposed scheme is compared with some existing schemes with respect to following features such as network model, confidentiality, authentication, non-repudiation, impersonation attack, centralized administration, key escrow, and network traffic. As discussed in the literature, MANET has limited resources and is infrastructureless, and therefore, it cannot be administered by a centralized authority failing which violates the characteristics of MANET (Table 3).

6 Conclusion

The traditional security mechanisms such as public key cryptosystem, identity-based cryptosystem, and certificateless public key cryptosystem could not satisfy the attributes of self-organizing, autonomous wireless nature of communication due to lack of security infrastructure. We propose a certificateless signcryption scheme for MANET which removes the central dependency of partial key generation on KGC by self-generating public/private key pair for each node. To demonstrate the availability, a signcryption with public verifiability scheme is presented to control the malicious behavior of sender node. The simulation result is generated using widely accepted AVISPA tool for formal security specification. The proposed scheme is proved to be secured against active and passive attacks by an adversary in an open channel.

References

1. I. Chlamtac, M. Conti, J.J.N. Liu, Mobile ad hoc networking: imperatives and challenges. Ad Hoc Netw. **1**(1), 13–64 (2003)
2. P.M. Jawandhiya et al , A survey of mobile ad hoc network attacks. Int. J. Eng. Sci. Technol. **2.9**, 4063–4071 (2010)

3. A. Shamir, Identity-based cryptosystems and signature schemes, in *Advances in cryptology* (Springer, Berlin Heidelberg, 1985)
4. Y. Zhang et al., Securing mobile ad hoc networks with certificateless public keys. Dependable Secure Comput. IEEE Trans. **3**(**4**), 386–399 (2006)
5. Z. Zhang, W. Susilo, R. Raad, Mobile ad-hoc network key management with certificateless cryptography, in *2nd International Conference on Signal Processing and Communication Systems, 2008. ICSPCS 2008* (IEEE, 2008)
6. X. Boyen, Multipurpose identity-based signcryption, in *Advances in Cryptology-CRYPTO 2003* (Springer, Berlin Heidelberg, 2003), pp. 383–399
7. L. Chen, J. Malone-Lee, Improved identity-based signcryption, in *Public Key Cryptography-PKC 2005* (Springer, Berlin Heidelberg, 2005), pp. 362–379
8. S.S. Al-Riyami, K.G. Paterson, Certificateless public key cryptography, in *Advances in Cryptology-ASIACRYPT 2003* (Springer, Berlin Heidelberg, 2003), pp. 452–473
9. Y. Zheng, Digital signcryption or how to achieve cost (signature & encryption) $\ll$ cost (signature) + cost (encryption), in *Advances in Cryptology—CRYPTO'97* (Springer, Berlin Heidelberg, 1997), pp. 165–179
10. X. Lv, H. Li, B. Wang, Virtual private key generator based escrow-free certificateless public key cryptosystem for mobile ad hoc networks. Secur. Commun. Netw. **6**(1), 49–57 (2013)
11. M. Barbosa, P. Farshim, Certificateless signcryption, in *Proceedings of the 2008 ACM symposium on Information, computer and communications security* (ACM, 2008)
12. W.-H. Liu, C.-X. Xu, Certificateless signcryption scheme without bilinear pairing. Ruanjian Xuebao J. Softw. **22**(8), 1918–1926 (2011)
13. W. Xie, Z. Zhang, Efficient and provably secure certificateless signcryption from bilinear maps, in *IEEE International Conference on Wireless Communications, Networking and Information Security (WCNIS), 2010* (IEEE, 2010)
14. D-B. He, Security analysis of a certificateless signcryption scheme. Ruanjian Xuebao J. Softw. **24**(3), 618–622 (2013)
15. W. Shi et al., Cryptanalysis and improvement of a certificateless signcryption scheme without bilinear pairing. Front. Comput. Sci. **8**(4), 656–666 (2014)
16. B. Wang, J. Li, (*t*, *n*) Threshold signature scheme without a trusted party. Chin. J. Comput. **26** (11), 1581–1584 (2003). (Chinese Edition)
17. L.-C. Li, R.-S. Liu, Securing cluster-based ad hoc networks with distributed authorities. Wirel. Commun. IEEE Trans. **9**(10), 3072–3081 (2010)
18. H. Lee et al., ID-based key management scheme using threshold decryption for OPMD environment, in *IEEE International Conference on Consumer Electronics (ICCE), 2012* (IEEE, 2012)
19. F. Miao et al., Fully self-Organized Key management scheme in MANET and its applications, in *Computer Networks & Communications (NetCom)* (Springer, New York, 2013), pp. 381–391
20. AVISPA, Automated validation of internet security protocols and applications. http://www.avispa-project.org/. Accessed on Sept 2015
21. L. Vigan, Automated security protocol analysis with the AVISPA tool. Electron. Notes Theor. Comput. Sci. **155**, 61–86 (2006)

Impact of Sidewall Spacer Layers on the Analog/RF Performance of Nanoscale Double-Gate Junctionless Transistors

Debapriya Roy and Abhijit Biswas

Abstract Using extensive numerical device simulation, we investigate the influence of sidewall spacers on the analog/RF performance of double-gate junctionless transistors at channel length of 30 nm. Our findings reveal that peak transconductance and peak intrinsic gain increase by 5.2 and 71.3% for spacer dielectric constant $k = 30$ as compared to the respective values for $k = 3.9$, while peak unity gain cut-off frequency increases by 37% for $k = 3.9$ compared with the value for $k = 30$. The transconductance generation factor is found to be less sensitive to the variation in k. With increasing k the output conductance becomes less for low gate overdrive voltage V_{GT} while it shows a reverse trend for higher V_{GT}. It is evident from our studies that peak transconductance, peak transconductance generation factor, peak gain, and peak cut-off frequency increase by 13, 10, 27, and 20%, respectively, for spacer length of 5 nm compared with the corresponding values for spacer length of 15 nm. However, with a larger spacer length, the output conductance exhibits reduced value for lower V_{GT}, while it becomes comparable with the values for smaller spacer lengths as V_{GT} increases.

Keywords Analog/RF performance · Double-gate MOSFET · Gain
Junctionless transistor · Spacer layer · Unity gain cut-off frequency

1 Introduction

The remarkable performance improvement of integrated circuits (ICs) has been achieved by miniaturization of MOSFETs. As the device feature size falls in the sub-100 nm regime, short-channel effects (SCEs) come into play, which degrade device performance. Furthermore, to maintain super-steep doping profiles at the source-channel and drain-channel junctions of an extremely scaled transistor

D. Roy · A. Biswas (✉)
Institute of Radio Physics and Electronics, University of Calcutta,
92, Acharya Prafulla Chandra Road, Kolkata 700009, India
e-mail: abiswas5@rediffmail.com

V. Nath (ed.), *Proceedings of the International Conference on Microelectronics, Computing & Communication Systems*, Lecture Notes in Electrical Engineering 453,
https://doi.org/10.1007/978-981-10-5565-2_8

becomes really challenging. Recently, junctionless transistors (JLTs) [1–10] have been demonstrated in which no source-channel and drain-channel junctions exist. In contrast to inversion mode devices, such devices do not require any heavy source/drain implants. Additionally, due to the absence of source/drain charge sharing, JLTs turn out to be more immune to SCEs. While earlier findings demonstrate that the use of sidewall spacers on both sides of the gate of MOSFETs reduces SCEs through gate-bias-dependent effective channel length [11], a longer spacer length lowers the ON-current (I_{ON}) owing to longer effective channel length L_{eff} [11, 12]. Moreover, the use of a high-k spacer is beneficial for reduction of OFF-state current (I_{OFF}) because of the lateral extension of depletion width contributing to L_{eff} [13]. Clearly, it is expected that sidewall spacers influence the JL device parameters related to analog/RF circuit performance. Investigations in particular the design issues of spacer layers focusing on the improvement of device parameters associated with analog/RF performance have not yet been addressed.

In the present paper, we study the role of spacer layers for the improvement of analog/RF performance of double gate (DG) JLTs at channel length of 30 nm. In our investigation, we consider device parameters such as transconductance g_m, transconductance generation factor (TGF) g_m/I_D, output conductance g_{ds}, intrinsic gain A_v, and unity gain cut-off frequency f_T. Such parameters are obtained for various spacer dielectric constants ranging 3.9–30 and spacer lengths in the range 3–15 nm for various bias conditions. Furthermore, spacer dielectric constant and length have been identified for obtaining improved device parameters for analog/RF circuit applications.

2 Device Structure and Simulation Setup

In the present study, we consider an underlap double-gate junctionless transistor (DGJLT), as shown in Fig. 1. Various device design parameters are entered in Table 1. We employ numerical device simulator SILVACOATLAS [14] to simulate such devices. To capture the effects of different scattering mechanisms such as acoustic phonons, ionized impurities, surface scattering, vertical electric field on carrier mobility in the JLTs, the Lombardi CVT model is used. In order to include generation-recombination events in the device, Shockley-Read-Hall (SRH) recombination model is included. To capture bandgap narrowing due to heavy channel doping, bandgap narrowing model and to describe carrier distribution Fermi Dirac statistics are also activated. Furthermore, we energize the Selberherr impact ionization model [15] to take into account the impact ionization of carriers at the drain end of the channel particularly at high drain bias conditions. Drift-diffusion transport model is invoked with adjusted carrier velocity saturation as recommended in [16]. The source/drain resistance (S/D) which is crucial for extremely scaled multi-gate architecture is considered to be 180 Ω μm following [17]. We have not considered quantum-mechanical effects in our studies as such effects are not significant for our devices having Si body thickness of 5 nm [18].

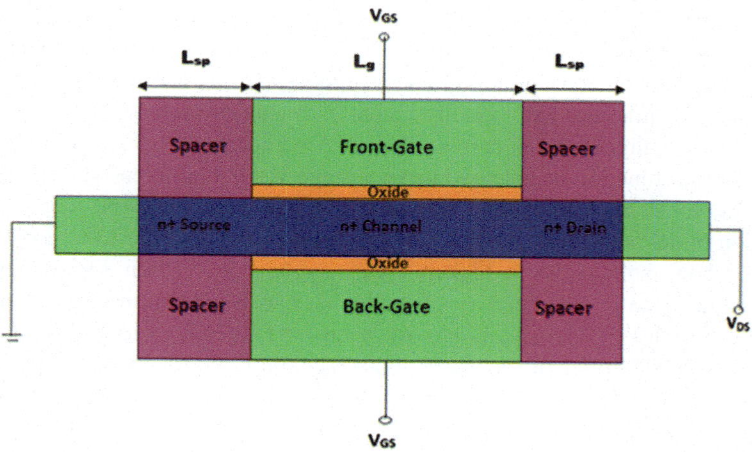

Fig. 1 Schematic diagram of a junctionless double-gate MOSFET with underlap spacer layers. Various insulators with different dielectric constants in the range 3.9–30 are used as spacer layers

Table 1 Various device design parameters of JLTs used in simulation

Sl. No.	Device design parameters	Value
1.	Gate length	30 nm
2.	Silicon channel thickness	5 nm
3.	Width	1 μm
4.	Donor-type channel doping concentration	10^{19} cm^{-3}
5.	Gate insulator SiO_2 thickness (EOT)	1 nm
6.	Source/drain resistance	180 Ω μm
7.	Spacer length	3, 5, 10, and 15 nm
8.	Spacer dielectric constant	3.9, 9.3, 22, and 30

3 Results and Discussion

3.1 Impact of Dielectric Constant of Spacer, k

Figure 2 shows the TCAD simulated transfer characteristics obtained for JLTs in both linear and logarithmic scales for two different sidewall spacer layers—SiO_2 with dielectric constant k of 3.9 and HfO_2 with $k = 22$ at drain to source voltage $V_{DS} = 0.5$ V. It is evident from Fig. 2 that the transfer characteristic curve for a spacer dielectric layer with a larger dielectric constant is steeper as compared to that for a spacer layer with a lower k. At lower V_{GS} for a spacer layer with a larger k, the vertical electric field (E_y) remains almost constant whereas the lateral electric field (E_x) reduces due to increase in L_{eff} thereby increasing the ratio of E_y/E_x. As a result, SCEs reduce and hence I_D dominated by diffusion current at low gate biases ($I_D = I_{OFF}$, at $V_{GS} = 0$ V and $V_{DS} = 0.5$ V) lowers. However, at higher V_{GS} the drain current is dominated by the drift component which in turn depends on the carrier concentration and carrier drift velocity. For a higher spacer dielectric constant k, the electron concentration remains almost same as that obtained for a lower k (E_y same) while the drift velocity increases owing to increase in E_x eventually resulting in a higher drain current. Figure 3 compares the transconductance of DGJLTs as a function of gate overdrive voltage V_{GT} for different values of spacer dielectric constant in the range 3.9–30 at channel length L_g = 30 nm. Since the variation of drain current with V_{GS} is steeper particularly for gate voltage exceeding the threshold voltage for a higher k as may be observed in Fig. 2, the peak transconductance g_m also increases for a higher k as is evident in Fig. 3. However, for lower values of V_{GT}, g_m is almost insensitive to change in k. The transconductance generation factor (g_m/I_D) is almost same for all spacers with a peak value of about 35 V^{-1} near the threshold region as shown in Fig. 4. Such variation is expected as a higher spacer k brings in both larger I_D and g_m, and reverse is true for a lower k. The dependence of output conductance g_{ds} with V_{GT} is demonstrated in Fig. 5 for four different values of spacer dielectric constant ranging 3.9–30. The dependence of g_{ds} with k can be explained by examining the variation of E_y/E_x with k. At a lower V_{GT}, the ratio of E_y/E_x increases thus lowering SCEs with increasing k as explained earlier. This leads to a lower drain-induced barrier lowering (DIBL) at a higher V_{DS} (= 0.5 V) resulting in a lower g_{ds} for a higher k. However, for moderate and higher values of V_{GT} the ratio shows an opposite behavior, thus leading to an increased g_{ds} for a higher k. Figure 6 compares the intrinsic gain ($A_v = g_m/g_{ds}$) as a function of V_{GT} for different values of spacer dielectric constant. The gain exhibits a peak value around V_{GT} = 0 V for all values of k with a maximum value of 60 at k = 30. At a lower V_{GT}, the gain increases due largely to reduction in g_{ds} with increase in k with g_m remaining almost constant for all k. On the contrary as V_{GT} increases beyond 0.2 V, both g_m and g_{ds} increase with increasing k resulting in a lower gain for a higher k. Such a variation occurs since the increase in g_{ds} plays a dominant role as compared to the increase in g_m with a higher k. The variation of cut-off frequency $f_T \left(= \frac{g_m}{2\Pi(C_{gs}+C_{gd})}\right)$ with V_{GT} is shown in Fig. 7. Our simulation results show that as k increases from 3.9 to 30, peak value of f_T reduces from 238.4 to 154.4 GHz (Fig. 7). Such a reduction in f_T with increasing k is explained in the following. In a JLT, there is no junction capacitances and the parasitic capacitance mainly comprises the inner fringing capacitance (C_{if}) and the outer fringing capacitance (C_{of}). With increase in k, the outer fringing capacitance of the gate sidewall increases thus leading to higher values of C_{gs} and C_{gd}. Hence, at a higher spacer k despite a larger value of g_m the peak f_T reduces due to larger values of C_{gs} and C_{gd}. Our findings show that peak g_m, peak TGF, and peak A_v increase marginally as k increases from 22 to 30, while f_T changes to some extent for the same increase in k.

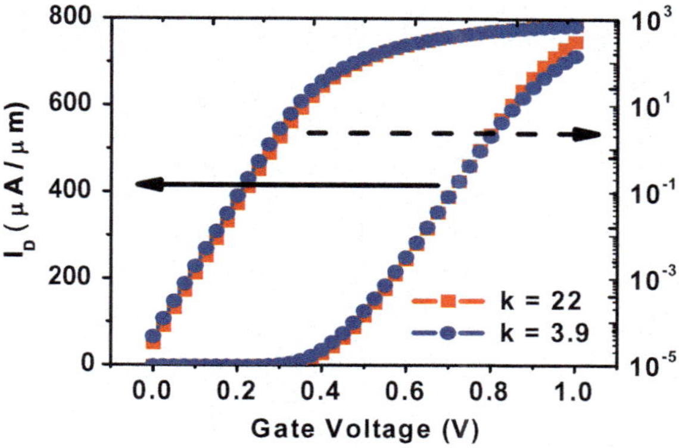

Fig. 2 Linear and logarithmic transfer characteristics of a DG junctionless MOSFET with channel length of 30 nm, channel thickness of 5 nm, equivalent oxide thickness (EOT) of 1 nm, and channel doping concentration of 10^{19} cm^{-3} for $V_{DS} = 0.5$ V. SiO_2 and HfO_2 with dielectric constants 3.9 and 22, respectively, have been used as spacer dielectric layers having spacer length of 5 nm

Fig. 3 Comparison of transconductance as a function of gate overdrive voltage, V_{GT}, for junctionless DG MOSFETs for different values of spacer dielectric constant, k

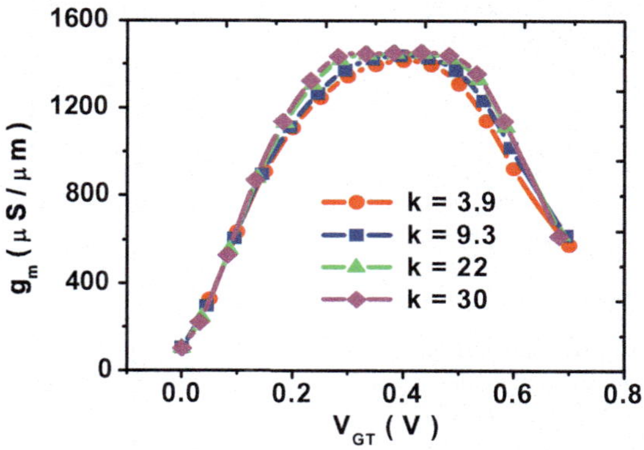

Fig. 4 Variation of transconductance generation factor with gate overdrive voltage, V_{GT}, for junctionless DG MOSFETs for different values of spacer dielectric constant, k

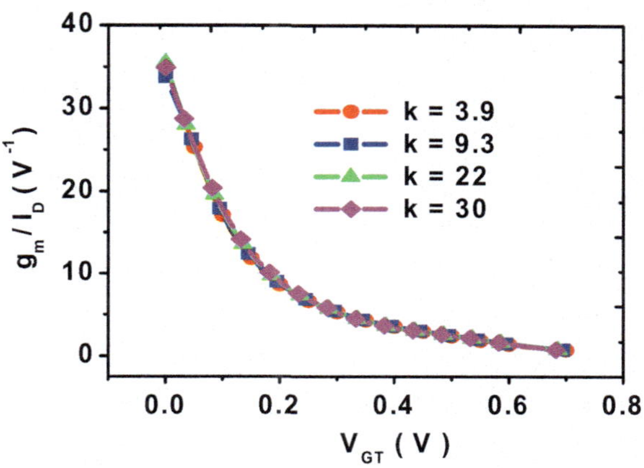

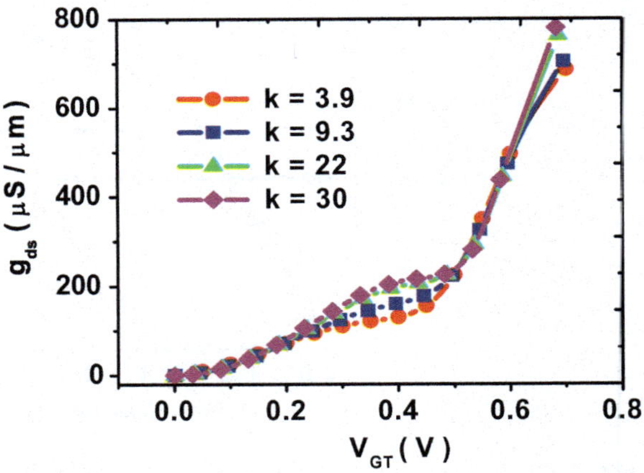

Fig. 5 Plot of output conductance as a function of gate overdrive voltage, V_{GT}, for junctionless DG MOSFETs for different values of spacer dielectric constant, *k*

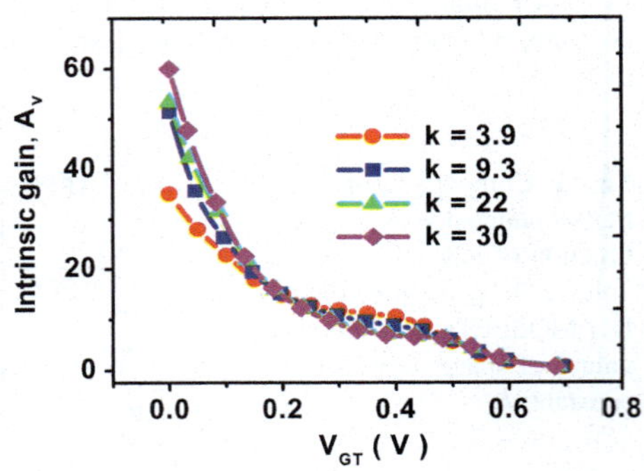

Fig. 6 Variation of Intrinsic gain with gate overdrive voltage, V_{GT}, for junctionless DG MOSFETs for different values of spacer dielectric constant, *k*

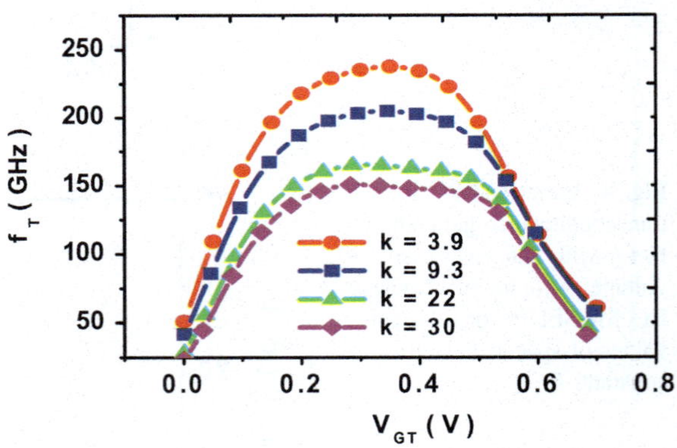

Fig. 7 Comparison of unity gain cut-off frequency with gate overdrive voltage, V_{GT}, for junctionless DG MOSFETs for different values of spacer dielectric constant, *k*

3.2 Impact of Underlap Length, L_{sp}

The variation of transconductance as a function of gate overdrive voltage V_{GT} is demonstrated in Fig. 8 for four different values of spacer dielectric length L_{sp} in the range 3–15 nm. The transconductance exhibits enhanced values over a considerable span of V_{GT} for the lowest value of L_{sp} viz., 3 nm. The thinnest L_{sp} results in lowest L_{eff} which in turn produces the highest I_D. The transconductance attains the highest value for the thinnest L_{sp} as g_m is inversely proportional to effective channel length L_{eff} being linearly dependent on L_{sp}. The variation of transconductance generation factor (TGF) as a function of V_{GT} is demonstrated in Fig. 9 for different values of L_{sp} ranging 3–15 nm. The TGF versus V_{GT} plots for different values of L_{sp} almost overlap each other for larger values of V_{GT} (> 0.2 V) with the highest value near the threshold condition as expected. The highest value of TGF is observed for L_{sp} = 3 nm which is very close the value obtained at L_{sp} = 5 nm. For instance, the peak value of g_m/I_D is 37 and 32 V^{-1} for L_{sp} = 3 nm and 15 nm, respectively. It is observed that at lower V_{GT} though I_D shows almost identical values, g_m increases with reduction in spacer length, which ultimately leads to a higher g_m/I_D ratio for thinnest spacer length i.e., at L_{sp} = 3 nm. Figure 10 compares the variation of g_{ds} with V_{GT} for various values of L_{sp}. At all values of V_{GT}, the ratio of E_y/E_x increases with larger L_{sp} thus lowering SCEs such as DIBL at higher V_{DS} (= 0.5 V). As a result g_{ds} exhibits lower value for larger L_{sp} as may be observed in Fig. 10. However, some crossovers of g_{ds} versus V_{GT} curves for different values of L_{sp} at V_{GT} ~ 0.5 V are noticed. This feature may be attributed to the impact ionizations at the drain end. The dependence of intrinsic gain of the transistor as a function of V_{GT} is compared in Fig. 11 for different values of L_{sp}. The intrinsic gain of the device depends on both g_m and g_{ds}. As evident from Fig. 11, the peak gain attains its highest value closer to the threshold condition at spacer length of 5 nm. With a larger L_{sp} though g_{ds} reduces, g_m also reduces thus reducing the ratio g_m/g_{ds} for higher L_{sp} at lower V_{GT} except at L_{sp} = 3 nm for which the increase in g_{ds} becomes the dominating factor. However, at V_{GT} in the range 0.2–0.45 V, the rate of reduction of g_{ds} is more as compared to the rate of reduction of g_m for a higher L_{sp} thereby leading to a higher gain for larger L_{sp}. For V_{GT} exceeding 0.45 V, the gain increases for a lower L_{sp} due primarily to the increase in g_m despite partial compensation of larger g_{ds}. The variation of unity gain cut-off frequency with V_{GT} is shown in Fig. 12 for different values of L_{sp}. The peak value of f_T increases from 138.7 to 179.6 GHz as the spacer length is reduced from 15 to 3 nm. It is worth noting that both C_{gs} and C_{gd} increase marginally with increasing L_{sp} owing largely to the increase in outer fringing capacitance despite partial compensation for the reduced inner fringing capacitance whereas g_m reduces with increasing L_{sp}. The interplay of these two effects eventually results in a higher value of cut-off frequency for a lower spacer length. The peak g_m, peak A_v, peak TGF, and peak f_T are entered in Table 2 for two different values of spacer dielectric constant k with spacer length L_{sp} = 5 nm. Furthermore, we enter the peak value of all the four analog/RF performance metrics reported elsewhere [3]. Clearly, it is found from

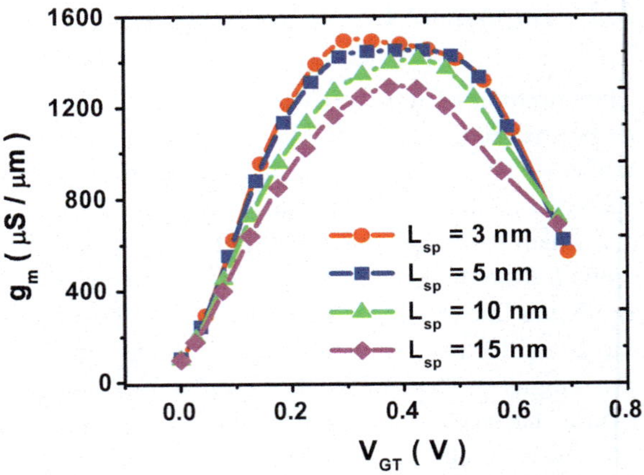

Fig. 8 Comparison of transconductance as a function of gate overdrive voltage, V_{GT}, for junctionless DG MOSFETs for different values of spacer length, L_{sp}

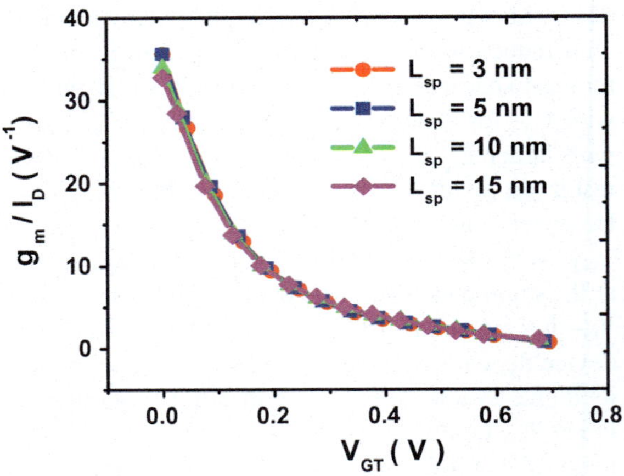

Fig. 9 Variation of transconductance generation factor with gate overdrive voltage, V_{GT}, for junctionless DG MOSFETs for different values of spacer length, L_{sp}

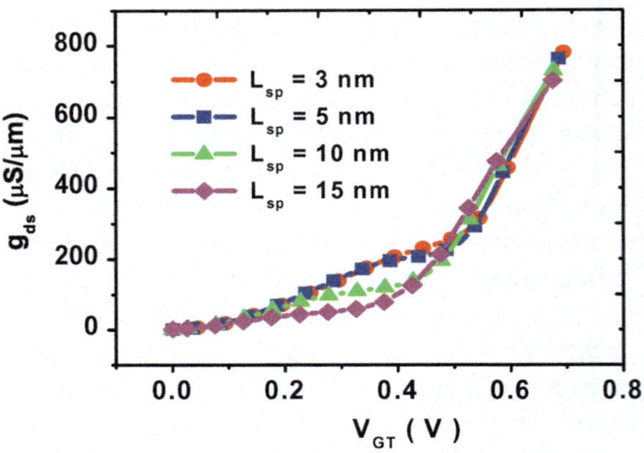

Fig. 10 Plot of output conductance as a function of gate overdrive voltage, V_{GT}, for junctionless DG MOSFETs for different values of spacer length, L_{sp}

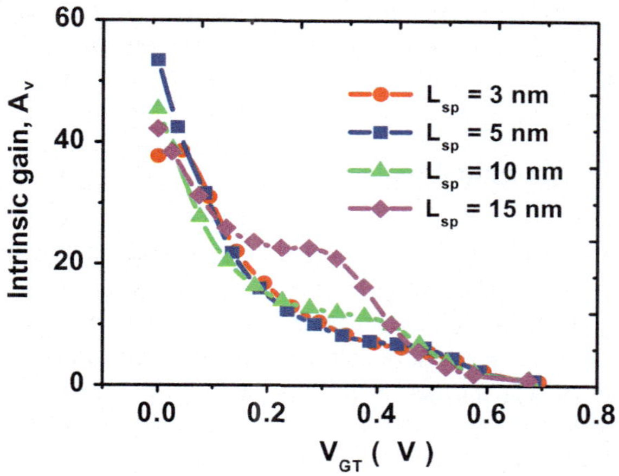

Fig. 11 Variation of intrinsic gain as a function of gate overdrive voltage, V_{GT}, for junctionless DG MOSFETs for different values of spacer length, L_{sp}

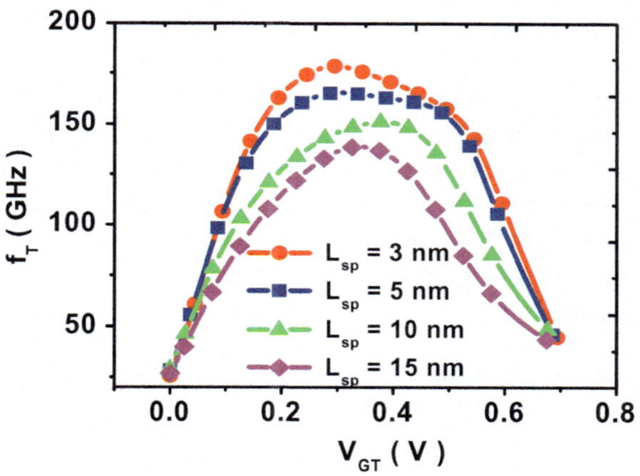

Fig. 12 Comparison of unity gain cut-off frequency with gate overdrive voltage, V_{GT}, for junctionless DG MOSFETs for different values of spacer length, L_{sp}

Table 2 Comparison of analog/RF performance metrics of JLTs with L_{sp} = 5 nm

Peak value of parameters	Spacer dielectric constant, k = 3.9	Spacer dielectric constant, k = 22	Reported data [3]
g_m (μS/μm)	1411.7	1448	350
TGF (V^{-1})	34	36	30
A_v	35.4	53.5	18
f_T (GHz)	238.4	166.9	145

Table 2 that our proposed JLT device with spacer length of 5 nm outperforms the reported JLT [3].

Our proposed JLT device with spacer dielectric constant of 22 and spacer length of 5 nm yields 313, 20, and 197.2% improvement of peak g_m, peak TGF, and peak A_v, respectively, as compared to the reported device [3]. In addition our JLT with spacer dielectric constant of 3.9 and spacer length of 5 nm produces 64.4% enhancement in peak f_T relative to the reported peak f_T in [3].

4 Conclusion

In this paper, we have presented systematically the impact of sidewall spacer layers on the analog/RF performance of nanoscale double-gate junctionless transistors. The analog/RF circuit performance has been evaluated in terms of device parameters such as transconductance, transconductance generation factor, output conductance, intrinsic gain, and unity gain cut-off frequency for different values of spacer dielectric constant ranging 3.9–30 and spacer length in the range 3–15 nm. Our analysis reveals that transconductance and intrinsic gain improve for spacer layer with a larger dielectric constant whereas the cut-off frequency degrades with increasing spacer dielectric constant. For a higher value of spacer dielectric constant, the output conductance exhibits lower and higher values for V_{GT} below and exceeding 0.6 V, respectively. However, transconductance generation factor exhibits less sensitivity to the variation in spacer dielectric constant. Furthermore, it is evident from our findings that all the parameters such as peak transconductance, peak transconductance generation factor, peak intrinsic gain, and peak cut-off frequency improve for spacer thickness of about ~5 nm. On the contrary, the output conductance is found to assume lower values for increasing spacer length. Hence, by choosing spacer dielectric constant and spacer length properly, analog/RF performance of the device can be improved significantly.

Acknowledgements The first author acknowledges TEQIP Phase II, University of Calcutta for providing financial support.

References

1. C.W. Lee, A. Afzalian, N.D. Akhavan, R. Yan, I. Ferain, J.-P. Colinge, Junctionless multigate field-effect transistor. Appl. Phys. Lett. **94**, 0535111–0535112 (2009)
2. J.-P. Colinge, C.-W. Lee, A. Afzalian, N.D. Akhavan, R. Yan, I. Ferain, P. Razavi, B. O'Neill, A. Blake, M. White, A.-M. Kelleher, B. McCarthy, R. Murphy, Nanowire transistors without junctions. Nat. Nanotechnol. **5**(3), 225–229 (2010)
3. D. Ghosh, M.S. Parihar, G.A. Armstrong, A. Kranti, High performance junctionless MOSFETs for ultra low power analog/RF applications. IEEE Electron Dev. Lett. **33**(10), 1477–1479 (2012)
4. J. Hur, D.-I.I. Moon, J.-M. Choi, M.-L. Seol, U.-S. Jeong, C.-H. Jeon, Y.-K. Choi, A core compact model for multiple-gate junctionless FETs. IEEE Trans. Electron Dev. **62**(7), 2285–2291 (2015)
5. K. Wei, L. Zeng, J. Wang, G. Du, X. Liu, Physically based evaluation of electron mobility in ultrathin-body double-gate junctionless transistors. IEEE Electron Dev. Lett. **35**(8), 817–819 (2014)

6. S.-J. Choi, D.-I.I. Moon, S. Kim, J.P. Duarte, Y.-K. Choi, Sensitivity of threshold voltage to nanowire width variation in junctionless transistors. IEEE Electron Dev. Lett. **32**(2), 125–127 (2011)
7. R.K. Baruah, R.P. Paily, The effect of high-k gate dielectrics on device and circuit performances of a junctionless transistor. J. Comput. Electron. **14**, 492–499 (2015)
8. Y. Chen, M. Mohamed, M. Jo, U. Ravaioli, R. Xu, Junctionless MOSFETs with laterally graded-doping channel for analog/RF applications. J. Comput. Electron. **12**, 757–764 (2013)
9. S.I. Amin, R.K. Sarin, Analog performance investigation of misaligned double gate junctionless transistor. J. Comput. Electron. **14**, 675–685 (2015)
10. X. Liu, M. Wu, X. Jin, R. Chuai, J.-H. Lee, Simulation study on deep nanoscale short channel junctionless SOI FinFETs with triple-gate or double-gate structures. J. Comput. Electron. **13**, 509–514 (2014)
11. J.G. Fossum, M.M. Chowdhury, V.P. Trivedi, T.-J. King, Y.-K. Choi, J. An, B. Yu, Physical insights on design and modeling of nanoscale FinFETs. IEEE Int. Electron Devi. Meet. Tech. Dig. 679–680 (2003)
12. V. Kilchytska, A. Nève, L. Vancaillie, D. Levacq, S. Student, H. van Adriaensen, K. De Meer, C. Meyer, M. Raynaud, J.-P. Dehan, D.Flandre Raskin, Influence of device engineering on the analog and RF performances of SOI MOSFETs. IEEE Trans. Electron Dev. **50**(3), 577–588 (2003)
13. S. Gundapaneni, S. Ganguly, A. Kottantharayil, Enhanced electrostatic integrity of short-channel junctionless transistor with high-κ spacers. IEEE Electron Dev. Lett. **32**(10), 1325–1327 (2011)
14. ATLAS: Users' Manual, Silvaco Santa Clara, CA, USA (2012). Available: www.silvaco.com
15. S. Selberherr, *Analysis and simulation of semiconductor devices* (Springer, Wien, New York, 1984)
16. J.D. Bude, MOSFET modeling into the ballistic regime, in: *Proceedings International Conference Simulation of Semiconductor Process Devices*, (2000) pp. 23–26
17. International Technology Roadmap for Semiconductors. (2008) [Online]. Available: http://www.itrs.net
18. J.-P. Colinge, J.C. Alderman, W. Xiong, C.R. Cleavelin, Quantum-Mechanical effects in trigate SOI MOSFETs. IEEE Trans. Electron Dev. **53**(5), 1131–1136 (2006)

A Novel Data Encryption Approach in the Grid-Structured Binary Image

Ram Ch. Barik, Sitanshu S. Sahu, Subhendu P. Bhoi and Suvamoy Changder

Abstract Data hiding from external malicious access is an important and timely issue. Cryptography is the backbone of information or processed data security. The existing cryptography techniques provide good security; however, its computational complexity is also very high. Hence, there is a need of an efficient as well simple cryptography approach. In this context, the paper proposes a novel technique for cryptography in the form of binary textures. The binary textures provide a form of security corresponding to the original message. The binary textures are generated, reshuffled, and arranged in an image form to make it robust from malicious access. The reliability of the proposed approach has been illustrated with some empirical case studies. The overall cryptography process in a digital image makes it a simple, low-cost, and effective methodology for the secure communication.

Keywords Grid structured · Cryptography · Binary image · Texture Shuffling pattern

R. Ch. Barik (✉)
Department of Computer Science and Engineering,
Vikash Institute of Technology, Bargarh 768028, Odisha, India
e-mail: ramchbarik@gmail.com

S. S. Sahu
Department of Electronics and Communication Engineering,
Birla Institute of Technology, Mesra 835215, Jharkhand, India
e-mail: sitanshusekhar@gmail.com

S. P. Bhoi
Department of Computer Science, Rajendra (Auto) College,
Bolangir 767002, Odisha, India
e-mail: bhoi.sprakash@gmail.com

S. Changder
Department of Computer Science and Engineering, National Institute of Technology,
Durgapur 713 209, West Bengal, India
e-mail: suvamoy.nitdgp@gmail.com

V. Nath (ed.), *Proceedings of the International Conference on Microelectronics, Computing & Communication Systems*, Lecture Notes in Electrical Engineering 453,
https://doi.org/10.1007/978-981-10-5565-2_9

1 Introduction

Data hiding from external malicious access is an ancient art. Cryptography is a procedure of data or information hiding applied to broad areas of information technology arena such as various types of authentication including text based or biometric authentication in both standalone and online-based order and payments. More precisely e-mail security is a major application area of today's cryptography. It plays a significant role of abstracting data in modern era of information technology. Data in the form of meaningful information is the centroid in today's computational world. In every dimension where information technology is being utilized as a tool of automation, there data is encapsulated and it is abstracted from malicious access so that only intended recipient can decode and understand it. Cryptography is an art of encrypting crucial information stored on different information carrier devices applicable for both standalone computational devices which is to be transmitted to unsecured network devices into no interpretable format. At receiver end, the cipher message or encoded information is processed by computational machine or human to decipher or decode it.

In the literature, many methods have been proposed to encrypt the data in a secure way. Islam et al. [1] proposed an encryption technique to embed or encode message in edges of the host image. Sukalyan et al. [2] proposed a chaos-based encryption over gray scale images by decomposing it into eight binary bit planes using tent map-based pseudorandom binary number generator (PRBNG). The four significant bit planes, determined by 5% level of significance on contribution of a bit-plane in determination of a pixel value, are encrypted using keys which are obtained by applying the recurrence relation of tent map based PRBNG. The four insignificant bit planes along with encrypted significant bit planes are combined to form the final cipher image gives optimistic and efficient security level [2]. As this literature is cited from Sukalyan Som et al. which clearly narrates among eight bit planes 4 are significant encrypted bit plane and 4 are insignificant bit planes. Yicong et al. [3] proposed a novel encryption algorithm using a bit plane of a source image as the security key to encrypt images. Chung and Yu [4] proposed an encryption scheme which is better as compared to its predecessors but it is still vulnerable to attacks if it uses the same key to encrypt different images. But again Chang gives a different approach to show that their scheme can be broken with some pairs of plain image and cipher image for secure communication. Wang et al. [5] presented a new method of optical image encryption with binary Fourier transform computer-generated hologram (CGH) and pixel-scrambling technology. The orders of the pixel scrambling and the encrypted image are used as the keys to decrypt the original image. Therefore, higher security is achieved. Furthermore, the encrypted image is binary, so it is easy to be fabricated and robust against noise and distortion. Lin [6] proposed a new type of encoding methods to encrypt hidden (covert) information in host images. The encrypted information can be plot, fax, word, or network, and it must be encoded with binary codes. Chung et al. [7] proposed an approach for encrypting binary images by putting different scan patterns at the same level in the scan tree structure and employing a two-dimensional run-encoding (2DRE) technique, it can encrypt images with higher security and good compression ratio. Zhang et al. [8] presented a new image encryption scheme based on DNA sequence addition operation and chaos which achieves good encryption; at the same time it can also resist exhaustive attack, statistical attack, and differential attack. Li et al. [9] consider the algorithm as a typical binary image scrambling/permutation algorithm exerting on plain text of fixed size and propose a novel optimal method to break it with some known/chosen plaintexts. Review given by Kocarev [10] a chaos-based image encryption processes that go through two processes, i.e., chaotic confusion and pixel diffusion. In chaotic confusion process, the pixels of plain image are permuted with a two-dimensional chaotic map, whereas the pixel diffusion alternates the value (gray level) of each pixel in a sequential manner. Yang and Kot [11] proposed a new binary image authentications based on interpixel relationship by connectivity of pixels in a local neighborhood which uses blind data hiding method.

Yu et al. [12] proposed a pattern substitution-based reversible data hiding method by calculating the frequency occurrence of patterns then quantifying the frequency occurrence from pattern to pattern. In the extraction stage, these patterns are reversed to their original forms and rebuild an undistorted cover image. Wang et al. [13] propose block pattern-based data hiding scheme to authenticate and annotate scanned images which gives a significantly improved embedding capacity. Jung et al. [14] propose block masking-based data hiding scheme for binary images by distributing keys to two parts and then authenticating the right authorized part. Tsai et al. [15] propose pair-wise logical computation (PWLC)-based data hiding technique which takes benefits of reversible hidden data extraction and lossless reconstruction of host image. Phan and Kim [16] proposed an improved matrix encoding (IME) scheme for hiding data into a two-color binary image.

Conventional security schemes are computationally expensive and introduce inherent difficulty in the processing of data. Cryptography applied to a digital image is a new dimension to information security field. Thus, in this paper, we introduce an efficient data encryption methodology for secure communication through binary image. Accessing account details or cash from an automatic teller machine (ATM) or banking system requires a numerical or text secret code to get login into it in a secure manner. Generation of secret numeric code and to make that code unpredictable from malicious access is major challenge today. The proposed approach is simple, low cost, and effective for secure communication of numeric strings.

2 Materials and Method

The proposed encryption method is based on the grid-structured binary image which makes it a unique approach. The message is concealed in a binary image having certain number of grids or blocks. Basic overview and empirical analysis is described below.

First, a matrix of $N \times N$ pixels with black background is created; i.e., the intensity is zero which can be realized as an image having black foreground as well as background. Over the black background, the numerical digits are represented as foreground object by white symbolic lines. The texture of the numerical digits are represented in the form of binary images of $M \times M$ pixels and stored in both sender and receiver side. The original binary matrix of $N \times N$ pixels is divided into small blocks/grids of the size of the sub-images ($M \times M$). All these sub-images are to be contained within the $N \times N$ in a judicial way. The numerical message is converted to a cipher numeric code by using a key. Now the cipher numerical digits are kept in the $N \times N$ binary matrix in random manner and again the index of the location of each digit also stored in the same matrix by shuffling the rest of the locations. The shuffling pattern can be of many types. Then the binary matrix is saved as an image (binary image) and sent to the receiver through any medium.

At the receiver end, the image is read and all the blocks ($M \times M$ pixels) are dismantled and arranged in an array. Read all the locations of the blocks and their shuffling type (arrangement pattern). Each sub-images or blocks are matched with the numerical textures from where the coded numerical message is identified and then the actual numerical string is extracted by using the key. The flow graph of both encryption and decryption process is presented in Figs. 1 and 2, respectively.

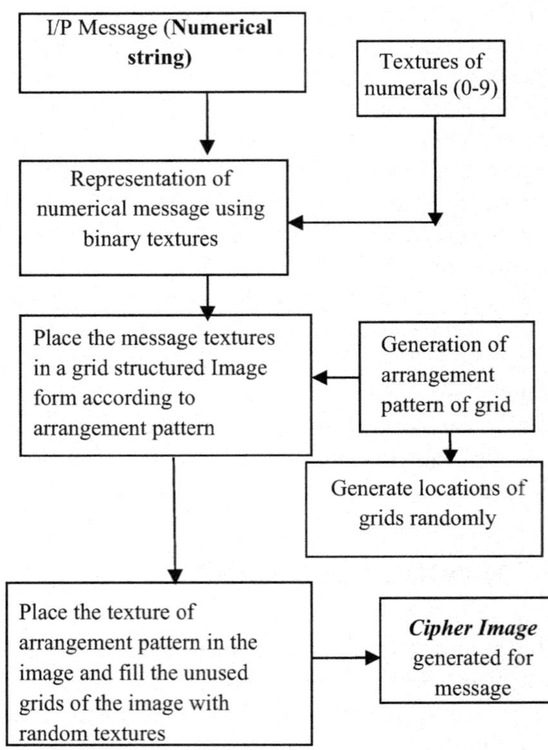

Fig. 1 Flow graph of encryption process

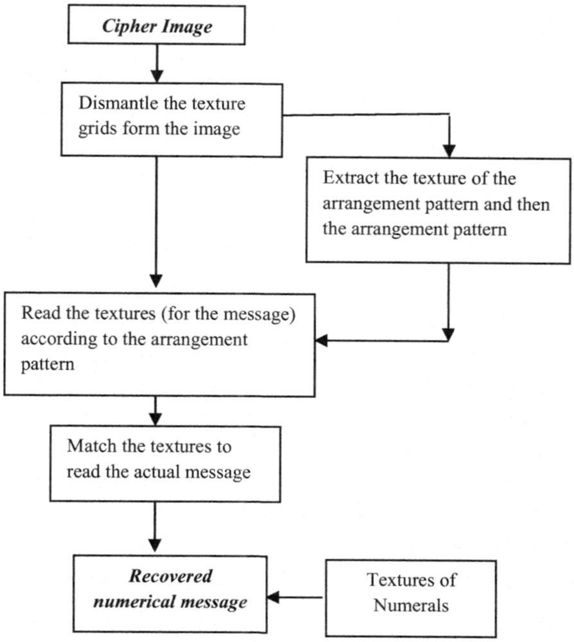

Fig. 2 Flow graph of decryption process

2.1 *Empirical Analysis of Proposed Method*

A matrix of size 500 × 500 pixels is created. Assign zeros to the whole matrix to form a black background image. Divide this matrix into 100 grids of 50 × 50 sized grids or sub-matrices. Further, create 10 binary matrix (realized as image) of 50 × 50 pixels to represent texture of the numeric digits, 0, 1, 2, 3, 4, 5, 6, 7, 8, and 9 as symbolic objects. Figure 3 describes the symbolic notation of the texture of the numeric digits.

The actual texture representation of the numeric digit ‘8’ represented in a small scale (10 × 10 grids) is shown in Fig. 4.

Input any numeric number of 16 digits as the plain text. The key used to make the cipher image of the numeral string is 10. Subtract each digit from the key to make it a cipher numeric code. If it has any digit as zero (0) then it will not be subtracted from 10 rather remains as same zero (0). Create 16

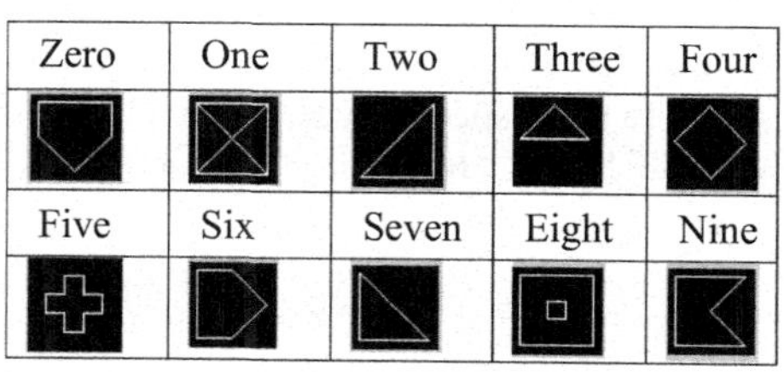

Fig. 3 Texture representation of numeric digits in the form of binary image

Fig. 4 Actual binary texture of digit Eight (8)

0	0	0	0	0	0	0	0	0	0
0	1	1	1	1	1	1	1	1	0
0	1	0	0	0	0	0	0	1	0
0	1	0	1	1	1	1	0	1	0
0	1	0	1	0	0	1	0	1	0
0	1	0	1	0	0	1	0	1	0
0	1	0	1	1	1	1	0	1	0
0	1	0	0	0	0	0	0	1	0
0	1	1	1	1	1	1	1	1	0
0	0	0	0	0	0	0	0	0	0

computer-program-generated pseudorandom numbers from 1 to 50 ($1 \leq$ number ≤ 50) and assign each cipher numeric digit image to the random-generated location of grid. The pseudorandom numbers are generated using the linear congruential method.

For an example, the generated random numbers for the cipher digit string are 41, 46, 07, 32, 05, 14, 28, 48, 49, 08, 25, 22, 40, 33, 02, and 43. This 16 cipher numeric code is to be inserted to each grid location referenced by the random numbers horizontally. If the random number generated is of two digits then first dismantle these two digits and then assign these two digits in two grid locations from 51st to 99th grid representing the pair. For example, 41 is the location where the first digit of the cipher number is placed. Then it is isolating as 4 and 1 and stored the texture binary image of digit 4 in 61st grid and 1 in 99th grid. Likewise, keep on assigning other random number location in the grid. After assigning these locations to the corresponding grids which are left vacant are padded with the digits texture randomly. Then the whole ($N * N$) matrix is converted to an image which is binary in nature and transmit to the receiver.

The above arrangement of numerical digits could be done in many different ways. Suppose there are 10 different types of arrangements. All these 10 different shuffling patterns are shown in Fig. 5. Receivers do have the knowledge of these 10 different arrangements. The 100th grid of the binary image is embedded with the particular shuffling arrangement by the corresponding geometric shape of the numerals.

At the receiver, read the image and dismantle all the blocks and arrange in array of matrices from 1 to 100. Read 100th block/grid by pattern matching and find the shuffling type (arrangement of numbers). Receiver already knows the key that is the number 10. Also it knows the 10 sub-images having the texture (Zero, One, Two, Three, Four, Five, Six, Seven, Eight, and Nine) in binary format as another key. After reading the image as data matrices of size 50×50, read to the location of the numeric messages (32 cells). Arrange the location pattern, for example, 4 and 6 as 46 go to the respective cell and read the texture emblem and decode the digits of the number by matching the matrixes to predict the respective digit. After getting the cipher digits, subtract it from 10, and upshot of this will give the actual message.

Fig. 5 Ten different arrangement patterns from 0 to 9 (this pattern number is placed in the 100th grid in the binary image)

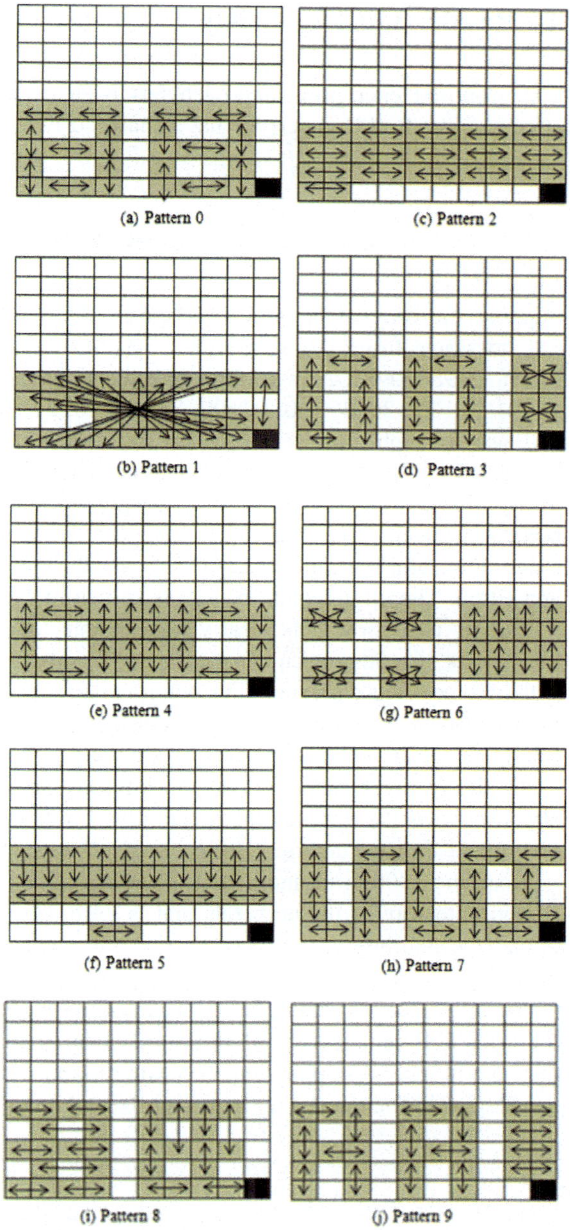

3 Results and Discussion

The proposed encryption approach has been extensively studied to validate its effectiveness. In this paper, we have analyzed two different case studies.

3.1 *Case Study I: Arrangement Pattern 1*

Suppose the original numeric message is: **3456278912230129**. After subtracting this message from the key value, i.e., 10 the cipher message becomes as: 7654832198870981. Now, the random locations generated between 1 and 50 are: 46, 01, 24, 22, 39, 17, 40, 02, 09, 37, 08, 18, 31, 10, 13, and 14.

Place the cipher message as a matrix of size 50 × 50 into the respective grid between 1 to 50th location. Then we have to store the grid locations in the 49 grids in the image from 51st to 99th. 100th grid location is reserved for the shuffling pattern. Consider the shuffling pattern for this numeric string is 1. Hence, the random location for numeric digit '7' is 46, then store 4 in 61st cell, 6 in 99th cell, and so on. The arrangement pattern is shown in Fig. 6.

Out of 100 grids, it is clearly noticed that total 49 grids are reserved for all location as 16 grids for cipher message, 32 grids for random location, and one for pattern type. The corresponding binary image is shown in Fig. 7. Still 51 grids remain vacant which needs to be padded with the digit textures randomly. After filling the vacant cells of the image, the generated image is shown in Fig. 8.

At the receiver side, retrieve the binary image sent by the sender. Receiver already knows the first key that is the lowest two-digit number, i.e., 10. Again the textures of the 10 numerical digits are also known to receiver.

Read the 100th grid and match the symbolic texture of the 100th grid with the symbolic texture of the numerals that is Zero, One … Nine. After matching, recognize the shuffling pattern between 0 and 9. Then the pattern is decoded in reverse order.

Read the received binary image as data matrix of 100 block of size 50 × 50. All the grids between 61st and 99th are matched against the ten sub-images, and the matched numbers are paired up according to the shuffling pattern. In this example, the 61st grid is matched with numeric texture 4 and 99th grid matched with numeric

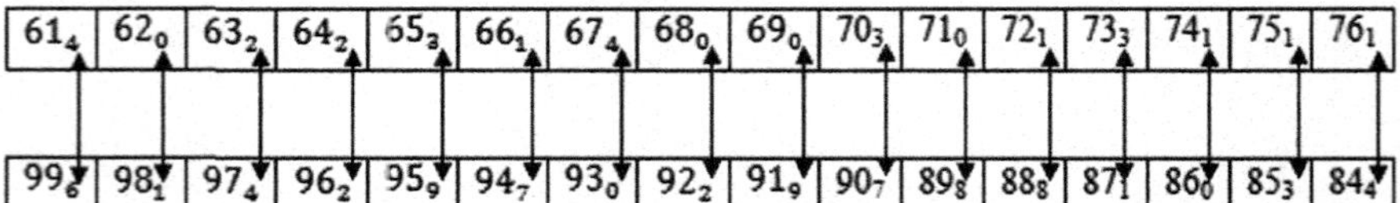

Fig. 6 Shuffling pattern for placing the random locations for pattern 1. Subscript refers to the numeral message digit

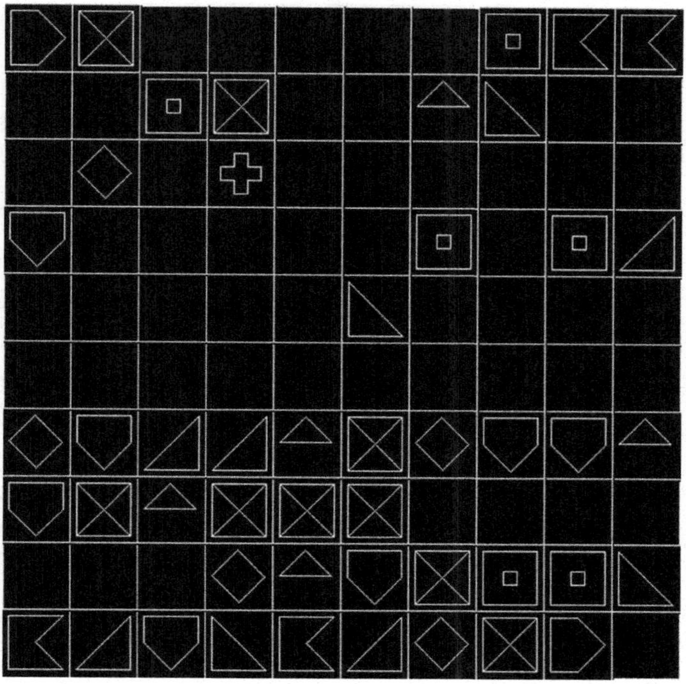

Fig. 7 Embed of cipher message and its location in 49 grids following pattern 1

texture 6. Thus, it is paired up to form the location 46. Again the 46th grid is matched against the ten sub-images and identified the stored numerical digits as 7. Then it is subtracted from the first key, i.e., ten to get back the original numerical message as 3. Similarly, the remaining 15 numeric messages are identified as **3456278912230129** which is the original numeric string that has been transmitted.

3.2 *Case Study II: Arrangement Pattern 2*

Suppose the original message is: **6667658909843221**. After subtracting this message from the first key value, i.e., 10 we got the cipher numeric message as: **4443452101267889**.

Now the random number locations generated as: 39 32 47 49 10 7 35 05 27 44 25 20 34 38 18 and 08. Hence, in 39th and 32nd locations of the image is embedded with the block texture image of Four (4).

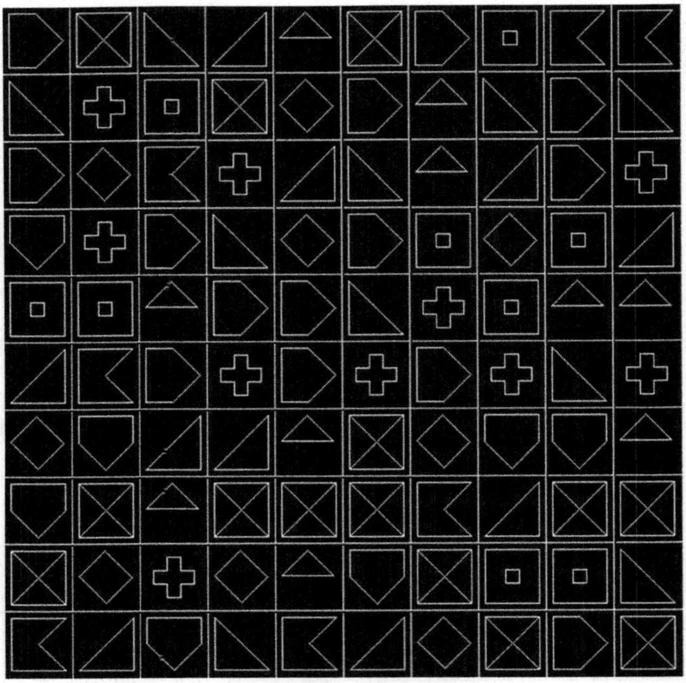

Fig. 8 Binary image of 500 × 500 size containing message of 16 digit text using pattern 1

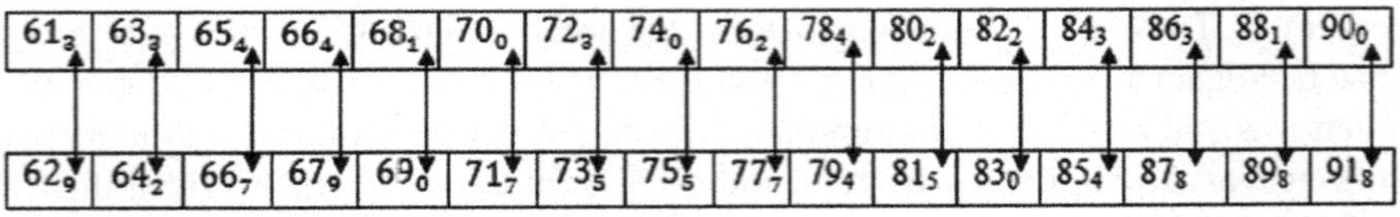

Fig. 9 Shuffling pattern for placing the random locations for pattern 2. Subscript refers to the numeral message digit

Again the grid locations are stored in grids from 61st to 99th. The 100th grid location is reserved for the pairing pattern. Suppose the shuffling pattern is 2. Figure 9 shows the shuffling pattern 2.

After placing the numeric textures in the image, the corresponding binary image is shown in Fig. 10 and rest of the grids are filled with random textures and the generated binary encrypted image is shown in Fig. 11. At the receiver side, the

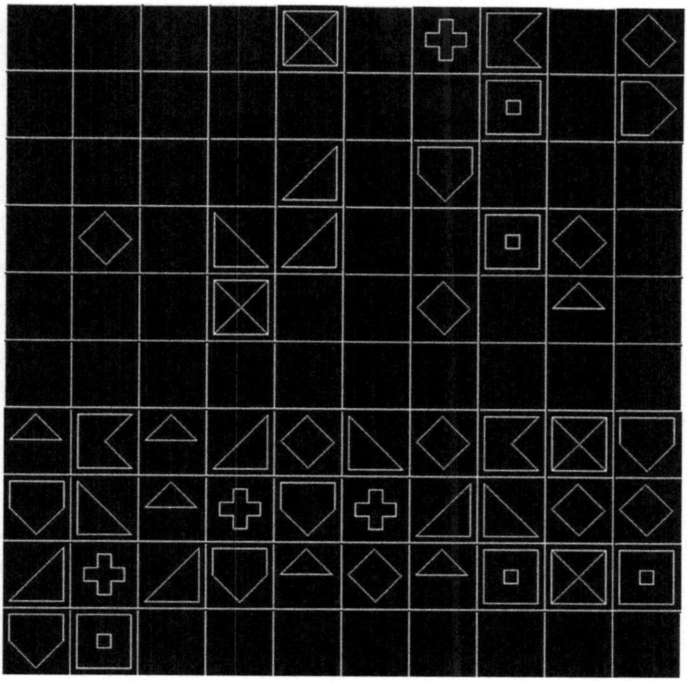

Fig. 10 Embed of cipher message and its location in 49 grids of pattern 2

encrypted message is read in the similar way as in case study 1, and the numeric string is recovered as **6667658909843221**.

The performance of the proposed method is compared with the existing standard cryptography methods such as DES, RSA, and AES. The comparison has been assessed through the computational time required for encryption of a 16-byte numerical data. The simulation studies have been done with a 64-bit core I3 Intel processor and listed in Table 1. Although our proposed approach takes bit more time, it is simple and provides equivalent security.

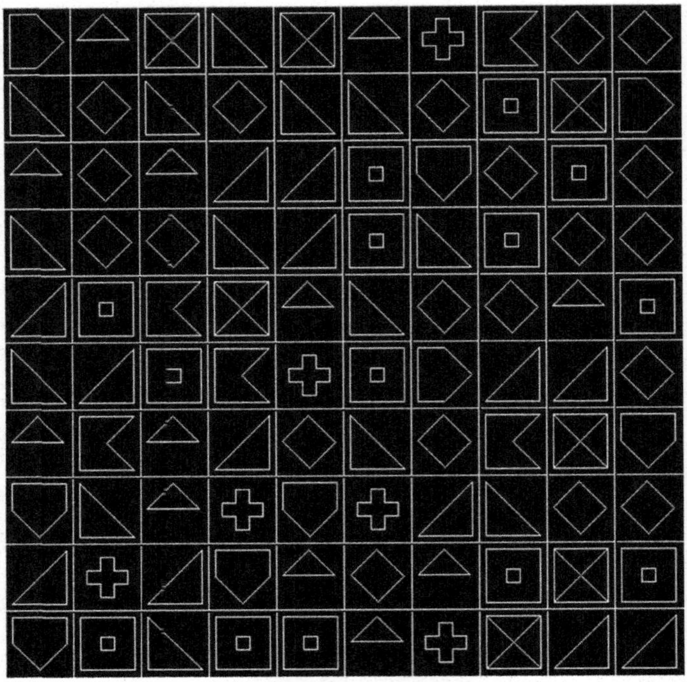

Fig. 11 Binary image of 500 × 500 size containing message of 16 digit text using pattern 2

Table 1 Performance comparison of the proposed method with the existing methods

Packet size: 16 Byte	DES	RSA	AES	Proposed method
Time consumption (s)	0.967543	0.823152	0.512341	1.106669

4 Conclusion

In this paper, a simple, low-cost, and efficient method of encryption is proposed for secure numeric data communication in terms of space and time complexity. The data is encrypted within a binary image and decrypted at the receiver side with a secure way. The effectiveness of the proposed approach has been demonstrated with suitable example. As the data is encrypted in the form of an image, its bit difficult to be cracked the original data. Usage of binary image is robust against noise and distortion. The current work can be extended for entire ASCII character for secure communication by increasing the grid numbers.

References

1. S. Islam, M.R. Modi, P. Gupta, Edge-based image steganography. EURASIP J. Inf. Secur. **2014**(8), 1–14 (2014)
2. S. Sukalyan, S. Sayani, A non-adaptive partial encryption of grayscale images based on Chaos. In: *International conference on computational intelligence: modeling techniques and applications*, Procedia Technology, Vol. 10 (2013), pp. 663–671
3. Yicong Zhou, Weijia Cao, C.L. Philip Chen, Image encryption using binary bitplane. Sig. Process. **100**, 197–207 (2014)
4. Chin-Chen Chang, Yu. Tai-Xing, Cryptanalysis of an encryption scheme for binary images. Pattern Recogn. Lett. **23**(14), 1847–1852 (2002)
5. Y.-Y. Wang, Y.-R. Wang, Y. Wang, H.-J. Li, W.-J. Sun, Optical image encryption based on binary Fourier transform computer-generated hologram and pixel scrambling technology. Opt. Lasers Eng. **45**(7), 761–765 (2007)
6. K.T. Lin, Digital information encrypted in an image using binary encoding. Opt. Commun. **281**(13), 3447–3453 (2008)
7. K.-L. Chung, L.-C. Chang, Large encrypting binary images with higher security. Pattern Recogn. Lett. **19**(5–6), 461–468 (1998)
8. Q. Zhang, L. Guo, X. Wei, Image encryption using DNA addition combining with chaotic maps. Math. Comput. Model. Elsevier **52**(11–12), 2028–2035 (2010)
9. C. Li, K.-T. Lo, Optimal quantitative cryptanalysis of permutation-only multimedia ciphers against plaintext attacks. Sig. Process. **91**(4), 949–954 (2011)
10. L. Kocarev, Chaos-Based Cryptography: A Brief Overview. Circuits Syst. Mag. IEEE **1**(3), 1531–1636 (2002)
11. H. Yang, A.C. Kot, Pattern-based data hiding for binary image authentication by connectivity-preserving. IEEE Trans. Multimed. **9**(3), 475–486 (2007)
12. Y.-A. Hoa, Y.-K. Chanb, H.-C. Wuc, Y.-P. Wuc, High-capacity reversible data hiding in binary images using pattern substitution. Comput. Stand. Interfaces **31**(4), 787–794 (2009)
13. C.-C. Wang, Y.-F. Chang, C.-C. Chang, J.-K. Jan, C.-C. Lin, A high capacity data hiding scheme for binary images based on block patterns. J. Syst. Softw. **93**, 152–162 (2014)
14. K.-H. Jung, K.-Y. Yoo, Data hiding method in binary images based on block masking for key authentication. Inf. Sci. **277**, 188–196 (2014)
15. C.-L. Tsai, H.-F. Chiangb, K.-C. Fanb, C.-D. Chungc, Reversible data hiding and lossless reconstruction of binary images using pair-wise logical computation mechanism. Pattern Recogn. **38**(11), 1993–2006 (2005)
16. P.T. Huy, C. Kim, Binary Image Data Hiding Using Matrix Encoding Technique in Sensors. Int. J. Distrib. Sensor Netw. **2013**, Article ID 340963, (2013)

Balanced Wrapper Design to Test the Embedded Core Partitioned into Multiple Layer for 3D SOC Targeting Power and Number of TSVs

Niranjan Raj and Indranil Sen Gupta

Abstract Manufacturing of three-dimensional (3D) IC chips is become executable nowadays with the furtherance in fabrication engineering. However, the process of designing and testing of tools in this regards are even if non-autumnal. One of the main challenges is to reduce the total time for testing of such chips. In order to make a reduction in the test application time, the wrapper design must be balanced such that all scan chain lengths are almost of equal length. Minimization of the scan test time is possible with the help of above-proposed work with the available numeral of through silicon vias (TSVs). The Verilog coding intended for the proposed implementation has been done using Cadence tool to analyze power and delay.

Keywords Scan chain · Wrapper · 3D IC · Test access mechanism · TSV

1 Introduction

A steep electrical connectedness going entirely by the way of wafer of silicon or die for link up of multiple layers of three-dimensional integrated circuit chips is termed as through silicon via (TSV). We can use TSVs [1] are as an option to wire-bond and flip chips to form 3D packages and 3D integrated circuits. The reason to accomplish broad interconnectedness and also to attain larger space efficiencies in compared to the flip chip peeling and wire-bounding it is used there.

N. Raj (✉)
Department of Electronics & Communication Engineering,
National Institute of Technology, Shillong 793003, Meghalaya, India
e-mail: niranjan1990nitm@gmail.com

I. S. Gupta
Department of Computer Science & Engineering,
Indian Institute of Technology, Kharagpur 721302, India
e-mail: isg@iitkgp.ac.in

V. Nath (ed.), *Proceedings of the International Conference on Microelectronics, Computing & Communication Systems*, Lecture Notes in Electrical Engineering 453,
https://doi.org/10.1007/978-981-10-5565-2_10

In system-on-chip (SOC) designs, the storage cells that are present in the embedded cores of a chip are typically interconnected in such a way that the formation of more than one shift registers is become possible which are termed as scan chains. Now, the input test patterns are applied at the inputs of these registers and the resultant is sensed through a suitable test access mechanism (TAM). When the connection made between the internal scan chain and the wrapper input–output cell, the wrapper is formed which is a thick shell associated close to the core. The number of wires in the TAM, called TAM width, must always be equal to the number of wrapper chains. The internal scan chain (ISC) and wrapper input–output cell are distributed to these wrapper chains in such a way that each wrapper chain is almost of equal length. By doing so, we get balanced sized wrapper chains.

The 3D IC design can be carried out through circuit partitioning at two layers of granularity, namely,

1. Coarse granularity
2. Fine granularity.

Considering fine granularity partitioning, where cores in the SoC chips are divided into multiple layers, the scan chain design will traverse all around the layers using vertical interconnection through silicon via (TSVs) which will result in the reduction in the gross wire length for the wrapper design technique. Location for scan chain terminals is on any of the levels, and interface must be done between the test wrapper and input/output cell and the scan chain for each particular core can now traverse with multiple levels.

The wrapper elements like ISC and input/output cell must be distributed to multiple layers. Availability of all the pins of the chip can be found at the last level. So, a scan chain wrapper starts and stops at the lowest level. For connecting the wrapper elements, TSV is required which goes through multiple layers. A wrapper chain is complete when it contains at least one ISC, one input cell and one output cell. There is no any requisite of TSV to create a wrapper chain if the presence of all the wrapper elements is found in the last layer. Maximum numeral of TSVs of any wrapper chain will be twice the numeral of levels present in the core when all the wrapper chain evolves an equal number of TSVs.

For example, if we consider a 4-layer core, in any wrapper chain the employment of maximum number of TSV there is 8.

Figure 1 shows a schematic showing the interconnection of input cell, output cell, and internal scan chains. It also shows via the connections through multiple layers. In the figure, 1 indicates the connection between layer-1 and layer-3. For this, TSV required is 1. Similarly, 2 indicates the connection between input cell and ISC, and for this TSVs required is 1. Also, 3 indicates the connection between ISC and ISC. Since TSVs internal to the scan chain is not counted, so TSVs required is 0. And, 4 indicates the connection between ISC and output cell, for which TSVs required, is 1. Lastly, 5 indicates the connection between layer-3 and layer-1, for this TSVs required is 1.

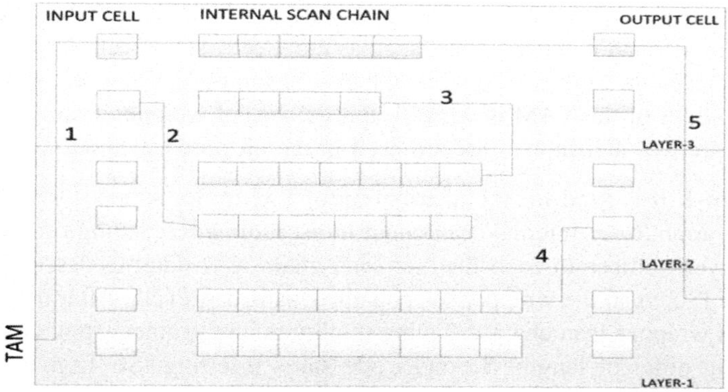

Fig. 1 TSV analysis

2 Work Done Previously on Wrapper Design

In optimization process of wrapper, a number of works for 3D SoC have been proposed, but all these works have considered only ideal condition where all internal scan chains are of equal length. Also, the number of ISC and wrapper input/output cell is taken such that all wrapper scan chain lengths become equal. Some of these wrapper optimization techniques include heuristic algorithm [2], ILP [3], genetic algorithm [6], simulated annealing to optimize the post-bond and pre-bond test time, and some other methods. For the testing of 3D ICs, some works have been possible. Design of TAM for 3D SOC is rendered in [3] where for the partition of TAM wires into many test buses and designate the cores to test buses for reducing the test time below TSV restraint integer linear programming (ILP) is used. For reduction of weighted test cost with restraint on the width of test pin in post-bond and pre-bond, a method is applicable which is known as trial-and-error method.

3 Proposed Methodology

To equal the distribution of the wrapper component for all the wrapper scan chains, thereby reducing the test time is the main motive of the proposed algorithm. The scan test time is minimized since the length of the longest wrapper scan chain is almost equal to all other wrapper scan chain.

The conceptualization for the above problem can be given by utilization of core with contrary functional parameters such as input cell, output cell, level, a group of internal scan chain having unequal length also including TAM width, and TSVmax. Here the width of the TAM is equal to the number of wrapper scan chain present.

There are two algorithms that are used in the proposed methodology.

Algorithm-1 It is used for the distribution of ISC to the wrapper scan chain. For doing so, arrange the internal scan chain in ascending/descending order of length. Assign layer number to each internal scan chain. Considering descending order, insert one ISC to all the wrapper scan chain. This is first cycle. After the end of first cycle, last wrapper scan chain is having minimum length since we have considered descending order of length. Second cycle starts inserting ISC from last wrapper scan chain toward the first one. After the end of the second cycle, if there exist some more ISC, then start inserting from the first toward the last one. At the end of every cycle, calculate the size of each scan chain wrapper. During the insertion of ISC to the wrapper scan chain, there is no need to calculate the number of TSVs, since internal TSVs present in the scan chain are not measured.

Algorithm-2 It is used for the distribution of input/output cell and calculation of TSVs. Firstly, one input cell and one output cell is inserted into all the wrapper scan chain. Assign layer number to these input/output cells. Now, calculate the length of each wrapper chain as well as a number of TSVs required.

Assume the wrapper chain with minimum length; compare its length with second minimum length wrapper chain. Add input/output cell to minimum length wrapper chain such that its length is equal to the second minimum length wrapper chain. Simultaneously, assign the layer number to these input/output cells. Now, these two wrapper chains are of equal length. Calculate the TSVs required. Compare its length with the third minimum length wrapper chain and add input/output cell to the wrapper chain till their length becomes equal. Do the same with all wrapper chain. If at any point, the number of TSVs required becomes equal to the TSVmax, then all the remaining input/output cells must be placed in the lowest layer since no TSVs are required to connect the input/output cells which are inserted in the lowest layer. In this way, optimization in the numeral of TSVs becomes possible. When the length of all the wrapper chain becomes equal, and there exist some more input/output cells then randomly add one input/output cell to any scan chain wrapper. Again calculate the length of each wrapper chain and arrange it in ascending/descending order. Again compare the length and do the same as above.

Algorithm-1: Distribution of ISC to wrapper chain

```
Input: ISC, TAM_width, no_of_layer
Output: wrapper chain
Number of wrapper chains = TAM_width
begin
Arrange ISC in ascending/descending order of length
Assign layer no. to each ISC
for wrapper chain=1 to TAM_width
    Wrapper chain(1)=ISC(1)
    Wrapper chain(2)=ISC(2)
    -
    -
    Wrapper chain(TAM_width)=ISC(TAM_width)

  if ISC(n) > ISC(TAM_width)
  for wrapper chain=TAM_width to 1
     Wrapper chain(TAM_width)=ISC(TAM_width+1)
     Wrapper chain(TAM_width+1)=ISC(TAM_width+2)
     -
     -
     Wrapper chain(1)-ISC(2*TAM_width)
  end
```

Algorithm-2: Distribution of input/output cell and calculation of TSVs

```
Input: input cell, output cell, TSVmax, no_of_layer
Output: wrapper chain design
for Input cell = 1 to n
for Output cell = 1 to n
  for wrapper chain i=1 to TAM_width
     Wrapper chain(i) = input cell(1)
     Wrapper chain(i) = output cell(1)
  end
 Calculate number of TSVs required
 Calculate length of wrapper chain
 Arrange wrapper chain in ascending order of length
 for i=1 to TAM_width
  if wrapper_chain_length(i) < wrapper_chain_length(i+1)
     wrapper_chain(i) <= input/output cell
 end
Now wrapper_chain_length(i)=wrapper_chain_length(i+1)
if wrapper_chain_length(i& i+1) < wrapper_chain_length(i+2)
   Wrapper_chain(i) <= input/output cell
   Wrapper_chain(i+1) <= input/output cell
end
-
-
end
Calculate TSVs at every instant
 if TSVs=TSVmax
   Place remaining input/output cell in lowest layer
 end
```

A. Illustrative example for creating wrapper chain

Consider 6 internal scan chain having length 8, 7, 8, 6, 4, 5, respectively, and 12 input cells and 12 output cells. For creating the wrapper chain having the width is 4, we have to take the TAM width equal to 4. According to the Algorithm-1, the ISC is arranged in descending order of length as 8, 8, 7, 6, 5, 4. The number of ISC is greater than the TAM width, so two cycles are needed for the distribution of ISC to wrapper scan chain. In the first cycle, one ISC is inserted in each wrapper scan chain. The remaining 2 ISC is inserted in the last two wrappers according to the length as shown in Fig. 2a. SC indicates the internal scan chains which are represented by their length. I and O indicate the input and output cell which is having length 1. The wrapper scan chain is represented by W1, W2, W3, and W4.

In order to complete the wrapper chain, one input and one output cell is inserted to all wrapper scan chain as shown in Fig. 2b. Now calculate the length of each wrapper chain. Wrapper W1 and W2 is having length 10, and wrapper W3 and W4 is having length 13. So, 3 input or output cells or its combination must be inserted

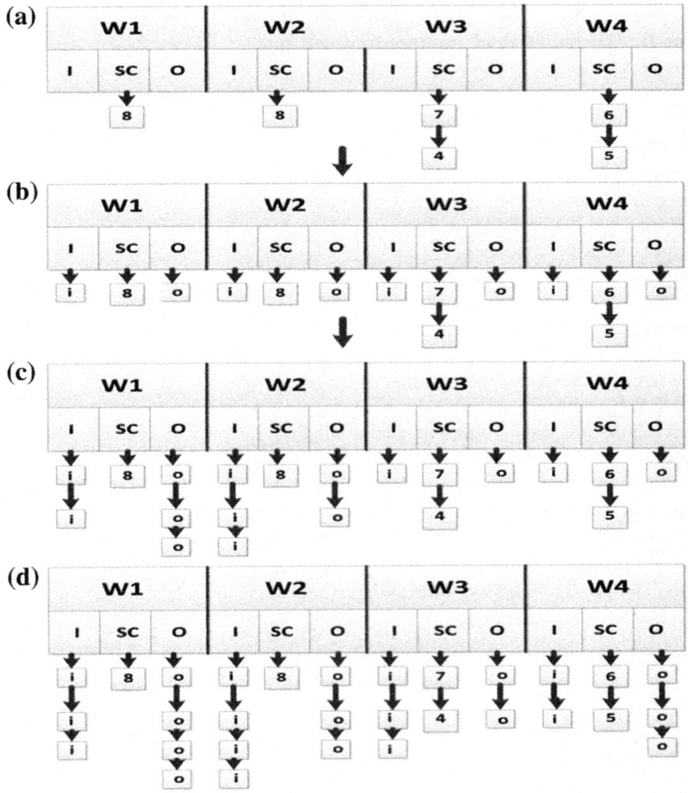

Fig. 2 Distribution of wrapper elements as per proposed algorithm

Table 1 Comparison of wrapper length with existing paper

TAM width	Wrapper	W1	W2	W3	W4
4	Wrapper length [our]	15	15	16	16
4	Wrapper length [5]	14	14	17	17

in W1 and W2 in order to make their length equal to W3 and W4 as shown in Fig. 2c where one input and two output cells are inserted in W1 and 2 input and 1 output cells are inserted in W2. The size of all wrapper scan chain is equal at that moment. The rest of 5 input cells and 5 output cells are distributed equally to all the wrapper chain. Now one input cell and one output cell is inserted to all wrapper scan chain as shown in Fig. 2d and calculate the length of each wrapper chain. At this point, each wrapper chain is having length 15. Still, one input and one output cell is remaining which is inserted into any wrapper scan chain randomly. Here, we have inserted one input cell in W3 and one output cell in W4 randomly as represented in Fig. 2d. Lastly, the size of the wrapper scan chain is 15, 15, 16, 16. That means all wrapper chain is almost of equal length resulting in balanced wrapper design. Table 1 shows the comparison of wrapper optimization of above example with [4].

4 Power and Delay Estimation

One major issue in testing a circuit is the optimization of test viewgraph similar to power, delay, and area. If the flip-flops present in the scan chain can be arranged in such a way that the number of transition within the flip-flops present in the scan chain is minimum, then the reduction in the test power can be achieved. When nearly all the time, the contiguous flip-flop stays in the unvaried state then it can be achieved. So, power consumption in testing a circuit is directly proportional to the number of transitions in the scan chain [1]. By using the proposed methodology, if we can reduce the size of the scan chain, then we can also made reduction in power consumption. With the help of using genetic algorithm (GA), the numeral of transition inside the scan chain can be reduced, which is a search and optimization algorithm. The numeral of transition present in the scan chain or chromosome is known as the fitness function in GA. The selection of the parent chromosome is done according to the lowest value of the fitness function. We can symbolize the result of GA by using chromosomes. Gene of the chromosome is represented by the numeral of flip-flops present inside the scan chain. The dependency of the size of chromosome is up to the numeral of scan flip-flops present in the circuit. During the phase of population generation mechanism for calculating respective chromosomes is provided by the fitness function.

With the help of the size of the scan chain, we can easily obtain the delay. So, there is automatically reduction in delay if the reduction made in the size of scan chain by using the above-proposed method. If the representation of the size of scan chain is given as '*L*' where *L* represents the Manhattan outstrip between the cells, then we can represent the delay as below:

Delay $\propto L^2$

So, the formation of scan chain is done in such a way that minimization in delay becomes possible as a result of which we can get reduction in power consumption.

5 Experimental Results

Under this section representation of the experimental results of the proposed methodology is completed. Initially, Verilog code is written for the above-mentioned two algorithms which are then imported to Cadence tool in order to analyze the power and delay. Based on the cores from ITC'02 SOC benchmarks, test circuit outcomes of simulation are rendered. Consistency for core 7 is 20 scan chains, 700 and 790 inputs and outputs, respectively for SOC d281. Also, consistency for core 4 is 23 scan chains, a number of inputs and outputs are 15 and 30, respectively for p93791. While core 5 of h953, consist of 4, 19, and 13 scan chains, inputs and outputs, respectively.

The scan test time consumed by this wrapper design is very much less as compared to the other wrapper design techniques like ILP, heuristic algorithm. *T* indicates the TAM width. Table 2 indicates the results on wrapper optimization for core 7 of SOC d281, Table 3 indicates the results on wrapper optimization for core 4 of p93791, and Table 4 indicates the results on wrapper optimization for core 5 of h953.

Table 2 Optimization results on wrapper for core 7 of d281

T	TSV max	Shortened WC length [our]	Shortened WC length [5]	Longest WC length [our]	Longest WC length [5]
3	18	709	710	710	816
4	18	532	532	532	724
5	22	425	426	426	618
6	22	354	355	355	473

Table 3 Optimization results on wrapper for core 4 of p93791

T	TSV max	Shortened WC length [our]	Shortened WC length [5]	Longest WC length [our]	Longest WC length [5]
3	18	51	51	52	55
4	18	38	39	39	56
5	20	30	35	31	51
6	20	25	28	26	40

Table 4 Optimization results on wrapper for core 5 of h953

T	TSV max	Shortened WC length [our]	Shortened WC length [5]	Longest WC length [our]
3	16	136	241	243
4	16	128	129	129
5	16	24	129	123
6	16	16	129	123

6 Conclusion

In this paper, we have presented a noble design methodology for designing a balanced wrapper for three-dimensional multilayer integrated circuits. The proposed methodology is applicable for any condition of internal scan chain whether all ISC are of equal length or every ISC is having different length. In every condition, the proposed methodology will generate balanced wrapper design. The number of TSVs also gets optimized. There is much improvement in the power, area, and scan test time. Hence, the proposed methodology provides an enhanced wrapper design for testing 3D ICs chips. Since the length of the longest wrapper scan chain is reduced, as a result power consumed to test the circuit also gets reduced.

References

1. C. Giri, S.K. Roy, B. Banerjee, H. Rahaman, *Scan Chain Design Targeting Power and Delay Optimization for 3D Integrated Circuits*, in Proceedings of IEEE International Conference on Advances in Computing, Control, and Telecommunication Technologies, India (2009), pp. 845–849
2. C. Giri, S.K. Roy, B. Banerjee, H. Rahaman, Test wrapper design for 3D system-on-chip using optimized number of TSVs, International Symposium on Electronic System Design(ISED) (2010), pp. 197–202
3. X. Wu, Y. Chen, K. Chakrabarty, Y. Xie, *Test Access Mechanism Optimization for Core-based Three-dimensional SOCs*, in IEEE International Conference on Computer Design (2008), pp. 212–218
4. S.K. Goel, E.J. Marinissen, "SOC test architecture design for efficient utilization of test bandwidth", ACM Trans. Des. Autom. Electron. Syst. **8**(4), 399–429 (2003)
5. B. Noia, K. Chakrabarty, Y. Xie, *Test Wrapper Optimization for Embedded Cores in TSV-based Three Dimensional SOCs*, in IEEE International Conference on Computer Design (2009), pp 70–77

Analysis of Electromagnetic Wave Using Explicit FDTD in TM Mode with Extrapolation

Bhattu HariPrasad Naik and Chandra Sekhar Paidimarry

Abstract In this paper, Explicit finite difference time domain (FDTD) method is used for electromagnetic wave analysis. Explicit method has computational simplicity in linear medium with superior stability by the CFL condition. The method is unstable with nonlinear mediums when materials have $\varepsilon_r > 1$. Here, a conventional Explicit FDTD method is used along with Interpolation and Extrapolation techniques for the EM wave analysis in TMz mode. Deriving the higher-order approximations from a lower-order approximation is called as Extrapolation. This technique is used to eliminate the second-order error $O(h^2)$ terms from first-order central difference approximation. The propagating wave Ez in TM mode is the summation of two triangular waves of Hx, Hy fields, which in turn form a 5-point wave stencil. The wave propagates through the grid using a cell-centred technique. The propagation speed of the wave depends on necessary parameters such as numerical dispersion of grid, time-step (Δt), x and y spatial step (Δx) and (Δy).

Keywords Finite difference time domain (FDTD) · CFL · Electro magnetic (EM) Central difference time domain (CDTD) · Explicit · Transverse magnetic (TM) Interpolation · Extrapolation

1 Introduction

The FDTD method is the most popular time domain method that is used extensively for solving Maxwell's equations. FDTD method [1] is a popular computational EM method due to its ease of implementation, flexibility and numerical efficiency. FDTD is a numerical analysis method that is used to find the approximate solutions to the system of partial differential equations. FDTD is a grid-based differential

---

B. HariPrasad Naik (✉) · C. S. Paidimarry
Department of ECE, UCE, Osmania University, Hyderabad, Telangana, India
e-mail: bhattu.hariprasadnaik@gmail.com

C. S. Paidimarry
e-mail: sekharpaidimarry@gmail.com

V. Nath (ed.), *Proceedings of the International Conference on Microelectronics, Computing & Communication Systems*, Lecture Notes in Electrical Engineering 453,
https://doi.org/10.1007/978-981-10-5565-2_11

numerical method. Using FDTD method, the time-dependent Maxwell's equations are discretized using the central difference time domain (CDTD) approximations with respect to forward in time and central in space. The resulting finite difference equations are solved in a leap-frog scheme [2]. FDTD method models many problems related to science and engineering that deals with EM waves and that are related with structures of the materials used.

For implementing a FDTD solution of the Maxwell's equation, requires a computation domain, i.e. a physical medium over which simulation is to be performed. Computational domain is discretized as rectangular numerical grid along with numerical dispersion of the medium. In the simulation phase, E and H fields are determined at each point in space and time. Types of the materials used as a medium for each cell of the grid in the computational domain need to be specified with their material constants (ε_r) during simulation phase.

The main approach is to develop a successful Explicit FDTD method with stability, accuracy and convergence of EM waves. The analysis of waves also involves linear and nonlinear mediums. Basic idea of this work is presented in [3]. In this paper, we provide a comprehensive description of the Explicit FDTD method with Interpolation and Extrapolation techniques to eliminate the effect of second-order error terms from the first-order central difference approximations which leads to increase in accuracy of the wave.

Method used to obtain a formula for $f'(x)$ of higher-order approximations from a formula of lower-order approximations is called Extrapolation.

The central difference approximation formula for an equation of first order is given as:

$$f'(x) = \frac{f(x+h) - f(x-h)}{2h} + E_{trunc}(f, h) \tag{1}$$

The local truncation error is given as

$$E_{trunc}(f, h) = \frac{-h^2 f^{(3)}(c)}{6} = \mathrm{O}(h^2) \tag{2}$$

The error function $E_{trunc}(f, h)$ makes the wave unstable with second-order error term $\mathrm{O}(h^2)$. So the error has to be removed in order to increase the accuracy of the wave.

2 Formulation

A conventional Explicit FDTD method formulation is presented here. Method uses first-order time-dependent Maxwell's equation in a TM mode for wave analysis. In the formulation of Explicit FDTD, Interpolation technique is used. Here a 2-D TM mode with wave propagation in Z direction is considered. The electromagnetic field components in TM mode Hx, Hy and Ez are arranged on the Yee's cell similar to that as conventional FDTD method [4]. The update value at $(n + 1)$ is determined at each point on the grid.

Maxwell's equations for the 2-D TM mode wave in a linear or lossless medium are expressed as:

$$\frac{\partial H_x}{\partial t} = \frac{-1}{\mu_0}\frac{\partial E_z}{\partial y} \tag{3}$$

$$\frac{\partial H_y}{\partial t} = \frac{1}{\mu_0}\frac{\partial E_z}{\partial x} \tag{4}$$

$$\frac{\partial E_z}{\partial t} = \frac{1}{\varepsilon_0}\left(\frac{\partial H_y}{\partial x} - \frac{\partial H_x}{\partial y}\right) \tag{5}$$

Where μ_0 and ε_0 are the permittivity and permeability of the medium, respectively.

An Explicit FDTD method is a forward difference in time and first-order central difference for space at a point. Explicit method [5] is numerically stable and convergent if and only if $\text{CFL} \le \frac{1}{2}$. It is the condition for the stability of the Explicit FDTD method, without which system is unstable [6]. Numerical errors $O(h^2)$ are proportional to time-step and square of space-step.

An Explicit FDTD method that is used to update the field Hx, Hy and Ez components at time level $(n + 1)$ with Interpolation [3] is shown in Fig. 1.

The node point at time level $j + 1$ is determined using $i - 1, i, i + 1$ node points. Wave obtained using $j + 1$ time level is a triangular wave.

$$H_x^{n+1}(i,j) = H_x^n(i,j) - \frac{\Delta t}{2 \cdot \mu \cdot \Delta y}\left[E_z^n(i,j+1) - E_z^n(i,j-1)\right] \tag{6}$$

$$H_y^{n+1}(i,j) = H_y^n(i,j) - \frac{\Delta t}{2 \cdot \mu \cdot \Delta x}\left[E_z^n(i+1,j) - E_z^n(i-1,j)\right] \tag{7}$$

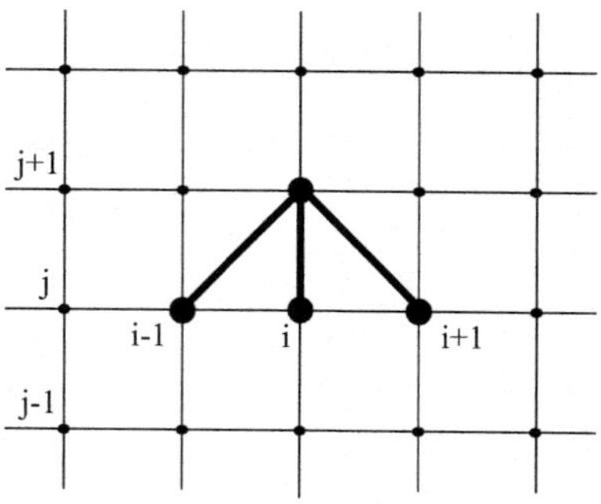

Fig. 1 Explicit FDTD with Interpolation

$$E_z^{n+1}(i,j) = E_z^{n}(i,j) + \frac{\Delta t}{2\cdot\varepsilon_0\cdot\Delta x}\left[H_y^n(i+1,j) - H_y^n(i-1,j)\right] + \frac{\Delta t}{2\cdot\varepsilon_0\cdot\Delta x}\left[H_x^n(i,j+1) - H_x^n(i,j-1)\right] \quad (8)$$

Where Δt is the time-step, n is the index of time level.

The magnetic field components Hx, Hy are the tri-diagonal system of equations. The electric filed component Ez is a combination of two tri-diagonal equations, when combined together form a 5-point stencil in wave propagation.

Figure 2 shows the triangular wave propagation due to either Hx or Hy field only in one direction x or y, respectively.

Figure 3 shows the wave propagation and scattering process in all the four directions using cell-centred technique.

The main difference between Explicit FDTD with Interpolation and Explicit [7] FDTD with Extrapolation is, in interpolation technique, local truncation error $O(h^2)$ exists as in (2). For removal of the local truncation error, Extrapolation is used, that is, to replace first-order central difference approximation with fourth-order approximations [1, 8] as shown in the Fig. 4.

For Extrapolation analysis, consider first time-step $\Delta t = h$,

$$f'(x) \approx D_1(h) - \frac{1^2\cdot h^2\cdot c}{6} \quad (9)$$

Second time-step $\Delta t = 2h$

$$f'(x) \approx D_1(2h) - \frac{2^2\cdot h^2\cdot c}{6} \quad (10)$$

$$4f'(x) \approx 4D_1(h) - \frac{4h^2c}{6}$$

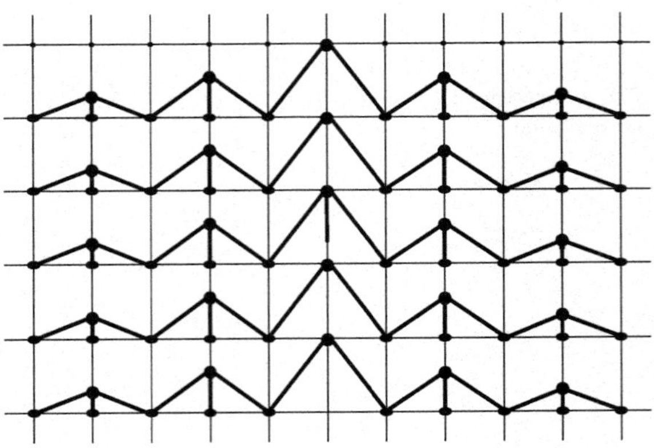

Fig. 2 Wave propagation with Hx field

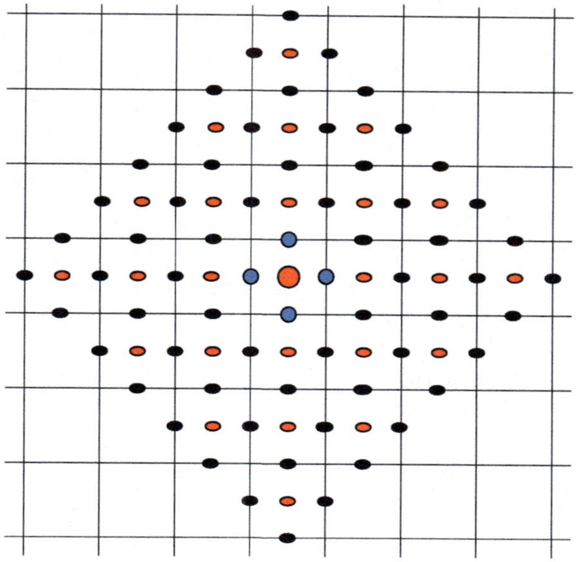

Fig. 3 Scattering process of the wave

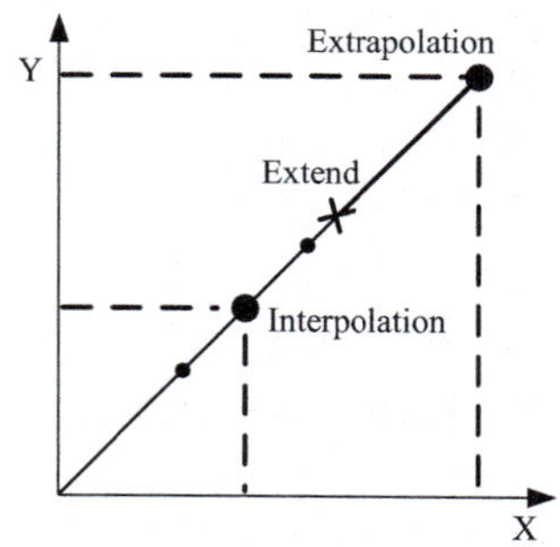

Fig. 4 Difference between Interpolation and Extrapolation

$$f'(x) \approx \frac{1}{3}[4 * D_1(h) - D_1(2h)]$$

$$f'(x) \approx \frac{1}{3}\left[4 * \left(\frac{f(x+h) - f(x-h)}{2h}\right) - \left(\frac{f(x+2h) - f(x-2h)}{4h}\right)\right]$$

$$f'(x) \approx \left[\left(\frac{-f(x+2h) + 8f(x+h) - 8f(x-h) + f(x-2h)}{12h}\right)\right] \quad (11)$$

Explicit FDTD algorithm that is used to update the field components Hx, Hy and Ez at time level $(n + 1)$ in linear medium with Extrapolation is shown in Fig. 5. The node point at time level $j + 1$ is determined using $i - 2$, $i - 1$, i, $i + 1$, $i + 2$ node points. Wave obtained using $j + 1$ time level is a triangular wave.

Fig. 5 Explicit FDTD with Extrapolation

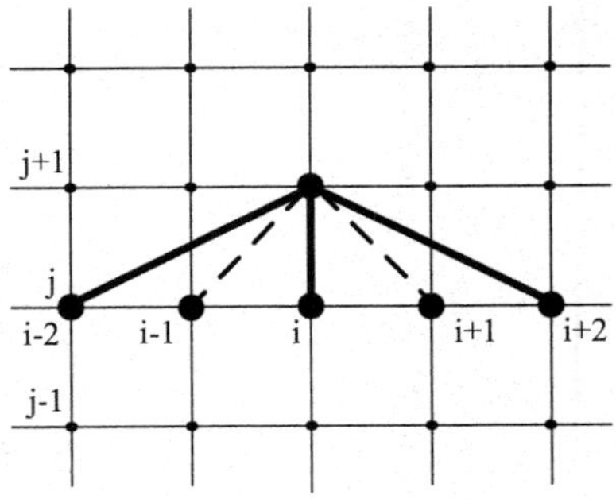

$$H_x^{n+1}(i,j) = H_x^n(i,j) - \frac{\Delta t}{2 \cdot \mu_C \cdot \Delta y}\left[-E_z^n(i,j+2) + 8 \cdot E_z^n(i,j+1) - 8 \cdot E_z^n(i,j-1) + E_z^n(i,j-2)\right] \tag{12}$$

$$H_y^{n+1}(i,j) = H_y^n(i,j) - \frac{\Delta t}{2 \cdot \mu_0 \cdot \Delta x}\left[-E_z^n(i+2,j) + 8 \cdot E_z^n(i+1,j) - 8 \cdot E_z^n(i-1,j) + E_z^n(i-2,j)\right] \tag{13}$$

$$\begin{aligned} E_z^{n+1}(i,j) = E_z^n(i,j) &+ \frac{\Delta t}{2 \cdot \varepsilon_0 \cdot \Delta x}\left[-H_y^n(i+2,j) + 8 \cdot H_y^n(i+1,j) - 8 \cdot H_y^n(i-1,j) + H_y^n(i-2,j)\right] \\ &+ \frac{\Delta t}{2 \cdot \varepsilon_0 \cdot \Delta y}\left[-H_x^n(i,j+2) + 8 \cdot H_x^n(i,j+1) - 8 \cdot H_x^n(i,j-1) + H_x^n(i,j-2)\right] \end{aligned} \tag{14}$$

3 Implementation of Explicit FDTD Method

3.1 *Explicit With Interpolation*

An implementation of Explicit FDTD [7, 9] method with Interpolation is presented here. Consider a linear or lossless medium of 100 × 100 m dimension with the free space parameters are:

$\varepsilon_0 = 8.854 * 10^{-12}$ F/m, $\mu_0 = 4\Pi * 10^{-7}$ H/m, $\Delta t = 1.667 * 10^{-9}$ s, $f_{\max} = 30$ MHz, $Pulse = \sin(2\Pi * f_{\max} * n * \Delta t)$, $\Delta x = \Delta y = 1$ m, $Velocity = \frac{\Delta t}{2 \cdot \varepsilon_0 \cdot \Delta x} = 0.3536$ m/s.

$$Ez(y,x) = Ez(y,x) + Pulse \tag{15}$$

For EM wave analysis, a soft source (15) is used. A sinusoidal signal of frequency 30 MHz is used as an input at the source point (50, 50) as shown in Fig. 6. The wave propagates and scatters in all four directions of equal values, and output is observed at destination point (70, 70) as shown in Fig. 7 (Fig. 8).

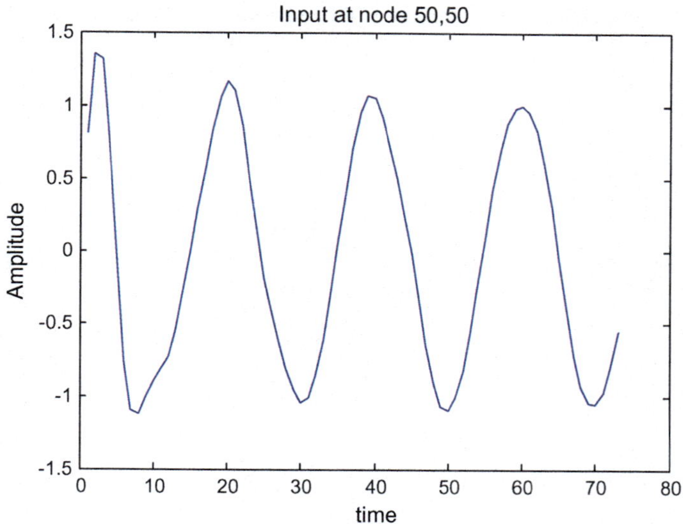

Fig. 6 Soft source input at node (50, 50)

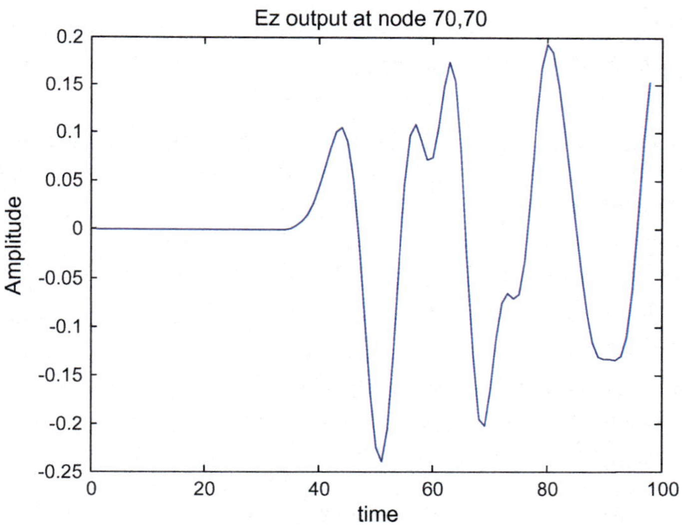

Fig. 7 Ez wave output at node (70, 70)

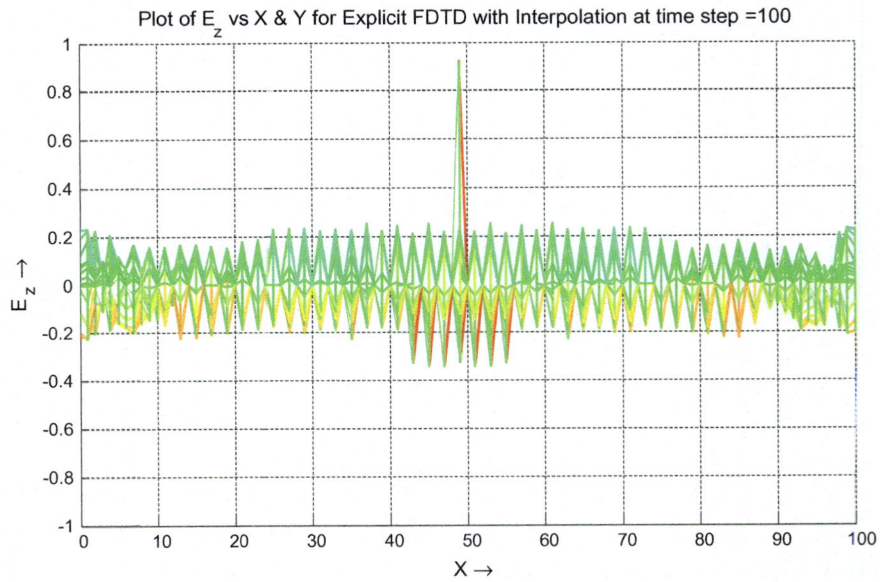

Fig. 8 Ez wave propagation with Interpolation

3.2 Explicit With Extrapolation

An implementation of Explicit FDTD method with Extrapolation is presented here. Consider a linear or lossless medium of 100 × 100 m dimension with the free space parameters are:

$\varepsilon_0 = 8.854 * 10^{-12}$ F/m, $\mu_0 = 4\Pi * 10^{-7}$ H/m, $\Delta t = 1.667 * 10^{-9}$ s, $f_{\max} = 30\,\text{MHz}$, $Pulse = \sin(2\Pi * f_{\max} * n * \Delta t)$, $\Delta x = \Delta y = 1$ m, $Velocity = \frac{\Delta t}{2 \cdot \varepsilon_0 \cdot \Delta x} = 0.0295$ m/s

For EM wave analysis in TM mode, a soft source (15) is used. A sinusoidal signal of frequency 30 MHz is used as an input at the source point (50, 50) as shown in Fig. 9. Here, second-order error $O(h^2)$ is eliminated using Extrapolation. The wave propagates and scatters in all four directions with equal values.

Ez wave output is observed at destination node point (70, 70). The output is shown in Fig. 10, which shows that accuracy of the wave is increased and stable when compared to the output shown in Fig. 7 (Fig. 11).

Table 1 shows the comparison of time taken for the wave to travel from the source point (50, 50) to the destination point (70, 70). Comparison shows that Extrapolation takes approximately 1 s more when compared to the Interpolation, but wave velocity of Extrapolation is decreased by 10 times, i.e. $V = 0.0295$ m/s and for Interpolation $V = 0.3536$ m/s. The accuracy of the output wave due to Extrapolation is increased and stable than the Interpolation output wave.

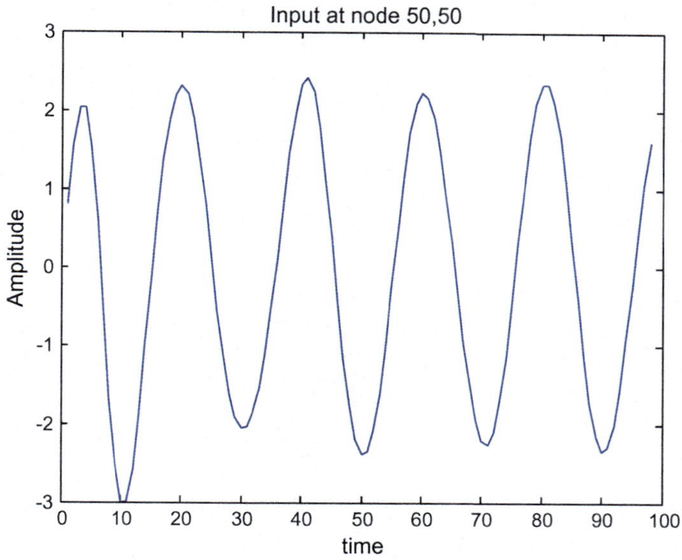

Fig. 9 Soft source input at node (50, 50)

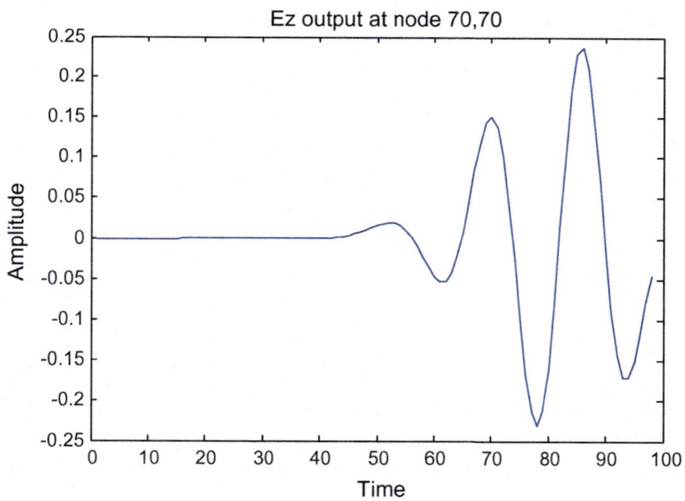

Fig. 10 Ez wave output at node (70, 70)

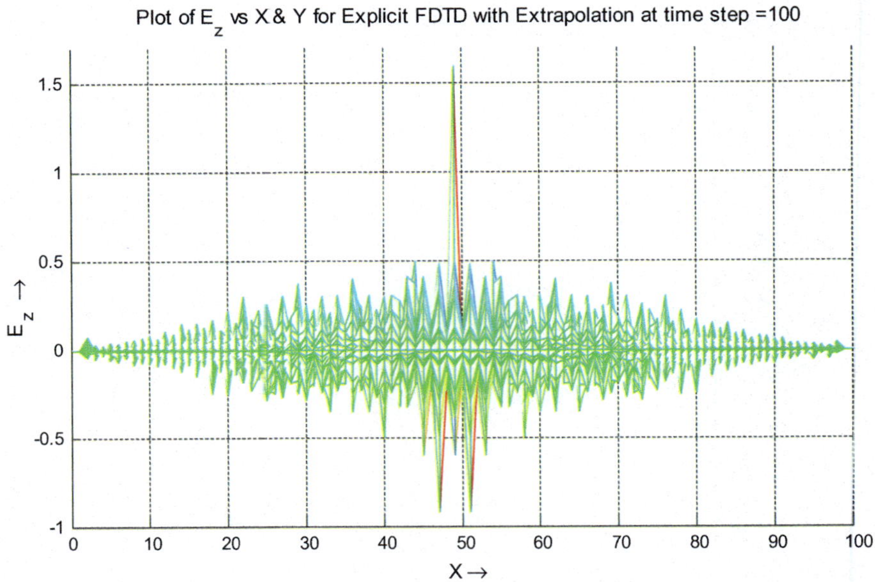

Fig. 11 Ez wave propagation with Extrapolation

Table 1 Comparison of time for Explicit FDTD

Method	Velocity of wave (m/s)	Elapsed time for wave (s)
Explicit with Interpolation	$V = \frac{\Delta t}{2 \cdot \varepsilon_0 \cdot \Delta x} = 0.3536$	8.891380
Explicit with Extrapolation	$V = \frac{\Delta t}{2*12 \cdot \varepsilon_0 \cdot \Delta x} = 0.0295$	9.856354

4 Conclusion

An alternative method to make Explicit FDTD method stable and accurate is developed. Here, we have adapted numerical method to eliminate the root cause of the accuracy of wave for a given time-step. Stability and applicability of the Explicit FDTD method with Interpolation and Extrapolation techniques are demonstrated. Here, the accuracy of the wave has been increased by eliminating the second-order error term $O(h^2)$ from the first-order finite difference approximation by using the Extrapolation. Numerical results have validated that the accuracy, efficiency and stability of the wave are increased with the proposed Extrapolation technique.

Acknowledgements The author would like to acknowledge UGC, Government of India, New Delhi, for providing the Fellowship.

References

1. J.H. Mathews, K.D. Fink, in *Numerical Methods Using MATLAB* (Prentice Hall, 1999)
2. S.T. Chu, S.K. Chaudhuri, in *Progress in Electromagnetic Research (PIER 1)* ed. by W. P. Huang. Finite-difference time-domain method for optical waveguide analysis, (1995), pp. 255–300
3. N.M. Nusi, M. Othman, Modified asymmetric explicit group (AEG) FDTD method for TM waves propagation, in *IEEE Asia-Pacific Conference On Applied Electromagnetics* (Apace 2010)
4. A. Taflove, S.C. Hagness, *Computational electrodynamics: the finite-difference time-domain method* (Artech House, Boston, MA, 2000)
5. M.N.O. Sadiku, in *Numerical Techniques in Electromagnetics* (CRC Press, 2009)
6. M. Othman, A.R. Abdullah, An efficient four points modified explicit group poisson solver. Int. J. Comput. Maths. **76**, 203–217 (2000)
7. M. Gaffar, D. Jiao, An explicit and unconditionally stable fdtd method for electromagnetic analysis. IEEE Trans. Micro. Theory Techn. **62**, 2538–2550 (2014)
8. D.J. Evans, Group explicit methods for the numerical solution of partial differential equations —(Topics in Computer Mathematics), Gordon and Breach Science Publishers, 1997
9. B.K. Huang, G. Wang, Y.S. Jiang, W.B. Wang, A hybrid implicit-explicit FDTD scheme with weakly conditional stability. Microw. Opt. Technol. Lett. **39**(2), 97–101 (2003)

A DEA-Based Evolutionary Computation Model for Stock Market Forecasting

S. S. Panigrahi, J. K. Mantri and P. Gahan

Abstract Stock market forecasting is used to draw attention of researcher since long and it will be. In this paper, a Data Envelop Analysis-based Gene Expression Programming model has been proposed and experimented with real data from BSE Sensex. The DEA has been used for filtering independent variables to be used as input variable of the GEP model. Different experiments have been made by first allowing all input variables to the GEP model directly without filtration by DEA and then allowing only those variables which are tested and marked as better variable to explain target variable. The result obtained from both the experiment has been put side by side and explained. From the analysis, it was noticed that the DEA-based GEP has better capabilities to forecast than the other one, even with less number of input variables.

Keywords Data Envelop Analysis · Gene Expression Programming · Stock market forecasting

1 Introduction

Stock market is a wonderful arena for 'risk management' and 'return generation.' It's a paradise for investors who love making money with certain level of risk. But there is an obvious question that, how much and how long a risk can be absorbed?

S. S. Panigrahi (✉) · J. K. Mantri
Department of Computer Science & Applications,
North Odisha University, Mayurbhanj, Odisha, India
e-mail: panigrahisasankasekhar@gmail.com

J. K. Mantri
e-mail: jkmantri@gmail.com

P. Gahan
Department of Business Administration,
Sambalpur University, Sambalpur, India
e-mail: pgahan7@gmail.com

V. Nath (ed.), *Proceedings of the International Conference on Microelectronics, Computing & Communication Systems*, Lecture Notes in Electrical Engineering 453,
https://doi.org/10.1007/978-981-10-5565-2_12

A cautious and intelligent move may lift the investor to become a gainer, whereas a silly mistake may also drug the investor to the line of looser. On the other hand, from the macroeconomic point of view, a stable financial market is one of the basic indicators of a stable economy. In general, a stable financial market is one where the volatility or the market fluctuation is not so erratic. Hence, there is always demand exists for forecasting the stock market behavior to creating better opportunity for investors. As the risk factors are associated with the higher expected returns, the area of stock market forecasting equally draws attention of researchers, economists, financial analysts, managerial decision makers, policy makers, and investors.

The need for better market predictability in right way and right time paved the way for experimentation with several models to forecast the stock market. The efficiency of various models has been tested with empirical data and analyzed. Further, availability of quality data at different warehouses along with handy data mining tools and analytical technologies has made statistical analysis an integral part of investment and managerial decision making. The unending process of such experiment and research in stock market forecasting is going on. Thousands of contributors are putting their efforts to make the field richer and richer day by day. The present work is a little effort in this direction.

Evolutionary computational models made a revolutionary change in the history of forecasting. It proves its efficiency through various empirical applications ranging from finance to computer science, engineering, biology, health science, etc. It gained wide popularity across the world as a forecasting tool due to its distinguished features over the existing methodologies as:

- Simple in approach.
- Change in approach to counter problem set.
- Different approaches to counter changing circumstance.
- More flexible.
- Capability to catch weak signals of the problem zone.
- Fruitful even where heuristic approaches failed.
- Better analytic power.
- Better predictability.

Here two established technique, viz. Data Envelop Analysis (DEA) and Gene Expression Programming (GEP) has been used for stock market forecasting. In the proposed model, the DEA is deployed for filtering attributes to be used in GEP. After experiment, it was shown that the DEA-based GEP has better predictability with less number of variables under consideration.

2 Review of Related Works

Application of evolutionary models adds a new dimension in soft computing approaches for forecasting financial market. A few landmark contributions related to the work has been reviewed for an understanding of the literature.

The year 2001 brings a new technique of evolutionary computing developed by Cândida Ferreira [1] over Genetic Algorithm and Genetic Programming known as Gene Expression Programming, which is a genotype/phenotype Genetic Algorithm for the creation of computer programs. In its basics, it uses character linear chromosomes composed of genes structurally organized in a head and a tail manner. In 2006, Ajith Abraham et al. [2] made a review of the common variants of genetic programming including GEP such as linear genetic programming (LGP), multi-expression programming (MEP), Cartesian genetic programming (CGP), traceless genetic programming (TGP). In the same year, Sam Mahfoud and Ganesh Mani [3] proposed a new genetic-algorithm-based system for predicting the future performances of individual stocks. They claimed to extend GAs from their traditional domain of optimization to inductive machine learning or classification. The year 2011 witnesses an introduction and experimental evaluation of neuro-genetic system for short-term stock index prediction by Jacek Ma_ndziuk and Marcin Jaruszewicz [4]. They tried to find out an optimal set of input variables for a one-day prediction via GA from a pool of input variables. In the same year, a new hybrid of genetic algorithm and particle swarm optimization model with perturbation term is proposed by Tarek Aboueldahab and Mahumod Fakhreldin to overcome the problem of local search restriction in standard hybrid of GA and PSO models. Alina Barbulescu and Iulia Ilie [5] proposed a hybrid model of ARMA and GEP for time series forecasting in 2012. They had tried to extract the ARMA for identification of linear trend and GEP to capture nonlinear patterns from data. In 2013, Alaa Sheta, Hossam Fari, and Mouhammd Alkasassbeh [6] compared GP with some other soft computing techniques in stock market modeling. They used S & P index for empirical studies and conclude in favor of GP. The same year a model named Genetically optimized Neural Network (GNN) has been proposed by Tuhin Mukherjee and Aritra Banerjee [7]. The model was based on Artificial Neural Network and Genetic Algorithm. The proposed model was tested and found superior over traditional ARCH/GARCH models. Chih-Ming Hsu [8] proposed an integrated procedure using DEA, Ant Colony Optimization (ACO), and GEP for stock trend prediction in 2014. The work claimed to provide a strategy to the investors for making profits even though the overall stock market suffers a loss. Further, it claimed to provide the right time of transaction to minimize risk by offering certain transaction rules for the investors. In the same year, Cheng-Han Lee et al. [9] tried to find out some good trading strategies from the historical series from Taiwan stock market and then trained the profitable strategies by the Gene Expression Programming, which involves some technical indicators as features. Another work in the same year by Dharamveer Sisodia et al. [2] tried to combine techniques of classification and GA. The classification of stock market is used to provide classification among different values of the stock market and then Genetic Algorithm is used to provide a close among these values. In the year 2015, Monir Mahmoudi et al. [10] proposed a prediction method to forecast the descending/ ascending direction of the future stock price by using GA on Istanbul Stock Exchange index. Their experiment shows that the proposed system works well in both upward and downward trends.

3 Objectives

The prediction capability of Gene Expression Programming in the area of stock market forecasting is proven by several studies. Now there is a necessity to converge some other proven models with GEP to enhance its efficacy toward better predictability. Here, we made an effort to combine DEA with GEP for better forecasting. The basic objectives of the study are bulleted as.

- As the financial market is one of the vital indicators of an economy, its non/improper predictability in right manner and right time has far-reaching consequences.
- There is a need for experimentation with the established computing approaches to be merged for creating hybrids which can perform better.
- As the GEP is an extension of GA and GP, further improvement of its predictability may be enhanced by making a hybrid with some other models.

4 Preparation of Data

A data set for empirical experiment needs to be a strong representative of the universe. In our case, we opt for the BSE Sensex data and related technical indicators for empirical study of the proposed model. As the stocks that constitute the BSE Sensex are drawn from almost all the market sectors of India, it is supposed that it could capable of tracking the movements of the market accurately. The historical data from January 2008 to July 2015 are treated for empirical study. The data have been collected on daily basis and standardized to run through the software for analysis. Historical data of daily open, high, low, close, and volume together with 12 numbers of selected technical indicators as represented in Table 1 constitute the base of the study. All the variables used are continuous in nature.

Table 1 Attributes for experiment

Attribute	Remarks	Attribute	Remarks
Open	Current day open price of the stock	EMA	The last 100 days exponential moving average of Close
High	Current day maximum price of the stock	PPO	The price oscillator in percentage
Low	Current day minimum price of the stock	PAIN	The price action indicator calculated for the current day
Close	Current day close price of the stock	MACD	The difference between 9 and 26 days EMA
Volume	Volume of transaction for current day	RSI	Calculated from 14 days avg. max and avg. min.
Highest High	The last 21 days highest of high	Momentum	The difference between previous and current close
Lowest low	The last 21 days lowest of low	% K	The stochastic which represents the percentage alert line
SMA	The last 21 days simple moving average of close	% D	The stochastic which represents the percentage definite line
WMA	The last 65 days weighted moving average of close	–	–

5 Methodologies

After collection of data set for the period mentioned in previous section, the technical indicators based on these data in Table 1 are calculated and except 'close' all are used as predicators. 'Close' is used as the target variable of the model for prediction. The methodologies are described in stepwise as follows:

Step 1 Collection of Historical data set.
Step 2 Selection and calculation of technical indicators.
Step 3 Selection of target variable.
Step 4 Run over the GEP model.
Step 5 Implementation of DEA over predicators of the GEP model.
Step 6 Selection of efficient predicators as per performance of DMUs.
Step 7 Run the GEP model over the predicators selected by DEA model keeping the target variable as usual.
Step 8 Compare the result obtained from step-4 and 7 on the basis of statistical parameters.
Step 9 Stop if step 7 gives better performance than step 4.
Step 10 Else move to step 5 and repeat the process up to step 8.

6 Models

The present work is built up over two well-established models—the Data Envelopment Analysis (DEA) and Gene Expression Programming (GEP). The DEA is a proven methodology to handle multiple inputs and multiple outputs with easy for measuring the efficiency of Decision Making Units (DMUs). On the other hand, GEP is a powerful evolutionary predictive method which can even trace the weak trends in a noise data set. A brief notes on both of the techniques used is followed.

A. *Data Envelopment Analysis*

Data Envelopment Analysis introduced by Charnes, Cooper, and Rhodes (CCR) is one of the latest additions to the operation research and management science which is used to develop efficiency scores for DMUs under consideration on a scale of zero to 1. The DMU scored closer to 1 is more efficient. It has been extensively applied for performance evaluation and benchmarking of nonprofitable units having qualitative data, e.g., educational institutions, hospitals, conservations of environment, banks, etc. It is generally used for multifactor productivity analysis measuring the relative efficiencies of a homogenous set of decision making units. The efficiency score of DMUs is calculated by the formula

$$\text{Efficiency} = \frac{\text{Weighted sum of outputs}}{\text{Weighted sum of inputs}}.$$

A major breakthrough was brought by Banker, Charnes, and Cooper (BCC) for extension of the CCR model to accommodate technologies that exhibit variable returns to scale.

B. *Gene Expression Programming*

The evolutionary computational models are constructed over certain common concepts drawn from the natural evolution of living beings and genetics. Most of the common models developed and used by the researchers under evolutionary computation include Genetic Algorithms, Genetic Programming, evolutionary programming, linear genetic programming, Gene Expression Programming, multi-expression programming, Cartesian genetic programming, traceless genetic programming.

Gene Expression Programming (GEP) is a technique which combines features from Genetic Programming (GP) and Genetic Algorithms [1]. Though all of them imitate natural, biological evolution yet the basic difference between them is in the way they deal with the individuals of a population of solutions. GEP sticks with the way Darwinian principle of the survival of the fittest works and uses populations of candidate solutions to a given problem in order to evolve new ones. The evolving populations passed through selective pressure and their individuals are submitted to genetic operators. Since proposed by Cândida Ferreira [1] in 2001, the Gene

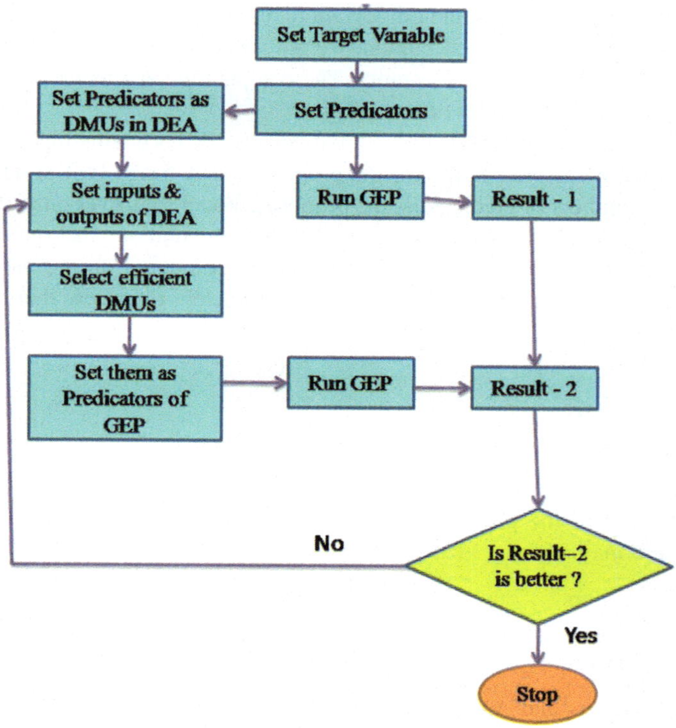

Fig. 1 Schematic view of the proposed model

Expression Programming (GEP) served to overcome the drawbacks of Genetic Algorithm and Genetic Programming for practical problems.

7 The Proposed Model

The proposed model is a hybrid of DEA and GEP. Though the GEP model has the ability to assignee weight to its input variables here, the DEA is used to filter input variables for the GEP. A schematic representation of the model is given in Fig. 1. The collected data of BSE Sensex and the technical indicators under study are put into the GEP model for taking a snapshot of prediction capability of the GEP model which can be compared with the proposed model. For both the model (GEP and DEA-based GEP), 'Close' is treated as target variable. Then the remaining variables are treated as predicators of the GEP model.

Each predicators of the GEP are taken as DMUs for DEA. To calculate the comparative efficiencies of the DMUs, certain statistical measures and parameters are taken as input and output of the DEA. The inputs of the DEA are mean, median, mode, range, variance, and standard deviation, and the z-score is taken as the single output. The inputs are assigned random weight where the sum of weights assigned to all inputs of a DMU is equal to unity. The comparative efficiencies of DMUs are calculated as per CCR model of the DEA.

Out of 16 numbers of DMUs, top 10 numbers of DMUs having higher scores are filtered to be used as predicators for the proposed DEA-based GEP model. Instead of taking 16 numbers of predicators, the proposed model runs over only 10 numbers of selected predicators. After its successful implementation, the outputs of the previous GEP model and the proposed model are compared on the basis of GEP parameters, analysis run time, and statistical parameters to review their respective performances.

8 Output Analyses

The efficiencies of the proposed model are basically compared with the original GEP in three major fronts, viz.

- GEP Parameters.
- Statistical Parameters.
- Analysis runs time.

GEP parameters and Analysis run time are reproduced in table.

From Table 2, it can be observed that though the complexity before simplification is 36 in both the cases the complexity of the proposed model comes down to 21 against 33 for the original GEP. Similarly, the generations required for simplification has drastically been reduced for the proposed model to reduce the analysis runtime. The analysis run time is quite less in case of proposed model. Further, the number of evolutions of the fitness function is also less for the proposed one. Hence in this front, the proposed model outperforms the existing one.

So far as the statistical parameters are concerned, we take the assistance of parameters, viz. R^2, CV, NMSE, RMSE, MAE. Table 3 can be referred for an easy grasp of the parameters compared. R^2 is a statistical parameter which gives us ideas about the proportion of variance explained by the model. Higher the R^2

Table 2 Comparative view of GEP parameters and analysis run time

Parameters	GEP	DEA-based GEP
Generations required to train model	1880	1964
Complexity of model before simplification	36	36
Complexity of model after simplification	33	21
Generations required for simplification	80	1
Number of evaluations of the fitness function	114,100	110,150
Number of execution threads used	4	4
Analysis run time	00:44:21	00:36:85

Table 3 Comparative view of statistical parameters

Parameters	GEP		DEA-based GEP	
	Training	Validation	Training	Validation
R^2	85.53%	–	85.75%	–
Coefficient of variation	0.093	0.0513	0.0923	0.0394
NMSE	0.1446	19.3636	0.1424	11.445
RMSE	1743.2792	1443.4133	1729.811	1109.718
MSE	3.039e+006	2.0834e+006	2.9922e+006	1.2315e+006
MAE	1432.5776	1421.3307	1457.679	1080.8393
MAPE	8.5818	5.06607	8.7013	3.8549

value better the model explained the proportion of variance. Here, we can notice that the proposed model bits the existing model. In case of the measurement of variation, i.e., the coefficient of variation, the proposed model outperforms the existing model in both training and validation. The remaining parameters in Table 3 measure the correctness of the prediction in terms of levels and the deviation between the actual and predicted values. The smaller the values are the better is the prediction. Here, we can observe that except for the MAE of the training data set the proposed model performs better than the existing model which proves its efficacy even with less number of predicators.

9 Findings and Concluding Remarks

The trend of financial time series data is very hard to be explained by a single method. Thanks to soft computing approaches for allowing hybrids to be made of existing methods to enhance prediction capabilities. The study concludes that the implementation of DEA for filtering predicators into the GEP certainly enhances the prediction capabilities of the GEP even if with less number of predicators. Due to comparatively less number of predictors are in use, the proposed model can run in less time by saving both analysis time and memory. The DEA model if used judiciously may improve the efficiencies of other existing forecasting models.

References

1. C.-H. Lee, C.-B. Yang, H.-H. Chen, Taiwan stock investment with gene expression programming, 18th International Conference on Knowledge-Based and Intelligent Information & Engineering Systems - KES2014. Procedia Comput. Sci. **35,** 137–146 (2014)
2. J. Ma_ndziuk, M. Jaruszewicz, Neuro-genetic system for stock index prediction. J. Intell. Fuzzy Syst. **22**(2), 93–123 (2011) (IOS Press)
3. T. Aboueldahab, M. Fakhreldin, Prediction of stock market indices using hybrid genetic algorithm/particle swarm optimization with perturbation term. Int. Conf. Swarm Intell. (2011)

4. M. Mahmoudi, R. Nouralsana, F. Rasouli, Estimating stock market index variations pattern using genetic algorithm and selected artificial intelligence methods. Indian J. Fundam. Appl. Life Sci. **5** (2015)
5. A. Bărbulescu, I. Ilie, Stock Market Indices Prediction Using Time Series Analysis. Math. Models Methods Appl. Sci. ISBN: 978–1-61804-098-5, BUBICHE-25, 2012
6. A. Sheta, F. Hossam, M. Alkasassbeh, A genetic programming model for S&P 500 stock market prediction. Int. J. Control Autom. **6**(5), 303–314 (2013)
7. https://in.finance.yahoo.com/
8. D. Sisodia, B. Kumar, J.K. Gupta, Efficient prediction of close value using genetic algorithm based horizontal partition decision tree in stock market. Int. J. Adv. Res. Comput. Sci. Manage. Stud. **2**(1) (2014)
9. C.-M. Hsu, An integrated procedure for resolving portfolio optimization problems using data envelopment analysis, ant colony optimization and gene expression programming. Int. J. Comput. Sci. Bus. Inform. **9**(1) (2014)
10. S. Mahfoud, G. Mani, Financial forecasting using genetic algorithms. Appl. Artif. Intell. **10**, 543–565 (1996)
11. A. Abraham, N. Nedjah, L. de Macedo Mourelle, in *Studies in Computational Intelligence (SCI)*, Evolutionary computation: from genetic algorithms to genetic programming, vol 13 Springer, Berlin, Heidelberg, 2006), pp. 1–20
12. T. Mukherjee, A. Banerjee, Prediction through genetic algorithm: a case study in Indian share market. Int. J. Emerg. Technol. Adv. Eng. **3**(6) (2013)

Investigations on the Logic Circuit Behaviour of Hybrid CMOSFETs Comprising InGaAs nMOS and Ge pMOS Devices with Barrier Layers

Suchismita Tewari, Abhijit Biswas and Abhijit Mallik

Abstract We investigate the logic circuit behaviour of hybrid CMOS devices made of InGaAs n-channel and Ge p-channel MOSFETs with Si and InP barrier layers, respectively, at channel length L_g = 20 and 30 nm. Rise and fall time, noise margin of hybrid CMOS inverters and frequency of oscillations, energy-delay product of 3-stage ring oscillators comprising hybrid CMOS inverters have been investigated to evaluate the performance of the proposed CMOS device. Our findings show a significant amount of reduction of 92.2 and 82.5% for rise and fall time, respectively, in case of proposed hybrid inverter, compared with the corresponding values for equivalent Si CMOS at L_g = 30 nm. Oscillation frequency of a 3-stage ring oscillator is found to be 264% higher when compared with its Si counterpart. Also there is an improvement of 17.8 and 77.4% in power-delay and energy-delay product, respectively, for hybrid CMOS inverters in comparison with their equivalent Si counterparts for a channel length of 30 nm. Similar trend is observed in case of channel length of 20 nm.

Keywords Hybrid CMOS · Logic performance · Noise margin
Rise time · Fall time · Frequency of oscillations

1 Introduction

To continue the historical performance improvement of complementary metal-oxide semiconductor (CMOS) devices for logic circuit applications beyond 32 nm channel length, the CMOS device made up of a p-Ge and n-InGaAs

S. Tewari · A. Biswas (✉)
Department of Radio Physics and Electronics, University of Calcutta,
92 Acharya Prafulla Chandra Road, Kolkata 700009, India
e-mail: abiswas5@rediffmail.com

A. Mallik
Department of Electronic Science, University of Calcutta,
92 Acharya Prafulla Chandra Road, Kolkata 700009, India

V. Nath (ed.), *Proceedings of the International Conference on Microelectronics, Computing & Communication Systems*, Lecture Notes in Electrical Engineering 453,
https://doi.org/10.1007/978-981-10-5565-2_13

MOSFETs has become a promising alternative to Si counterpart [1–8]. The choice of InGaAs channel is due to its outstanding electron transport properties such as mobility and injection velocity, while Ge is chosen owing to its record high hole mobility. Use of a barrier layer such as InP for InGaAs channel and a Si barrier for Ge channel confines carriers in the channel and improves carrier mobility further.

Several reports have been published in which the integration of a p-Ge and n-InGaAs MOSFETs was demonstrated to form the CMOS device on a common Si platform [9]. The task of co-integration of InGaAs nMOS and Ge pMOS on a common platform of Si offers various challenges; however, Takagi et al. have recently reported fabrication of both the devices on the same Si platform using direct wafer bonding technique [9]. Wang et al. have already reported in ref. [10] the direct growth of III–V and Ge in STI trenches of Si adopting aspect ratio trapping. Additionally, CMOS circuits comprising InGaAs nMOS and Ge pMOS devices, using the same gate stack, as demonstrated in [11], empower high-performance and cost-effective solution in the domain of circuit applications. Usually, Al_2O_3 is used as the gate dielectric for InGaAs nMOS device to achieve a lower value of D_{it}, and the same dielectric or SiO_2/HfO_2 is used for Ge pMOS devices [12].

In the paper, we report the logic circuit performance of a hybrid CMOS made of an InGaAs nMOS and Ge pMOS devices with InP and Si barrier layers, respectively. The logic performance is judged in terms of noise margin, fall and rise time of inverter circuit and frequency of oscillations and energy-delay product of a 3-stage ring oscillator built using hybrid CMOSFETs. Furthermore, the performance parameters for hybrid CMOS devices are compared with their corresponding Si values.

2 Device Structure and Simulation Framework

In our investigation, we have considered a hybrid CMOS device made of an InGaAs nMOS device and a Ge pMOS device, on a common platform of Si. SILVACOATLAS [13], a device simulator has been used for simulating each device followed by the use of MIXED-MODE circuit simulator for circuit-level analysis of inverters and 3-stage ring oscillator circuits. Figure 1 shows a schematic diagram of the CMOS inverter that we have simulated. In the Refs. [12, 14], process flows of the fabricated Ge and InGaAs devices are reported in detail. An interface-trap charge density in the range of $1 \times 10^{12} - 2 \times 10^{12}\ eV^{-1}\ cm^{-2}$ [12, 14] has been incorporated in our device simulation for realistic analysis. The dielectric constant (K) of InGaAs is computed using a linear interpolation technique, between the corresponding parameters of InAs and GaAs [15, 16]. The energy gap, valence and conduction band offsets have been calculated, incorporating the effect of compressive strain developed in between $In_xGa_{1-x}As$ channel and the InAlAs buffer layer [17]. The K-value and the E_g of buffer layer InAlAs and InP substrate are taken from the reported values [18]. The equivalent oxide

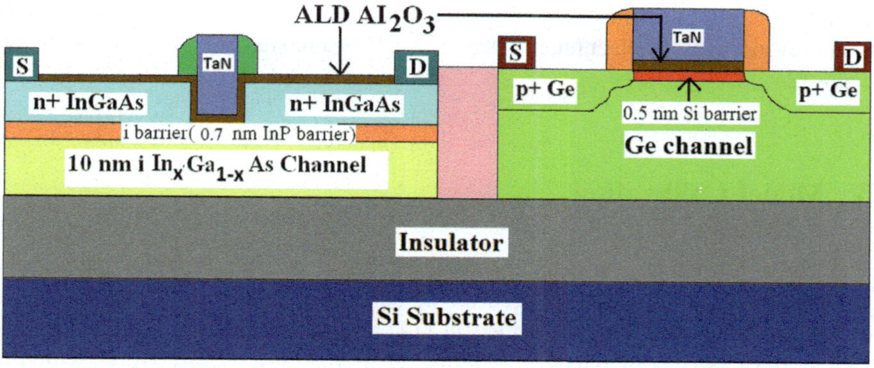

Fig. 1 Illustration of device structure of a CMOS made of a p-Ge and n-InGaAs MOSFET on the same Si platform. Each device has a width of 1 μm

thickness (EOT) for devices having L_g = 30 and 20 nm is chosen as 1 and 0.8 nm, respectively, according to ITRS road map [19]. The source/drain (S/D) contact resistances for Ge pMOSFET and InGaAs nMOSFET are considered as 93 and 82 Ω μm [20, 21], respectively. The work function of the metal gates for both the InGaAs and Ge devices are adjusted to set the threshold voltages (V_{th}) at −0.2 V and +0.2 V, respectively, and a common gate stack of TaN and HfO_2 has been used for both pMOSFETs and nMOSFETs.

The models used in our simulation are noted below. To simulate both the InGaAs nMOS and Ge pMOS devices, Lombardi CVT [12–14] model is invoked to take into account the dependency of mobility of electrons on carrier concentration, temperature and the vertical and horizontal fields. To capture the effect of velocity overshoot, which is important for InGaAs devices, the energy balance model has been used. This model utilizes the higher-order approximation of Boltzmann transport equation and models transport parameters, such as mobility as a function of local carrier temperature. To incorporate the tunnel effect, Hurkx band-to-band tunnelling models [12, 13] are incorporated in our simulation. Auger generation recombination and Shockley–Read–Hall (SRH) models are taken into account in the simulation for generation and recombination of the carriers. For carrier distribution, Fermi–Dirac (F-D) statistics have been taken into account during device simulation. We have taken into account the quantum-mechanical effect by using Schrodinger–Poisson equation coupled solver. For accurate estimation of leakage current, band-to-band (BTBT) [13, 22] and trap-assisted tunnelling (TAT) [13] are also incorporated in the simulation. Lombardi CVT model, SRH and Auger generation and recombination model, F-D statistics and the energy balance model have been invoked during the device simulation of Si p- and nMOS devices. Different material parameters of Si, e.g. dielectric constant, band gap, are taken from the default library values of SILVACOATLAS [13] with a D_{it} value of 1×10^{12} eV^{-1} cm^{-2} and S/D contact resistance as 30 Ω μm [23]. We have not, however, considered the quantum-mechanical effect while simulating Si MOS devices, as the thickness of channel of present devices is 10 nm and the quantum-mechanical effect takes place in case of Si channel having thickness below 5 nm [24].

3 Model Calibration

The device structures, various material combinations and different parameters, reported in [12, 14] have been used for model calibration. A comparison between experimental characteristics, reported in [14] and [12] for InGaAs nMOS and Ge pMOS devices, having channel lengths 40 and 30 nm, respectively, with our simulated characteristics is shown in Fig. 2a, b. From the Fig. 2a, b, it is obvious that our simulated curves show good agreement with the reported experimental curves [12, 14] which ensures the validation of our simulation scheme.

4 Results and Discussion

First, we have varied the barrier depth of InP layer in InGaAs nMOS device and have observed the electron confinement within the channel of the device for 2, 1.5, 1 and 0.7 nm barrier depths, having L_g = 30 and 20 nm, respectively. It is found that the carrier confinement achieves the maximum value for 0.7 nm barrier depth for both the channel lengths. We observe that the peak value of carrier concentration, i.e. 1.44×10^{18} cm^{-3} and 1.51×10^{18} cm^{-3}, has been attained at channel lengths L_g = 20 and 30 nm, respectively, for InGaAs channel nMOS devices with the InP single barrier of depth 0.7 nm. Hence in order to get an optimized

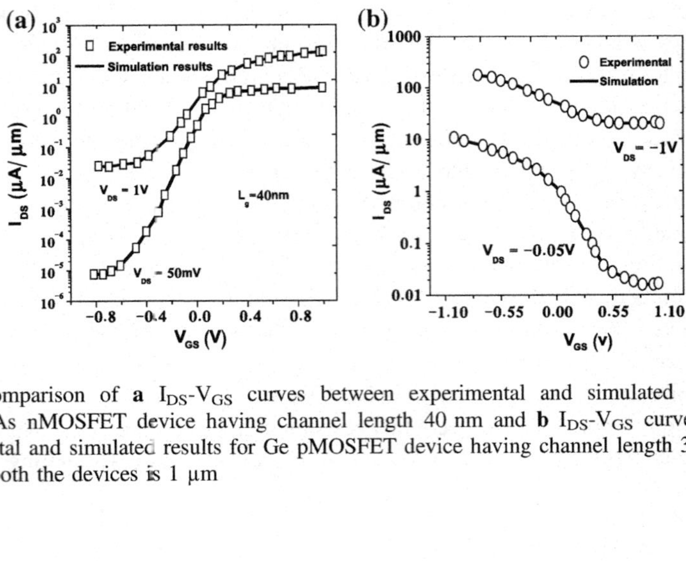

Fig. 2 Comparison of **a** I_{DS}-V_{GS} curves between experimental and simulated results for $In_{0.7}Ga_{0.3}As$ nMOSFET device having channel length 40 nm and **b** I_{DS}-V_{GS} curves between experimental and simulated results for Ge pMOSFET device having channel length 30 nm. The width of both the devices is 1 μm

performance we choose the nMOS device with barrier depth of 0.7 nm. For Ge device also we have varied the depth of the ultra-thin Si barrier layer and have studied the carrier confinement inside the channel of the device for 1.2, 1, 0.7 and 0.5 nm barrier depths, at channel lengths 30 and 20 nm, respectively. For such devices, the carrier concentration attains its maximum value of 6.43×10^{19} cm^{-3} and 5.44×10^{19} cm^{-3} for barrier depths of 0.5 nm at channel length L_g = 20 and 30 nm, respectively. We have found that the channel carrier confinement is maximum with Si barrier depth of 0.5 nm. Hence, we choose the Ge pMOS device having a barrier depth of 0.5 nm.

The two devices, chosen in the above-noted manner, are used to form the hybrid CMOS device for the circuit simulation using SILVACO MIXED-MODE simulator [13]. A schematic of circuit diagram of CMOS circuit with proper biasing is shown in Fig. 3. A family of VTC curves for various W_p/W_n ratio values of both Si and hybrid CMOS inverters, having channel lengths 30 and 20 nm, respectively, is being plotted in Fig. 4a–d. The sharpness of the VTC curves, which is nothing but the slope of the VTC curves is found to be much greater for hybrid CMOS inverter, compared to its equivalent Si counterpart. As the hole mobility in Ge and electron mobility in InGaAs are about three times more than that of in Si [23], the ON resistance of both Ge and InGaAs MOSFETs have been reduced due to the augmented carrier mobility, resulting more steep VTC curves for hybrid CMOS, made of n-InGaAs and p-Ge MOS devices. It is also observed from Fig. 4 that the VTC curve becomes steepest for W_p/W_n = 3. Hence, we construct a CMOS inverter using W_p/W_n = 3. The gain is essentially the slope of the VTC curve, and in each case, peak gain is maximum for W_p/W_n = 3. Hence, we have picked the CMOS device having W_p/W_n ratio equal to 3.

Following that, a hybrid push–pull inverter has been constructed comprising Ge PMOS and $In_{0.75}Ga_{0.25}As$ nMOS devices with a load capacitance C_L = 2.38 fF, which is determined by calculating the parallel combinations of the gate

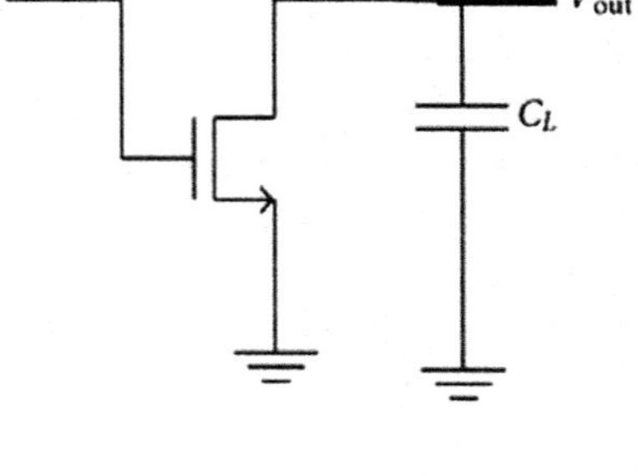

Fig. 3 Schematic of CMOS circuit diagram in push–pull configuration

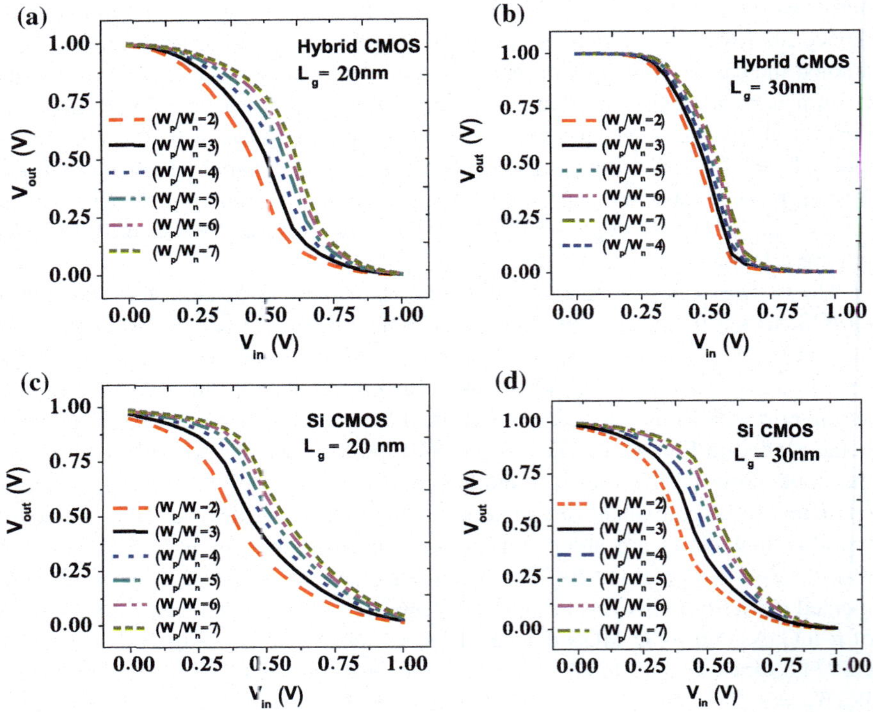

Fig. 4 Voltage transfer characteristics of hybrid CMOS at L_g = **a** 20 nm, **b** 30 nm and of Si CMOS at L_g = **c** 20 nm and **d** 30 nm having various W_p/W_n ratios

capacitances and neglecting the parasitic capacitances in present simulation. Next, we have applied a 5-GHz pulse train waveform having amplitude 1 V at the input of the constructed inverter, and on mixed-mode simulation platform, the output waveform is observed. Figure 5a–d shows a comparison of the output waveforms for the Si and Hybrid CMOS devices, for both the channel lengths. The rise time and fall time, extracted from the curves, are tabulated in Table 1. From Table 1, it has been observed that hybrid inverter is suffering from much lesser rise and fall time, compared to its equivalent Si inverters at both the channel lengths, which is due to the enhanced majority carrier mobilities in InGaAs and Ge channels, respectively. It is observed from Table 1 that the hybrid CMOS outperforms equivalent Si CMOS in terms of rise time as well as fall time by 92.16 and 82.47%, respectively, for L_g = 30 nm. Similar trend is observed for L_g = 20 nm.

Furthermore, we have calculated the off-state power consumption of hybrid and Si CMOS devices using the formula $P_{stat} = V_{DD} * (I_{offN} + I_{offP})/2$ for channel lengths 30 and 20 nm. The calculated values are incorporated in Table 1. It is observed from Table 1 that the off-state power consumption suffers greatly for Hybrid CMOS, which can be explained from the fact that off-state currents for both Si nMOS and pMOS devices are far more smaller than that of the InGaAs nMOS

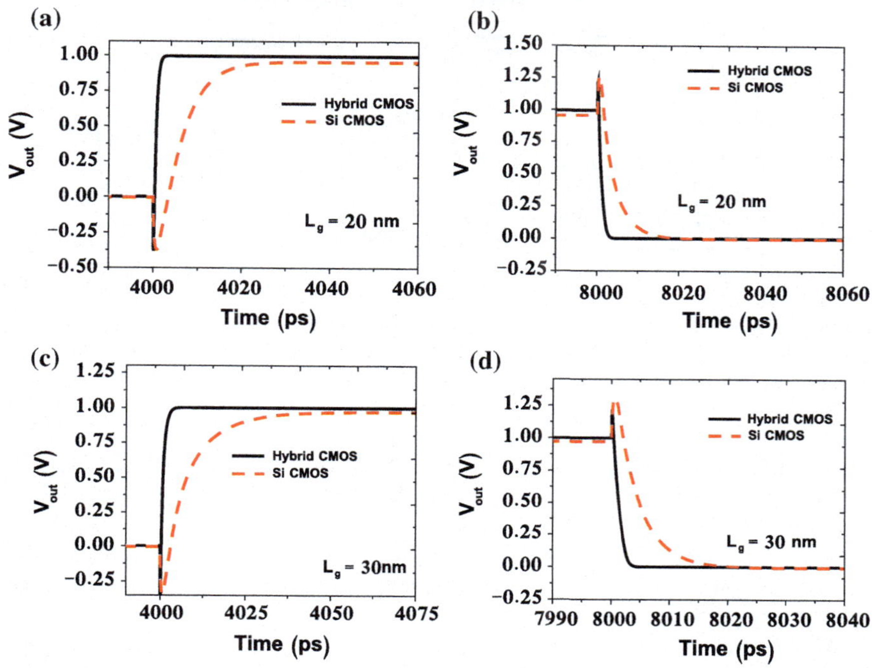

Fig. 5 Comparison of V_{out} as a function of time between Si and hybrid CMOS inverters. Plots show **a** rise time and **b** fall time at channel length = 20 nm, and also **c** rise time and **d** fall time at channel length = 30 nm

Table 1 Comparison of the Fall time and Rise time, PDP, EDP and off-state power consumption between Si and hybrid CMOS inverters for different L_g values

CMOS	L_g (nm)	Rise time per inverter (ps)	Fall time per inverter (ps)	Off-state power consumption (W/μm)
Hybrid	20	1.6	1.08	9.275×10^{-9}
Si		13.89	7.34	3.2×10^{-9}
Hybrid	30	1.8	1.5	2.975×10^{-9}
Si		22.98	8.56	1.4×10^{-9}

and Ge pMOS devices. This phenomenon occurs due to the larger short channel effects in InGaAs and Ge devices such as larger value of subthreshold slope.

Finally, we have made a 3-stage ring oscillator using such inverters. The comparison of the output waveforms from output of stage 3 between hybrid and Si devices for both the channel lengths is shown in Fig. 6a, b. The extracted delay per inverter and frequency of oscillations are entered in Table 2. The frequency of oscillation of the 3-stage ring oscillator comprising hybrid CMOS is showing 264 and 199% improvement over the equivalent Si CMOS-based 3-stage ring oscillator for L_g = 30 nm. For L_g = 20 nm, similar trend is followed. The considerable

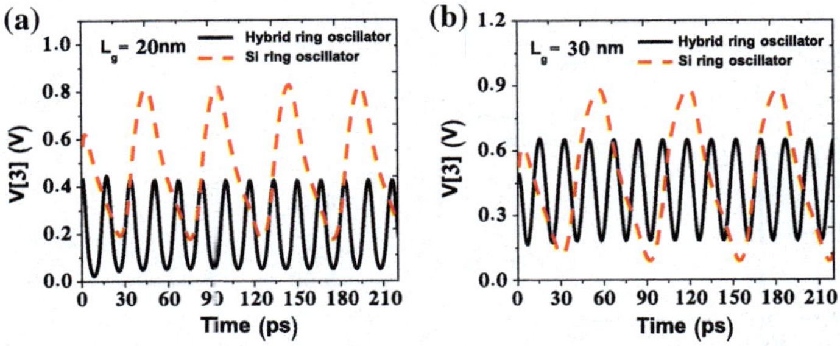

Fig. 6 Comparison of the output waveform V [3] obtained from final output stage of a 3-stage ring oscillator using Si CMOS and hybrid CMOS inverters at **a** channel length = 20 nm and **b** channel length = 30 nm

Table 2 Comparison between the time period and frequency of oscillations of 3-stage ring oscillators using hybrid and Si CMOS devices and also comparison between the delay per inverter for different channel length values

3-stage ring oscillator	L_g (nm)	Time period (ps)	Frequency (GHz)	Delay per inverter (ps)	Power-delay product (J/μm)	Energy-delay product (JS/μm)
Hybrid	20	16.6	60.31	5.52	5.37×10^{-15}	2.96×10^{-26}
Si		49.7	20.13	16.6	5.84×10^{-15}	9.69×10^{-26}
Hybrid	30	17.06	58.63	5.69	5.59×10^{-15}	3.18×10^{-26}
Si		62.05	16.11	20.68	6.8×10^{-15}	14.06×10^{-26}

enhancement in f_{osc} and the significant drop in propagation delay are the outcome of the superior hole and electron mobility in Ge and in InGaAs channel. The energy-delay product (EDP) and the power-delay product (PDP) of both the Si and hybrid CMOS inverters are calculated at both channel lengths, and the corresponding values are entered in Table 2. It is observed from Table 2 that both PDP and EDP improve for hybrid CMOS inverters by 17.79 and 77.38%, respectively, as compared to Si CMOS inverters for L_g = 30 nm. This trend is maintained for the device having L_g = 20 nm. This feature is attributed to the lower delay for the hybrid CMOS inverter. Figure 7a, b compares the high-to-low noise margins (NM_H) and the low-to-high noise margins (NM_L) as a function of W_p/W_n ratio for Si and hybrid CMOS inverters at channel lengths of 30 and 20 nm. It is obvious from Fig. 7 that for W_p/W_n = 3, both the Si and the hybrid CMOS inverters produce symmetrical VTC curve. Hence, for further analysis, we have chosen the W_p/W_n = 3 for both the devices. Figure 8a–d demonstrates the butterfly plot of VTC curve for both Si and hybrid CMOS inverters at channel lengths of 20 and 30 nm. From the butterfly plots, we have extracted symmetric noise margin (SNM), which is essentially the side of the largest square, fitted in the butterfly curve. From Fig. 9, it is clear that for each value of supply voltage, the SNM maintains a larger value

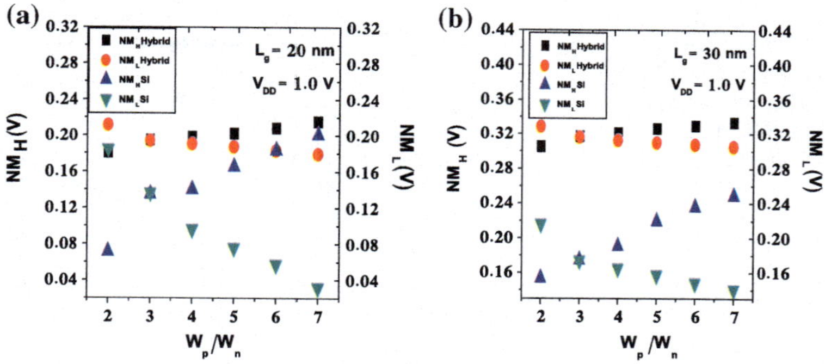

Fig. 7 Comparison of high (NM_H) and low (NM_L) noise margins for hybrid and equivalent Si CMOS inverters for channel length **a** 20 nm and **b** 30 nm

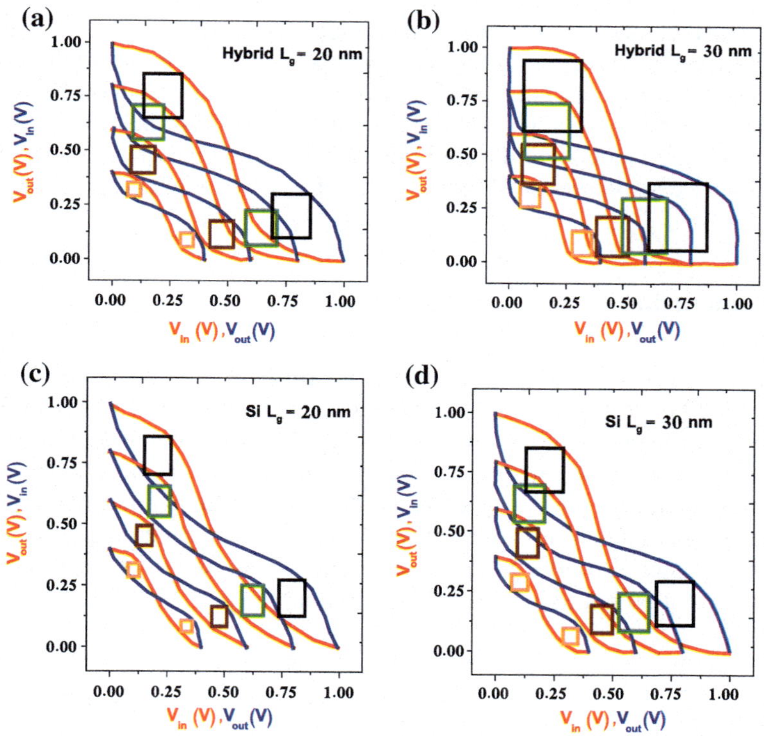

Fig. 8 Butterfly plot of VTC curves with supply voltage V_{DD} as a parameter, of hybrid CMOS at channel length of **a** 20 nm and **b** 30 nm and of Si CMOS at **c** L_g = 20 nm and **d** L_g = 30 nm for different V_{DD}

Fig. 9 Comparison of symmetric noise margins (SNM) as a function of supply voltage V_{DD} for hybrid and equivalent Si CMOS inverters at L_g = 20 and 30 nm

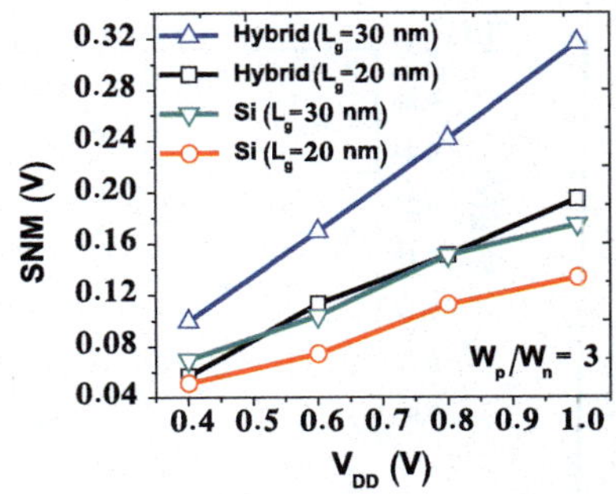

for hybrid CMOS device compared to its equivalent Si CMOS device for L_g = 20 and 30 nm. In summary, the hybrid CMOS inverter yields much lesser value of delay, rise time and fall time, and greater value of noise margins as compared to the equivalent Si CMOS device. Moreover, the f_{osc} for 3-stage inverters built using hybrid CMOS inverters is significantly higher as compared to the value for equivalent Si-based ring oscillator. Such improvement for hybrid CMOS devices is attributed to the superior hole mobility in Ge channel producing lower value of ON resistance of Ge pMOSFET during charging of load capacitance C_L and also superior electron mobility in InGaAs channel which in turn decrease down InGaAs nMOS device resistance value during the discharge of the load capacitance.

5 Conclusion

We have presented the logic behaviour of hybrid CMOS device made of InGaAs nMOS and Ge pMOS devices with barrier layers with channel length of 30 and 20 nm. Our investigations show that hybrid CMOS inverter outperforms its Si counterpart in terms of rise time and fall time and noise margin. Additionally, the oscillation frequency for a 3-stage ring oscillator built with hybrid CMOS inverters as components exhibits an improvement of 264% as compared to its equivalent Si inverters. Hence, hybrid CMOS devices turn out to be useful for future logic applications at 30 nm channel length and beyond.

References

1. S. Tewari, A. Biswas, A. Mallik, Study of InGaAs-channel MOSFETs for analog/mixed-signal system-on-chip applications. IEEE Electron Device Lett. **33**(3), 372–374 (2012)
2. S. Tewari, A. Biswas, A. Mallik, Investigation on high-performance CMOS with pGe and n-InGaAs MOSFETs for logic applications. IEEE Trans. Nanotechnol **14**(2), 275–281 (2015)

3. S. Tewari, A. Biswas, A. Mallik, Performance of CMOS with Si pMOS and asymmetric InP/InGaAs nMOS for analog circuit application. IEEE Trans. Electron Devices **62**(5), 1655–1658 (2015)
4. C. Mondal, A. Biswas, 2-D compact model for drain current of fully depleted nanoscale GeOI MOSFETs for improved analog circuit design. IEEE Trans. Electron Devices **60**(8), 2525–2531 (2013)
5. V. Palankovski, S. Selberherr, Micro materials modeling in MINIMOS-NT. Microsys. Technol. **7**(4), 183–187 (2001). Springer
6. S.K. Ray, R. Mahapatra, S. Maikap, High-k gate oxide for silicon heterostructure MOSFET devices. J. Mater. Sci. Mater. Electron. **17**(9), 689–710 (2006). Springer
7. K. Kalna, L. Yang, A. Asenov, Simulation of high performance III-V MOSFETs for digital apllications. J. Comput. Electron. **2**(2), 341–345 (2003). Springer
8. Y. Singh, M.S. Adhikari, Performance evaluation of a lateral trench-gate power MOSFET on InGaAs. J. Comput. Electron. **13**(1), 155–160 (2013). Springer
9. S. Takagi, High mobility channel MOS device technologies toward nano-CMOS era, in IEEE Nanotechnology Materials and Devices Conference, 281–290, Oct 2011
10. G. Wang, E. Rosseel, R. Loo, P. Favia, H. Bender, M.M. Heyns, W. Vandervorst, High quality Ge epitaxial layers in narrow channels on Si (001) substrates. Appl. Phys. Lett. **96**(11), 111903-1-3 (2010)
11. D. Lin, G. Brammertz, S. Sioncke, C. Fleischmann, A. Delabie, K. Martens, H. Bender, T. Conard, W.H. Tseng, J.C. Lin, W.E. Wang, K. Temst, A. Vatomme, J. Mitard, M. Caymax, M. Meuris, M. Heyns, T. Hoffmann, Ënabling the high-performance InGaAs/Ge CMOS: a common gate stack solution. Proc. Int. Electron Devices Meeting **327–330**, 2009 (2009)
12. L. Hutin, C.L. Royer, J.F. Damlencourt, J.M. Hartmann, H. Grampeix, V. Mazzocchi, C. Tabone, B. Previtali, A. Pouydebasque, M. Vinet, O. Faynot, GeOI pMOSFETs scaled down to 30-nm gate length with record off-state current. IEEE Electron Device Lett. **31**(3), 234–236 (2010)
13. ATLAS User's Manual, *A Device Simulation Software Package* (SILVACO Int, Santa Clara, CA, 2012)
14. F. Xue, A. Jiang, H. Zhao, Y.T. Chen, Y. Wang, F. Zhou, J. Lee, Sub-50-nm $In_{0.7}Ga_{0.3}As$ MOSFETs with various barrier layer materials. IEEE Electron Device Lett. **33**(1), 32–34 (2012)
15. D.L. Cortie, R.A. Lewis, The importance of scattering, surface potential, and vanguard counter-potential in terahertz emission from gallium arsenide. Appl. Phys. Lett. **100**(26), 261601-1-3 (2012)
16. N. Bouarissa, M. Boucenna, Band parameters for AlAs, InAs and their ternary mixed crystals. Phys. Scr. **79**(01), 015701-1-7 (2009)
17. S. Tewari, A. Biswas, A. Mallik, Impact of different barrier layers and indium content of the channel on the analog performance of InGaAs MOSFETs. IEEE Trans. Electron Devices **60** (5), 1584–1589 (2012)
18. W.Q. Chen, S.K. Hark, Strain-induced effects in (111)-oriented InAs/InP, InGaAs/InP, and InGaAs/InAlAs quantum wells on InP substrates. J. Appl. Phys. **77**, 5747–5750 (1995)
19. (2012). The International Technology Roadmap for Semiconductors. [Online] Available: http://www.itrs.net
20. Y. Yonai, T. Kanazawaza, S. Ikeda, Y. Miyamoto, High drain current (>2A/mm) InGaAs channel MOSFET at $V_D = 0.5$ V with shrinkage of channel length by InP anisotropic etching. IEEE IEDM, 307–310 (2011)
21. M.J.H. van dal, G. Vellianitis, B. Duriez, G. Doornbos, C.H. Hsieh, B.H. Lee, K.M. Yin, M. Passlack, C.H. Diaz, Germanium p-channel FinFET Fabricated by Aspect ratio Trapping. IEEE Trans. Electron Devices **61**(2), 430–436 (2014)
22. G. Eneman, M. Wiot, A. Brugere, O.S.I. Casain, S. Sonde, D.P. Brunco, B.D. Jaeger, A. Satta, G. Hellings, K.D. Meyer, C. Claeys, M. Meuris, M.M. Heyns, E. Simoen, Impact of donor concentration, electric field, and temperature effects on the leakage in germanium in p+/n junctions. IEEE Trans. Electron Devices **55**(9), 2287–2296 (2008)

23. A. Agrawal, J. Lin, M. Barth, R. White, B. Zheng, S. Chopra, S. Gupta, K. Wang, J. Gelatos, S.E. Mohney, S. Dutta, Fermi level depinning and contact resistivity reduction using a reduced titania interlayer in n-silicon metal-insulator-semiconductor ohmic contacts. Appl. Phys. Lett. **104**(11), 112101-1-3 (2014)
24. R. Granzner, V.M. Polyakov, F. Schwierz, M. Kittler, R.J. Luyken, W. Rosner, M. Stadele, Simulation of Nanoscale MOSFETs using modified drift-diffusion and hydrodynamic models and comparison with Monte Carlo results. Microelectron. Eng. **83**(2), 241–246 (2006)

Electrical Equivalent Model for Gene Regulatory System

Monalisa Dutta and Soma Barman

Abstract An electrical network model is designed to represent the central dogma of molecular biology and simulate the response to study the behaviors of bacteria gene *E. coli*. The transcription and translation processes of a biological system are represented by differential equations. These equations are mapped into electrical domain, and an equivalent electrical circuit is realized. The electrical response of circuit is simulated in SPICE domain, and result shows the structural and repressor protein behaves like a toggle switch which truly matches with biological system.

Keywords Operon · Genetic switch · Electrical model · Gene regulation
Central dogma · Ordinary differential equation (ODE)

1 Introduction

Central dogma is the process, which deals with unidirectional flow of information from DNA to protein within a cell. The transcription and translation mechanisms are the significant parts of central dogma (Fig. 1). Genetic regulation is the process of interaction between DNA–protein and protein–protein in an organism which is important to control gene expression level. In order to understand the nature of cellular processes, it is necessary to study the behaviors of gene regulatory system [1, 2]. The process of gene expression modeling and genetic regulatory system design are very significant research topics in the present day. In 1960, Jacob and Monod first predicted the existence of repressor protein in bacterial gene *E. coli,* and a general model of gene regulation is described [3]. The process of gene expression produces a product (protein, mRNA) using information contained within

M. Dutta · S. Barman (✉)
Institute of Radio Physics and Electronics, University of Calcutta,
Kolkata, West Bengal, India
e-mail: barmanmandal@gmail.com

M. Dutta
e-mail: dutta.monalisa01@gmail.com

V. Nath (ed.), *Proceedings of the International Conference on Microelectronics, Computing & Communication Systems*, Lecture Notes in Electrical Engineering 453,
https://doi.org/10.1007/978-981-10-5565-2_14

the genes [4–7]. Researchers from different field proposed various modeling and simulation techniques to describe the gene expression [8]. Model based on kinetic equations, stochastic formulation, and piecewise-linear differential equations also was proposed by researchers [9–11]. Harley and Adam proposed a stochastic mechanism to describe a protein product formation by a gene which also regulates the expression of another gene [12]. In order to describe gene expression process, some mathematical, algorithms, and Boolean network-based models are used [13–17]. Cells are complex processors of information and the processors respond to many signals in random manner. The complexity of cellular network can be simplified by considering the complex network as a set of simpler components which are interconnected through input–output signals. This network characteristic can be realized by electrical circuits, and an artificial gene network will be formed using electronic or electrical components [2].

In this paper, we proposed an equivalent electrical model of a biological system where mapping of biological processes is done in electrical domain and study the response of electrical system to correlate with the biological system.

The paper is divided into different sections; brief descriptions about the biological processes are illustrated in introduction. In Sect. 2, kinetic equations are used to describe the gene expression [9]. Equivalent differential equations for the kinetic equations are also developed, and solutions of the differential equations are mapped in electrical domain. In Sect. 3, the simulation results of the process are depicted. In Sect. 4, conclusion is drawn based on electrical response of biological

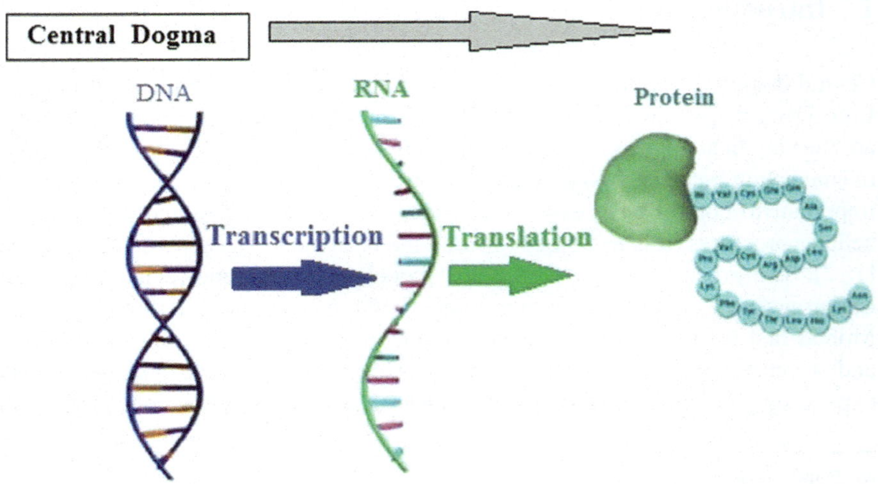

Fig. 1 Steps of central dogma

system.

2 Mathematical Modeling of Transcription, Translation, and Dimerization of Protein

A small functioning unit of prokaryotes DNA is known as operon was first described by Jacob and Monod in *E. coli*. An operon contains a group of structural genes, a repressor gene, and a promoter. The structural genes are three types lacZ, lacY, and lacA. The genes are encoded beta-galactosidase, lactose permease, and galactoside O-acetyltransferase protein, respectively. The repressor gene inhibits the expression of structural genes which is named as lac I that encodes repressor protein. The promoter is the primarily control sites of an operon. The control mechanism within the operon is regulated by lactose. The lactose is generated by the structural protein. When the lactose is absent, the repressor protein added to the promoter site and inhibits transcriptional activation of the structural gene. Therefore, protein production by structural gene is blocked and the process starts again whenever the repressor protein is removed from promoter site. Finally, a switch like phenomena occurs within operon [1], shown in Figs. 2 and 3.

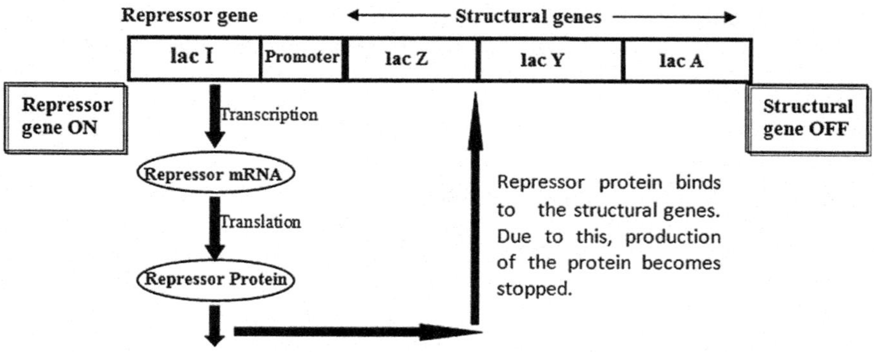

Fig. 2 Operon and the control path for structural protein

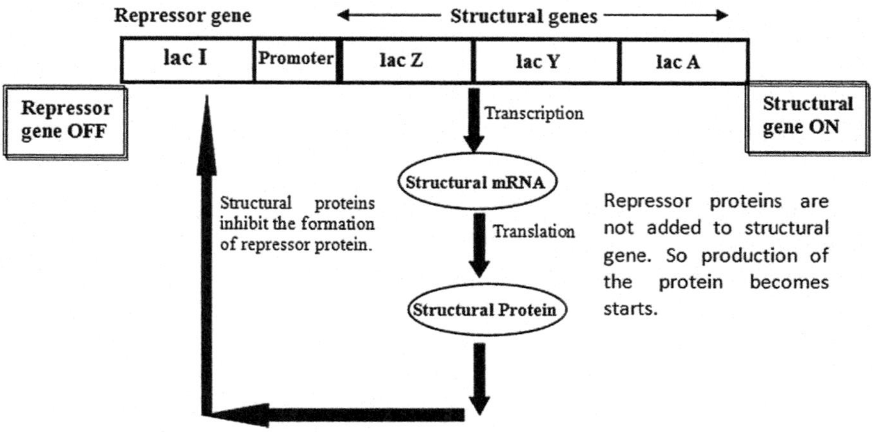

Fig. 3 Operon and the control path for repressor protein

2.1 Transcription from DNA to mRNA

In the transcription procedure, the genes in operon are converted into mRNA by the help of transcription factor. The transcription factor is sometimes called DNA-binding factor. It is a protein that binds with specific DNA sequence and controls the transcription process. Promoter is a sequence of base pairs in operon that initiates the process of transcription.

The rate of formation of mRNA from genes is zeroth order, and rate of decay is first order [18]. The corresponding equations for the transcription process are represented by Eqs. 1a–1e.

$$[P_{\mathrm{RS}}] + [T_{\mathrm{FS}}] \xrightarrow{K_{\mathrm{S1}}} [P_{\mathrm{RS}}] + [T_{\mathrm{FS}}] + [\mathrm{mRNA_S}] \tag{1a}$$

$$[\mathrm{mRNA_S}] \xrightarrow{K_{\mathrm{S2}}} \phi \tag{1b}$$

$[P_{\mathrm{RS}}]$, $[T_{\mathrm{FS}}]$, $[\mathrm{mRNA_S}]$, K_{S1}, K_{S2} represent concentration of promoter in operon, concentration of transfer factor protein, concentration of mRNA for structural protein, transcription initialization factor, mRNA degradation constant.

The above chemical equations can be rewritten as a differential equation.

$$\frac{\mathrm{d}}{\mathrm{d}t}[\mathrm{mRNA_S}] = K_{\mathrm{S1}}[P_{\mathrm{RS}}][T_{\mathrm{FS}}] - K_{\mathrm{S2}}[\mathrm{mRNA_S}] \tag{1c}$$

The solution of the differential equation is shown by Eqs. (1d).

$$[\mathrm{mRNA_S}] = \frac{K_{\mathrm{S1}}[P_{\mathrm{RS}}][T_{\mathrm{FS}}]}{K_{\mathrm{S2}}}[1 - e^{K_{\mathrm{S2}}t}] \tag{1d}$$

Similarly, the solution for repressor mRNA is given below

$$[\mathrm{mRNA_R}] = \frac{K_{\mathrm{R1}}[P_{\mathrm{RR}}][T_{\mathrm{FR}}]}{K_{\mathrm{R2}}}[1 - e^{K_{\mathrm{R2}}t}] \tag{1e}$$

2.2 Translation from mRNA to Protein

In the process of translation, mRNA is converted to protein. The rate of protein formation depends on the concentration of mRNA. The mRNA and protein both are time-dependent functions. The protein synthesis procedure is express by following Eqs. 2a–2e.

$$[\mathrm{mRNA_S}] \xrightarrow{K_{S3}} [m\mathrm{RNA_S}] + [P_S] \tag{2a}$$

$$[P_S] \xrightarrow{K_{S4}} \phi \tag{2b}$$

$[P_S]$, K_{S3}, K_{S4} represent concentration of structural protein, translation initialization factor, structural protein degradation constant.

The differential equation of above set of chemical reactions for structural protein is given below.

$$\frac{\mathrm{d}}{\mathrm{d}t}[P_S] = K_{S3}[\mathrm{mRNA_S}] - K_{S4}[P_S] \tag{2c}$$

The solution for structural protein is shown in Eqs. (2d).

$$[P_S] = \frac{K_{S3}[\mathrm{mRNA_S}]}{K_{S4}}[1 - \mathrm{e}^{K_{S4}\,t}] \tag{2d}$$

Similarly, the solutions for repressor protein are given below

$$[P_R] = \frac{K_{R3}[\mathrm{mRNA_R}]}{K_{R4}}[1 - \mathrm{e}^{K_{R4}\,t}] \tag{2e}$$

2.3 *Protein Dimerization*

The protein formed by translation of mRNA is monomer protein. This monomer converted to dimer protein. Dimer protein plays a significant role in regulation of another protein.

The necessary equations for the formation of dimer protein are represented Eqs. 3a–3f. The dimer protein and also the time-dependent functions depend on concentrations of monomer protein.

$$[P_S] + [P_S] \xrightarrow{K_{S5}} [PP_S] \tag{3a}$$

$$[PP_S] \xrightarrow{K_{S6}} [P_S] + [P_S] \tag{3b}$$

$$[PP_S] \xrightarrow{K_{S7}} \phi \tag{3c}$$

Where, $[PP_S]$, K_{S5}, K_{S6}, K_{S7} represent dimer concentration of structural protein, dimer association factor, dimer dissociation factor, dimer degradation factor.

The differential equation of above kinetic equations is given below.

$$\frac{d}{dt}[PP_S] = K_{S5}[P_S][P_S] - K_{S6}[PP_S] - K_{S7}[PP_S] \tag{3d}$$

The solution of the above differential equations for structural gene is shown below.

$$[PP_S] = \frac{K_{S5}[P_S][P_S]}{K_{S6} + K_{S7}}\left[1 - e^{(K_{S6} + K_{S7})t}\right] \tag{3e}$$

A similar solution for repressor dimer protein is represented by Eqs. (3f).

$$[PP_R] = \frac{K_{R5}[P_R][P_R]}{K_{R6} + K_{R7}}\left[1 - e^{(K_{R6} + K_{R7})t}\right] \tag{3f}$$

The biological processes are mapped into electrical domain and study the gene regulation of prokaryotic cell. The electrical equivalent parameters of biological phenomena are shown in Table 1. The value of biological entities and equivalent value of electrical entities are represented by Table 2. Table 3 represents the electrical equivalent of biological process based on transcription, translations, and dimerization. Based on above process, equivalent electrical circuits are realized (Fig. 4).

Table 1 The electrical equivalent parameters of biological phenomena

Biological parameter	Analogous electrical circuit parameter
$\frac{K_{S1}[P_{RS}][T_{FS}]}{K_{S2}}$ and $\frac{K_{R1}[P_{RR}][T_{FR}]}{K_{R2}}$	Input voltage for equivalent electrical mRNA concentration circuit of structural gene (V_{msi}) and repressor gene (V_{mri}), respectively
[mRNA$_S$] and [mRNA$_P$]	Output voltage V_{mso} and V_{mro} of mRNA circuit, respectively
$1/K_{S2}$ and $1/K_{R2}$	Product of resistance and capacitance $R_{ms2}C_{ms2}$ and $R_{mr2}C_{mr2}$, i.e., time constant of mRNA concentration circuit
$\frac{K_{S3}[\mathrm{mRNA}_S]}{K_{S4}}$ and $\frac{K_{R3}[\mathrm{m}RNA_R]}{K_{R4}}$	Input voltage of monomer structural (V_{psi}) and repressor(V_{pri}) protein formation circuit
$[P_S]$ and $[P_R]$	Output voltage of monomer protein formation circuit, Vpso and Vpro, respectively
$1/K_{S4}$ and $1/K_{R4}$	RC time constant. $R_{ps4}C_{ps4}$ and $R_{pr4}C_{pr4}$, respectively
$\frac{K_{S5}[P_S][P_S]}{K_{S6} + K_{S7}}$ and $\frac{K_{R5}[P_R][P_R]}{K_{R6} + K_{R7}}$	Input voltage of dimer structural (V_{ppsi}) and repressor protein (V_{ppri}) circuit
$[PP_S]$ and $[PP_R]$	Output voltage of dimer structural (V_{ppso}) and repressor protein (V_{ppro}) circuit
$1/(K_{S6} + K_{S7})$ and $1/(K_{R6} + K_{R7})$	RC time constant. $R_{pps6+pps7}C_{pps6+pps7}$ and $R_{ppr6+ppr7}C_{ppr6+ppr7}$, respectively

Table 2 The value of biological entities and equivalent value of electrical entities

Value of biological entities	Equivalents electrical values of biological entities	Electrical components
K_{S2}, K_{R2} = 0.023 $(100\ s)^{-1}$ [9]	τ_{ms2}, τ_{mr2} = (1/0.023) (100 s)	R_{ms2}, R_{mr2} = 435 kΩ, C_{ms2}, C_{mr2} = 100 μF
K_{S4}, K_{R4} = 0.077 $(100\ s)^{-1}$ [9]	τ_{ps4}, τ_{pr4} = (1/0.077) (100 s)	R_{ps4}, R_{pr4} = 130 kΩ, C_{ps2}, C_{pr2} = 100 μF
K_{S6}, K_{R6} = 0.023 $(100\ s)^{-1}$ [9] K_{S7}, K_{R7} = 0.058 $(100\ s)^{-1}$ [9]	τ_{pps4}, τ_{ppr4} = (1/ (0.023 + 0.058)) (100 s)	$R_{pps6+pps7}$, $R_{ppr6+ppr7}$ = 120 kΩ, $C_{pps6+pps7}$, $C_{pps6+pps7}$ = 100 μF

Table 3 The electrical equivalent of biological process based on transcription, translations, and dimerization

Biological process	Equations in biological field	Equations in electrical field
Transcription	$[\text{mRNA}_S] = \frac{K_{S1}[P_{RS}][T_{FS}]}{K_{S2}}[1 - e^{K_{S2}t}]$	$[V_{mso}] = [V_{msi}][1 - e^{t/\tau_{ms2}}]$
	$[\text{mRNA}_R] = \frac{K_{R1}[P_{RR}][T_{FR}]}{K_{R2}}[1 - e^{K_{R2}t}]$	$[V_{mro}] = [V_{mri}][1 - e^{t/\tau_{mr}}]$
Translation	$[P_S] = \frac{K_{S3}[\text{mRNA}_S]}{K_{S4}}[1 - e^{K_{S4}t}]$	$[V_{pso}] = [V_{psi}][1 - e^{t/\tau_{ps}}]$
	$[P_R] = \frac{K_{R3}[\text{mRNA}_R]}{K_{R4}}[1 - e^{K_{R4}t}]$	$[V_{\text{pro}}] = [V_{\text{pri}}][1 - e^{t/\tau_{\text{pr}}}]$
Dimerization	$[PP_S] = \frac{K_{S5}[P_S][P_S]}{K_{S6} + K_{S7}}[1 - e^{(K_{S6} + K_{S7})t}]$	$[V_{\text{ppso}}] = [V_{\text{ppsi}}][1 - e^{t/\tau_{\text{pps}}}]$
	$[PP_R] = \frac{K_{R5}[P_R][P_R]}{K_{R6} + K_{R7}}[1 - e^{(K_{R6} + K_{R7})t}]$	$[V_{\text{ppro}}] = [V_{\text{ppri}}][1 - e^{t/\tau_{\text{ppr}}}]$

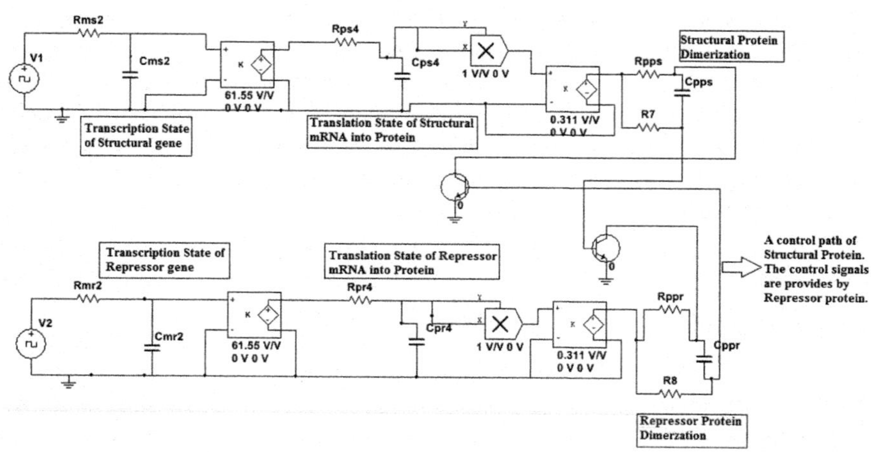

Fig. 4 Electrical equivalent circuit of gene regulatory system

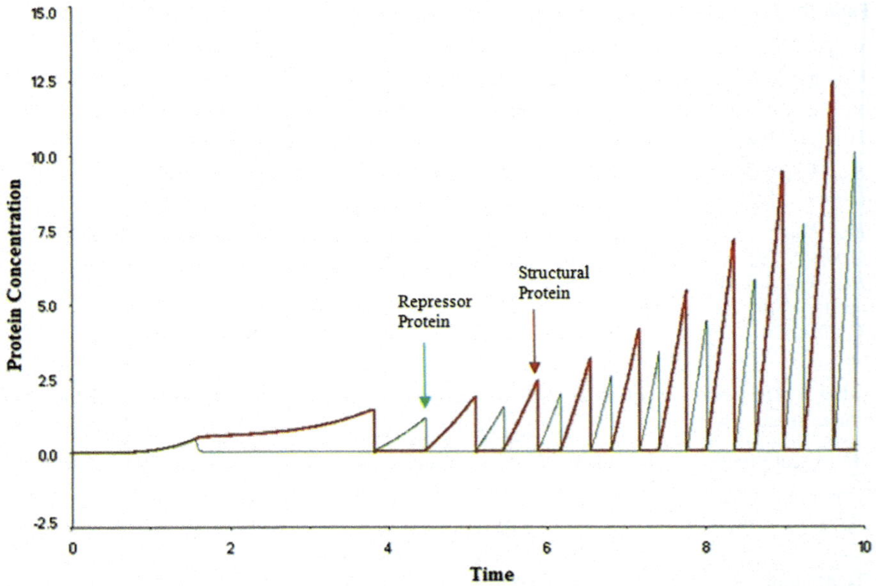

Fig. 5 Switch-like model of structural and repressor gene

3 Result and Discussion

The transcription, translation, and dimerization processes are represented by differential equations. The equations are modeled in electrical domain where the values of the electrical components are selected based on the solution of differential equations (Table 2). The model is tested in SPICE domain for a gene sample *E. coli* and simulates the concentration of structural and repressor protein with respect to time. The simulation results depict that when the repressor protein cross some certain concentration, it inhibits the production of structural gene. When repressor protein concentration goes below certain level, structural gene gets activated; therefore, the gene regulation in *E. coli* turning ON and OFF likes a switch [9]. The electrical response (Fig. 5) with respect to time truly matches the ODE of biological processes of prokaryotes gene (*E. coli*).

4 Conclusion

The gene regulation of *E. coli* gene is studied here in electrical domain. The protein concentration of the operon is represented by voltage response of electrical circuit. Simulation study in SPICE domain reflects that the structural and repressor protein behaves like a switch. The protein concentration in operon for *E. coli* is studied virtually without help of smelly laboratory and actual sample gene. Virtual gene network could provide deeper about the gene regulation of gene. The work helps in designing molecular devices or sensor development. It may further extend to regulate gene expression for other gene.

Acknowledgements The authors wish to thank DST, Science and Engineering Research Board (SERB/F/4504/2013–2014), Govt. of India for funding support of research work.

References

1. J. Hasty, D. McMillen, Engineered gene circuits. Nature 224–230 (2002). doi:10.1038/nature01257
2. M. Kaern, W.J. Blake, J.J. Collins, The engineering of gene regulatory networks. Annu. Rev. Biomed. Eng. 179–206 (2003). doi:10.1146/annurev.bioeng.5.040202.121553
3. F. Jacob, J. Monod, On the regulations of gene activity, in *Symposium on Cellular Regulatory Mechanisms* (Cold Spring Harbor laboratory, New York, 1961), pp 193–211
4. A. Becskei, Luis Serrano, Engineering stability in gene networks by autoregulation. Nature **405**, 590–593 (2000)
5. M.B. Elowitz, S. Leibler, A synthetic oscillatory network of transcriptional regulators. Nature **403**, 335–338 (2000)
6. S. Gardner, C.R. Cantor, J.J. Collins, Construction of a genetic toggle switch in *Escherichia coli.* Nature **403**, 339–342 (2000)
7. N.E. Buchler, U. Gerland, T. Hwa, Nonlinear protein degradation and the function of genetic circuits. Proc. Natl. Acad. Sci. U S A. **102**, 9559–9564 (2005)
8. H. De Jong, Modeling and simulation of genetic regulatory systems: a literature review. J. Comput. Biol. 67–103 (2002)
9. H. Kim, E. Gelenbe, Stochastic gene expression modeling with hill function for switch-like gene responses. IEEE Transac. Comput. Biol. Bioinform. **9**(X), 973–979 (2012)
10. H.H. McAdams, A. Arkin, Stochastic mechanisms in gene expression. Proc. Natl. Acad. Sci., USA **94**, 814–819 (1997)
11. T. Chen, H.L. He, Modeling gene expression with differential equations, in *Pacific Symposium of Biocomputing* (1999), pp. 17–28
12. H.H. McAdams, A. Arkin, Stochastic mechanisms in gene expression. Proc. Natl. Acad. Sci. **94**, 814–819 (1997)
13. I. Shmulevich et al., Probabilistic Boolean networks: a rule-based uncertainty model for gene regulatory networks. Bioinformatics **18**, 261–274 (2002)
14. C. Ferreira, in *Gene Expression Programming: Mathematical Modeling by an Artificial Intelligence*, vol 21, (Springer, 2006)
15. G. Bernot, et al., Modeling and analysis of gene regulatory networks, in *Modeling in Computational Biology and Biomedicine*, (Springer, Berlin, Heidelberg, 2013), pp. 47–80
16. S. Paul, D.A. Baxter, J.H. Byrne, Mathematical modeling of gene networks. Neuron **26**, 567–580 (2000)
17. N.R. Zabet, A.N.W. Hone, D.F. Chu, Design principles of transcriptional logic circuits. ALIFE, (2010), pp. 186–194
18. J.L. Hargrove, F.H. Schmidt, The role of mRNA and protein stability in gene expression. FASEB J. **3**, 2360–2370 (1989)

Antenna Path Loss Propagation in the Dehradun Region at 1800 MHz in L-Band

Ranjan Mishra, Piyush Kuchhal and Adesh Kumar

Abstract A proper and good coverage is an important parameter in the planning of cellular network. Path loss models are crucial in the planning of wireless network as they assist in interference estimations, frequency assignments, and evaluation of cell parameters. This paper reports the results of the propagation path loss to a fixed height of antenna at 1800 MHz in the outskirts of Dehradun city in the state of Uttarakhand, India. The results shown in the paper are for propagation path loss considering Okumura–Hata model and Walfisch–Ikegami model in the implementation of a digital cellular system in the region on the outskirt of Dehradun. An analysis of co-channel interference is also presented. 1800 MHz falls in the L-band of SHF, and after 900 MHz band, this is the most favorable frequency band for mobile communication.

Keywords Propagation model · Spectral efficiency · Co-channel interference

1 Introduction

Cellular communications is one of the fastest growing technologies today. The challenging aspect of it is to provide a seamless communication to all. A large percentage of all new telephone subscribers are cellular one nowadays, and there is a continuous increase in the number. Cellular digital technology [1] is going to be the universal way of communication in a long term. The rapid growth in the cellular mobile communications market has been driven by technological development and

R. Mishra (✉) · P. Kuchhal · A. Kumar
College of Engineering Studies, University of Petroleum and Energy Studies, Dehradun, India
e-mail: rmishra@ddn.upes.ac.in

P. Kuchhal
e-mail: pkuchhal@ddn.upes.ac.in

A. Kumar
e-mail: adeshmanav@gmail.com

V. Nath (ed.), *Proceedings of the International Conference on Microelectronics, Computing & Communication Systems*, Lecture Notes in Electrical Engineering 453,
https://doi.org/10.1007/978-981-10-5565-2_15

implementation of new systems. A major spread for cellular communication systems is enhanced coverage of the cell, which is fulfilled by adopting network planning early in the cycle and high spectrum efficiency later. A suitable implementation for good coverage, spectrally improved and quality cellular communication network is the need of hour. This is achieved by a proper network system planning by using the cell coverage area. The more the coverage is, the more the people will get benefitted with technology.

Antenna parameter is a major factor in the determination of cell coverage. Networking planning is done using the coverage of a cell. Cell coverage is dependent on antenna-defined parameters such as transmitting power, gain, location, and height of the base antenna. The geographical terrain parameters affect the radio frequency coverage. It includes environment, hills, valley, density of population, and building distribution. These are not pre-defined but they vary from one place of observation to another place. Also, the unpredictability is an issue there. This leads to the irregular practical cell in a multipath environment. Several prediction models [2–4] have been developed by different mobile planners in past years to develop a practical cell. The most efficient and dominating propagation models are Hata [5] and Okumura et al. [6] and Walfisch and Bertoni [7] and Ikegami et al. [8] propagation models. These two models are the basic foundation of many computer-aided propagation prediction tools for the network planning and mapping.

Propagation coverage area and the strength of the signal are the two prime factors in the designing of the cellular wireless systems. The coverage requirement of the network is fulfilled by the number of cells. Propagation models [9, 10] are used to calculate the total available number of cells in a cluster and its total cover area. Initially, the engineering design aspects focus on the cell coverage. A proper model is helpful [11] in the allocation of the traffic distribution. It also helps in off-loading from one cell site to another cell sites as part of a capacity enhancement. The propagation model is primarily useful in the determination of the location of the cell sites to get an optimized position [12] in the existing cluster. In this paper, we have presented Okumura–Hata model and Walfisch–Ikegami model at 1800 MHz in the outskirts of the Dehradun city and its valley region.

2 Sectorization

Co-channel interference (CCI) is a major threat in frequency reuse, as it leads to unwanted interference between channels in cell topology. The frequency reuse system is used frequently to increase the spectrum efficiency. But serious interference may occur for improper designed system. Optimum spectral efficiency [13] can be achieved by either reducing the size of the cell, or decreasing the factors influencing frequency reuse. In a cellular system, reusing each frequency at several regions of service area increases the capacity.

The interference hampers cell topology. It is primarily reduced by implementing directional antennas radiating in a specific sector. Using sectorization [14] techniques, co-channel interference is reduced and system performance increases. This technique in the reduction of co-channel interference by the use of specific antennas is termed as sectoring. It increases network system performance.

Generalized values of CCI under sectorization are modified to $(Q^{-n} + (Q + 0.7)^{-n})^{-1}$ for a three-sector case and $(Q + 0.7)^{n}$ for a six-sector case. In all these equations, n is the exponent value for the path loss.

Values of the different exponents assigned for the path loss exponent are provided in Table 1.

3 Experimental Work

A city like Dehradun has variation in the geographical terrain. The total valley area is fuzzy in nature with a combination of urban, semiurban, and greenery vegetation. RF propagation in this multipath environment is generally varied considerably due to irregular geographical terrain. The distribution of the building structure and its height also vary from one place to another and from one area to another. As a result of the irregularity, a precise propagation model is not suitably available. Although network operators have a well-advanced computer-aided propagation prediction tools, they too require terrain define factors. This lack of information in the prediction tool introduces impurities into the database and an inevitable error results. This equipment is a must in the planning of network and cellular implementation. After the deployment of the system, various radio values like verification, reduction of the interference reduction, and adjustments of handover parameters are carried out by means of an experimental survey.

Experimental work has been carried in the northwest part of Dehradun city. The area under test is in the outskirts of Dehradun city and is 10–15 km from the central location. The areas under observation are centrally located at Kandholi, Pondha, and Bhauwala of Dehradun. All the observations are measured up to 2.5 km because the population distribution is located within this range. Any two areas under observation are 5–6 km apart in distance.

Table 1 Exponent value range for path loss

Area under study *t*	Magnitude of exponent (*n*)
Space free transmission	1.9–2.1
Urban cell area	2.65–4.05
Vegetative cell area	2.9–4.9
Building sight in line	1.5–1.75
Obstruction between structures	2.9–3.94

4 Propagation Model

The propagation model helps in the determination of the exact location of the cell sites and its coverage to achieve an optimum planning [15] in the network. Urban propagation environments that enriched irregular terrains are mostly unpredictable in network. Here, the distribution of environment and terrain changes abruptly in place of observation. Therefore, it requires sharp characterization under dense urban, urban, suburban environment, etc. The planning under this propagation is done on the basis of propagation models such as Okumura and Hata model and Walfisch and Ikegami model. These two are the most widely acclaimed one. Recently, these models were exclusively used in Oman [16] and Egypt [17]. Calibrating the model to the actual propagation environment will be helpful in gaining confidence in the model. The purpose of the work is to validate the propagation model as per the geographical terrain.

5 Hata–Okumura Propagation Model

In the cellular mobile system, the Hata–Okumura propagation model is a deployed extension of Okumura work. This model is the most authenticated propagation model [9, 11] for predicting the system planning in the built-up areas. Okumura's report [6] is the most comprehensive propagation prediction methods for the deployment of cellular mobile radio system. An empirical formula is predicted by Hata taking the result based on Okumura's curves [5]. Hata propagation formulae are extensively utilized by the radio engineers in the calculation of the link budget for urban areas. This is dependent on ranges of frequency, and distance of observation.

6 Walfisch–Ikegami Model

This model [7, 8] is used for finding the losses in the urban area of observation and is best suited for the frequency range in higher UHF and lower SHF.

The model is valid from 800 to 2000 MHz and within a range of 0.02 to 5 km.

7 Results and Discussion

The various values of CCI losses, in the case of non-sectorization of the cells, under the different combinations of cell per cluster are plotted in Fig. 1. Here, both the measured results are plotted along with the theoretical values. A good proximity

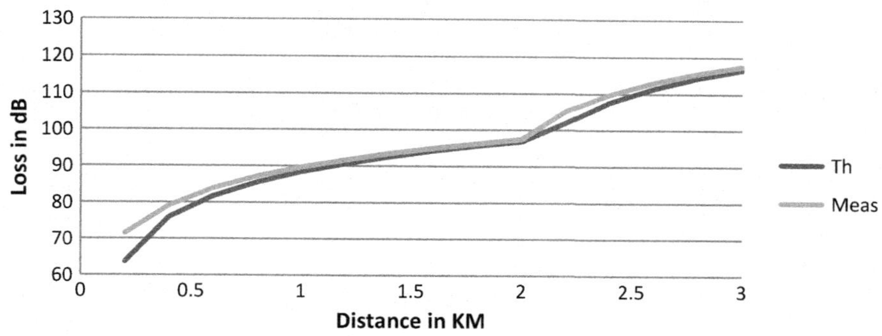

Fig. 1 CCI value for non-sectorization in suburban environment

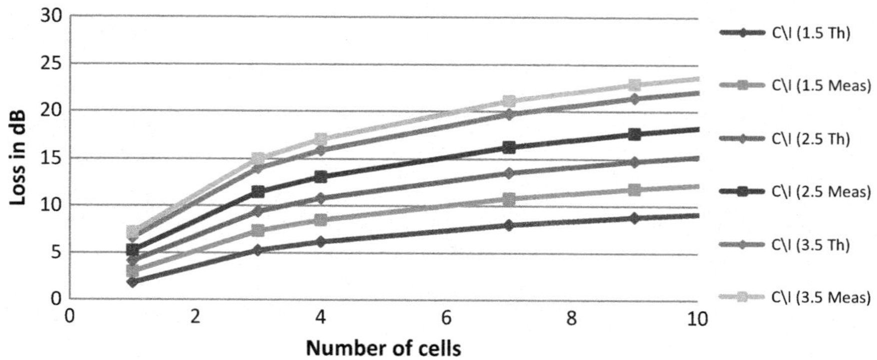

Fig. 2 CCI value for 3 sectorization for different terrain

between two has been obtained. For a distance in between 1 and 2 km, the best result is obtained. The size of omnidirectional antenna radiation is therefore confined to a radius of 2 km for the best result. This pattern is best suitable in rural environment.

The measured value of the CCI loss in the three-sector and six-sector cell patterns is plotted in Figs. 2 and 3. Along with the measured results, the theoretical values are also plotted as a means of comparing them. Three-sector cell structure is mostly applicable for suburban area, and six-sector configuration is used in dense urban environment. In these two figures, measured results are for three loss exponents. A comparative plot has been described in Fig. 4 between the results obtained from the three-antenna configuration. This result clearly demonstrated that the sectorization improves the spectral efficiency.

The measured propagation loss under the three distinct terrains in the outskirts of Dehradun has been plotted and compared with the Okumura–Hata model from Figs. 5, 6 and 7. Figure 5 shows the results measured in the urban environment. The observation measured in the semiurban environment is shown in Fig. 6. The

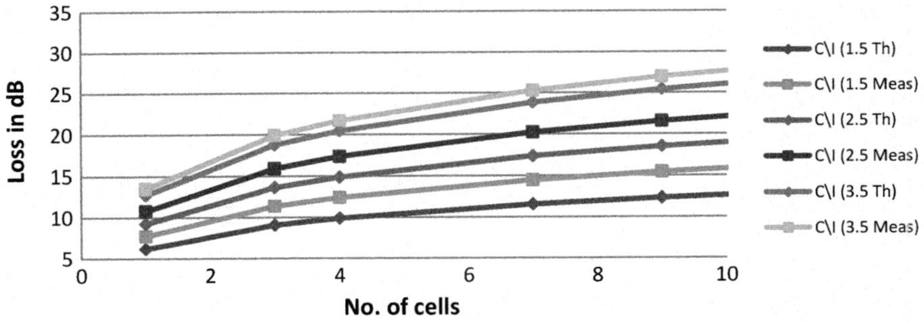

Fig. 3 CCI value for 6 sectorization for different terrain

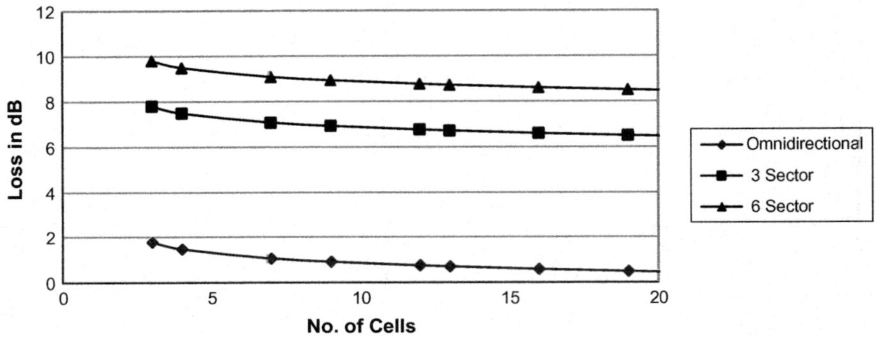

Fig. 4 Comparison of CCI value for different antenna sectorization

loss in urban environment and suburban environment reveals that the measured losses are very close to the results in the Okumura–Hata model. It is due to the proximity of the environment in the two cases. The measurement taken in the rural environment is shown in Fig. 7. These results in this region are mostly influenced with free space path loss. Most of the terrain in the outskirt of the city comes under semiurban.

Figure 8 shows the measured propagation path loss in the dense urban environment, and it is plotted against the Walfisch–Ikegami model. This environment is bounded with closely packed buildings and other high-density terrain. The close proximity as shown in Fig. 8 indicates that this model is best fit in the dense urban environment. When the distance between the cell sight antenna and the area under observation is small, the two results are very close to each another. As the distance goes beyond 1 km, the difference in the measured results and the theoretical values increases. The variation in the measured value with respect to the theoretical one depends on the building distribution in the dense urban environment.

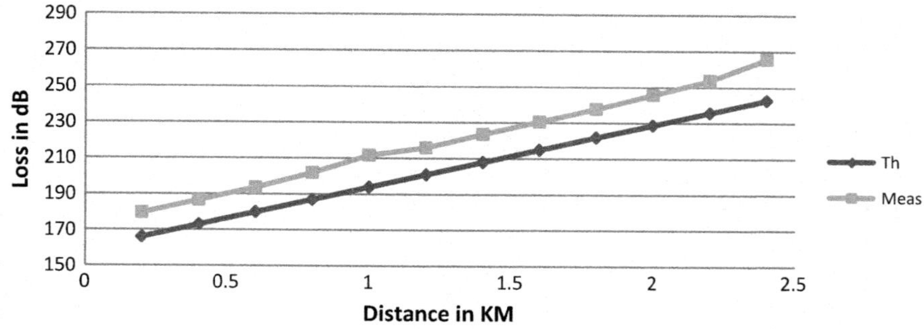

Fig. 5 Propagation loss in urban environment using Hata model

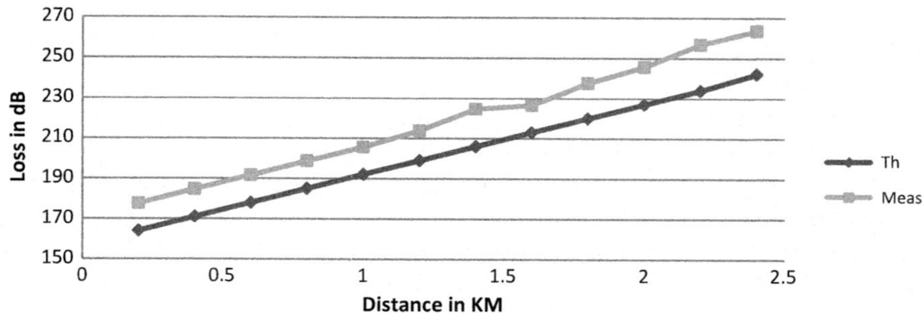

Fig. 6 Propagation loss in suburban environment using Okumura–Hata model

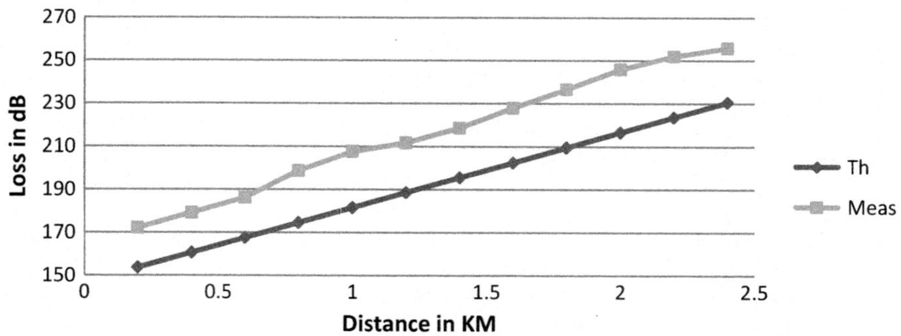

Fig. 7 Propagation loss in rural environment using Okumura–Hata model

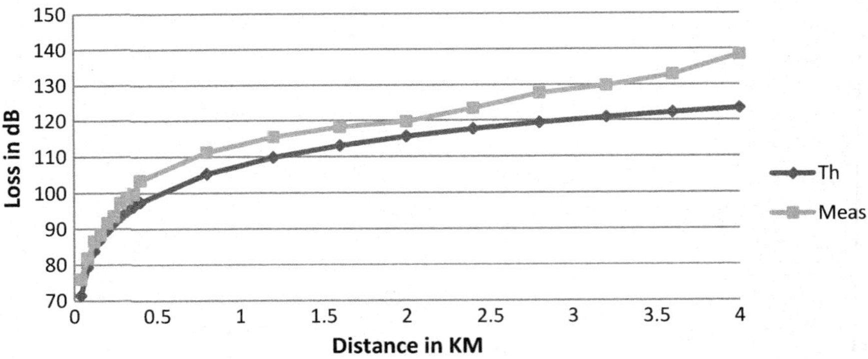

Fig. 8 Propagation loss in dense urban environment using Walfisch–Ikegami model

8 Conclusion

Dehradun area is vastly distributed due to the irregularity in the terrain environment. The cellular mobile system is a high-capacity system that can be designed to use spectrum efficiency while providing the highest quality services. The resulting *C*/*I* for three-sector cell pattern is found to be 24.5 dB. A good proximity of the theoretical and measured value is achieved as shown in the different plots.

References

1. A. Mehrotra, *Cellular Radio Performance Engineering* (Artech House, Norwood, MA, 1994)
2. T.S. Rappaport, Characterization of UHF multipath radio channels in factory buildings. IEEE Trans. Antenna Propag. **37**(8), 1058–1069 (1989)
3. R. Edwards, J. Durkin, Computer prediction 0f service area for VHF mobile radio networks. Proc. IEEE **116**(9), 1493–1500 (1969)
4. A.G. Longley, P.L. Rice, Prediction of tropospheric radio transmission loss over irregular terrain. ESA technical report (1968)
5. Hata, M., Empirical formula for propagation loss in land mobile radio services. IEEE Trans. Veh. Technol. **VT-29**, 317–325 (1980)
6. T. Okumura et al., Field strength and its variability in VHF and UHF land mobile service. Rev. Electr. Commun. Lab. **16**(9–10), 825–873 (1968)
7. J. Walfisch, H.L. Bertoni, A theoretical model of UHF propagation in urban environments. IEEE Trans. Antennas Propag. **36**, 1788–1796 (1988)
8. F. Ikegami, T. Takeuchi, S. Yoshida, Theoretical prediction of mean field strength for urban mobile radio. IEEE Trans. Antennas Propag. **39**(3) (1991)
9. T.K. Sarkar et al., A survey of various propagation models for mobile communications. IEEE Antennas Propag. Mag. **45**(3), 51–82 (2003)
10. Anderson, J.B., Rappaport, T.S., Yoshida, S., Propagation measurements and models for wireless communications channels. IEEE Commun. Mag., 42–49 (1995)

11. A. Medeisis, A. Kajackas, Modification and tuning of the universal Okumura-Hata model for radio wave propagation predictions, in *Asia-Pacific Microwave Conference 2007*, vol. 11–14 (2007), pp. 1–4
12. P.K. Sharma, R.K. Singh, Comparative analysis of propagation path loss models with field measured data. Int. J. Eng. Sci. Technol. **2**, 2008–2013 (2010)
13. G.K. Chan, Effects of sectorization on the spectrum efficiency of cellular radio systems. IEEE Trans. Veh. Technol. **41**(3) (1992)
14. Lee, W.C.Y., Spectrum efficiency in cellular. IEEE Trans. Veh. Technol. **38**(2) (1989)
15. M. Kumar, V. Kumar, S. Malik, Performance and analysis of propagation models for predicting RSS for efficient handoff. Int. J. Adv. Sci. Res. Technol. **1**(2) (2012)
16. Z. Nadir, M.I. Ahmad, Path loss determination using Okumura-Hata model and cubic regression for missing data for Oman. Proc. IMECS, vol. 2 (2010)
17. M. Farhoud et al., Empirical correction of the Okumura Hata model for 900 MHz band in Egypt, in Third IEEE International Conference on Communication and Information Technology (2013) pp. 386–390

Study of Strained-Si/SiGe Channel p-MOSFETs Using TCAD

Sanghamitra Das, Tara Prasanna Dash, Rajib Kumar Nanda and C. K. Maiti

Abstract A simulation study of strained-Si/SiGe channel heterostructure p-MOSFETs has been carried in order to enhance the performance of the experimentally reported such devices. Strained-Si channel device shows 40% mobility enhancement at 300 K and almost doubled at 200 K, when the results are compared with conventional Si-MOSFETs. The effects of low temperature operation on the performance of MOSFETs have been studied and discussed in terms of threshold voltage and output characteristics.

Keywords Heterostructure MOSFET · Hetero-FET · Device simulation

1 Introduction

Currently, the most mature device in SiGe is the SiGe heterojunction bipolar transistors (HBT). Due to the lower hole mobility of Si, p-MOSFET performance has always been limited. As far as speed and current drive capability are concerned, pMOS devices are mostly inferior to nMOS devices. Strained-SiGe and strained-Si grown over relaxed SiGe/Si layers can be used to benefit from the higher hole mobility in strained-SiGe [1]. As challenges to downscale MOSFETs are growing continuously, it is important to examine possible performance enhancements in strained-SiGe channel MOSFETs. Possible solutions are being sought to overcome the fundamental scaling limit via use of alternative channel materials, high-k/metal gate dielectrics, and non-classical device architectures. The development of a Si/SiGe field-effect transistors (FET) technology is also very interesting for the promises it offers to the current Si technology. On one hand, FETs are easier to fabricate than HBTs because they require fewer processing steps. In addition, FETs can offer lower noise at high frequencies; which is useful for MMIC applications [2–4].

S. Das · T. P. Dash (✉) · R. K. Nanda · C. K. Maiti
Department of Electronics and Communication Engineering, Siksha 'O' Anusandhan University, Bhubaneswar 751030, Odisha, India
e-mail: taradash@soauniversity.ac.in

V. Nath (ed.), *Proceedings of the International Conference on Microelectronics, Computing & Communication Systems*, Lecture Notes in Electrical Engineering 453,
https://doi.org/10.1007/978-981-10-5565-2_16

A wide variety of novel MOSFETs devices that are contenders for use in future high-speed and low-noise RF circuits have been evaluated [3]. The devices include (i) Strained-Si channel n- and p-MOSFETs on SiGe buffer layer, (ii) tri-gate FinFETs, and (iii) gate-all around FinFETs on buried oxide with reference to Si-MOSFETs [5–10].

When the trade-off between short channel effect, drive current and power consumption taken into consideration, FETs can be used in a complementary architecture which has significant advantages for circuit design and tremendous reduction for the power consumption of digital circuits. In this respect, Si/SiGe can offer much greater improvement compared to conventional Si CMOS. Beside matching the performance of the p- and n-type devices, the complementary heterojunction MOS (CHMOS) technology has the ability to have four times better power-delay product for CMOS [11].

The layers of SiGe has been grown using a wide variety of techniques. "Molecular beam epitaxy (MBE)", "ultra-high vacuum chemical vapor deposition (UHVCVD)", "atmospheric pressure chemical vapor deposition (APCVD)", "liquid phase epitaxy (LPE)", and "rapid thermal chemical vapor deposition (RTCVD)" are the most successful techniques for growing high-quality SiGe heterostructures. MBE provides the best control of the layer structure thickness and composition and allows growth at lower temperature [12]. When silicon is grown on a relaxed SiGe buffer results in tensile strain. It results in lowering of the heavy hole band and lifts the light hole band which leads to substantial improvement in the low field mobility of holes.

In this paper, we investigate the strained-Si/SiGe channel device performance issues associated with the heterostructure p-MOS devices. 2D simulation suite from Silvaco [13] has been used for the prediction of performance enhancement of the p-channel devices with strained-Si as channel layer. Basically a semi-analytical model based on physics including charge control model for SiGe quantum well was used for the simulation of the devices. Low-temperature hole mobility models are used from reported experimental data [14]. Briefly, in this work p-MOSFET with strained-Si channel is studied using the Silvaco-ATLAS simulation software.

2 Strained-Si Channel p-MOSFET

One of the main advantages of introducing SiGe to Si technology is the addition of other dimensions to the design of devices. Beside changing the geometry, implantation profiles, etc., one can also change the stacking, doping, and composition of the heterostructure. This can have significant effect on controlling parameters that are very hard to control for scaled Si MOSFET. Short channel effects in deep submicron FETs are one of the problems that heterostructures can help reduce. Biaxially strained-Si on relaxed $Si_{1-x}Ge_x$ buffer layer provides a possibility to improve hole mobility and thus increase p-MOSFETs drive current. Nayak et al. [14] reported of enhanced mobility $Si_{1-x}Ge_x$ p-MOSFET for the first time. Due to the difference in the energy band gap between Si (1.11 eV) and Ge (0.664 eV), the growth of SiGe with one Ge concentration on top of a SiGe layer with another Ge concentration results in a discontinuity in the valence and/or the conduction band. Nayak et al. [15] reported the fabrication of quantum well SiGe p-MOSFETs with higher channel mobility and saturation current.

p-channel MOSFET fabrication includes the growth of the buffer layer (step-graded) SiGe which is made possible by "gas source molecular beam epitaxy" (GSMBE) at 800°C. Long channel (100 μm) p-MOSFETs were fabricated following a standard process known "self-aligned n^+-poly Si process". The initial substrate was p-type having 3 in. diameter, of resistivity around 5–10 Ω-cm and orientation (100) Si. On top Si another buffer layer of Si having a thickness of 5000 Å, SiGe (step graded) buffer layer (Ge varies from 0 to 18% in 7 consecutive steps) of 2.1 μm thick, and buffer cap layer of 0.9 μm thick $Si_{0.82}Ge_{0.18}$ were subsequently grown. GSMBE (Daido Sanso VCE-S2020) was used to grow the SiGe buffer layer at 800 °C. The graded buffer (Ge concentration 0 to 0.18%) was made possible by following recipe. The ratio of silane to germane were maintained at (4.5/0–4.5/4.5) in 7 subsequent steps, each step of 5 min. The uniform buffer can was possible after 30 min of last step. Then, the strained-Si epitaxial layer of 180 Å thick was grown at 700 °C, with germane (99.99%) and disilane (99.99%).

For isolation LPCVD oxides were used. Devices were isolated with 7000 A LPCVD oxide. Conventional thermal oxidation was used to form the gate oxide. The silicon consumption during oxidation, resulted in a loss of around 0.44 times of oxide layer thickness of Si. This gate oxidation is the crucial step which determines the thickness of oxide layer and post oxidation remaining layer of the strained-Si channel. The strained-Si epitaxial layer (180 A) was undergone thermal oxidation for 140 min at 700 °C to develop a 100 A thick gate oxide. Both source and drain are maintained boron dose at 6×10^{14} cm^{-2} for implantation at 25 keV. Source/Drain implant activation can be achieved be following couple of steps: (i) 550 °C for 100 min and (ii) 700 °C for 60 min in nitrogen. Anneal at low-temperature helps epitaxial regrowth of damaged Si in the solid phase. However, the dopant atoms get activated by the high temperature. The metal contact Al is annealed for 20 min at 400 °C in forming gas. During the process the maximum temperature was maintained at 700 °C so that any degradation to the device can be avoided because of strain relaxation or inter diffusion. Generally p-MOSFETs were fabricated on two types of substrates: (i) unintentionally p-type doped (10^{16} cm^{-3}) 0.5 μm epitaxial silicon layer, grown on n^+ Si(100) substrate and (ii) film material of strained-Si, which can be grown entirely on relaxed buffer layer of $Si_{0.82}Ge_{0.18}$.

The structural representation of the experimental strained-Si p-channel MOSFET is depicted in Fig. 1a. With increase in gate bias, the number of charge carriers continuously increase in the surface channel. Finally a surface channel device results at large bias and due to the deteriorating action of the field in the surface channel the buring channel is almost removed. The $I_d - V_d$ characteristics of the p-MOSFET has been shown in Fig. 1b.

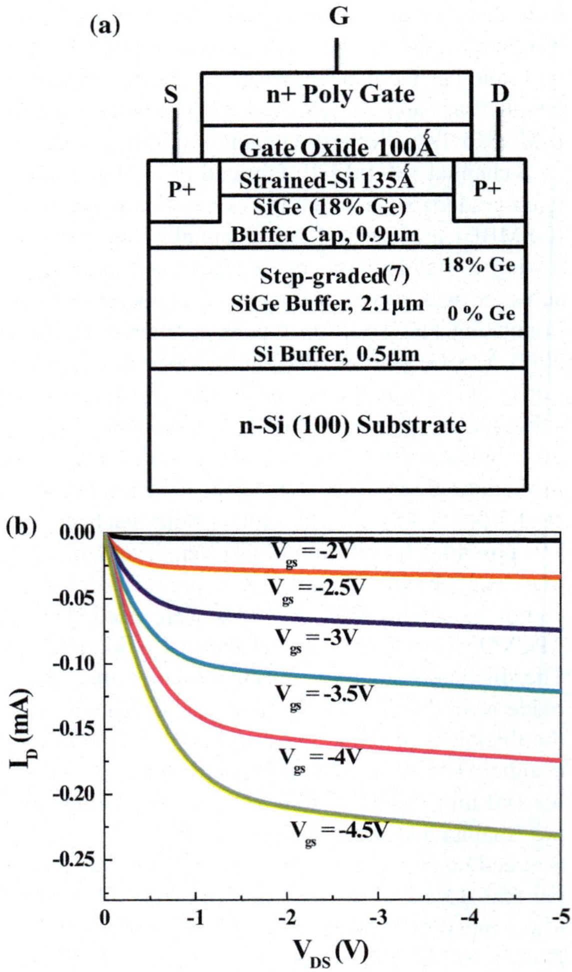

Fig. 1 **a** Structure of a p-channel MOSFET; **b** Typical $I_d - V_d$ characteristics of a strained-Si p-channel MOSFET with L/W = (300/100) μm

3 Strained-Si/SiGe p-MOSFET Simulation

A lot of research have been made on the fabrication of strained-SiGe channel MOSFETs due to the higher mobility of carriers. The main focus is to enhance the transport property of MOSFETs using heterostructures. Different groups of researchers interested in this area could successfully realize better performance with strained-SiGe compared to control-Si devices.

Few important parameters while designing strained-SiGe p-channel MOSFETs has to be considered such as the selection of the material for gate, Ge concentration, thickness of oxide layer and the strained-Si layer. So it is desirable to select the design considerations carefully for the thickness of the oxide layer, strained-SiGe layer and the graded SiGe layer, mole fraction of Ge and its profile, and the required substrate doping to achieve the desired threshold voltage, in order to obtain the optimal enactment. The densities of hole in the layer of SiGe and the parasitic channels of Si have been worked out using simulation. The electronic properties and material parameter models used in simulation for relaxed-SiGe and the strained-Si are based mainly on the work reported in the literature [16–18].

The basic structure used in simulation has been presented in Fig. 2 along with the band diagram.

Figure 3a, b show the Output Characteristics and transfer characteristics of strained-Si p-MOSFETs having effective channel length = 0.6 μm devices. The curves shows well agreement with long channel p-MOSFET characteristics. At 100 K the graph of transconductance depicts the clear transformation from buried to surface channel. At a low temperature of 200 K the improvement of drain current justifies the improvement in high-field transport phenomena and in mobility as well. With decrease in temperature, the mobility increases in strained-Si. The mobility increment factor can be determined by the scattering mechanisms at that particular operating gate voltage. The sharing of surface channel and parasitic buried channel and the bulk channel carriers is the real cause of the device performance. The carriers switch to the surface channel from the buried channel as the temperature is lowered. The important charge sharing controlling parameters can be stated as (a) width of the channel, (b) the gap between surface and the interface (c) the thermal energy and (d) Offset of valance band at the interface of strained-Si/SiGe (see Fig. 2).

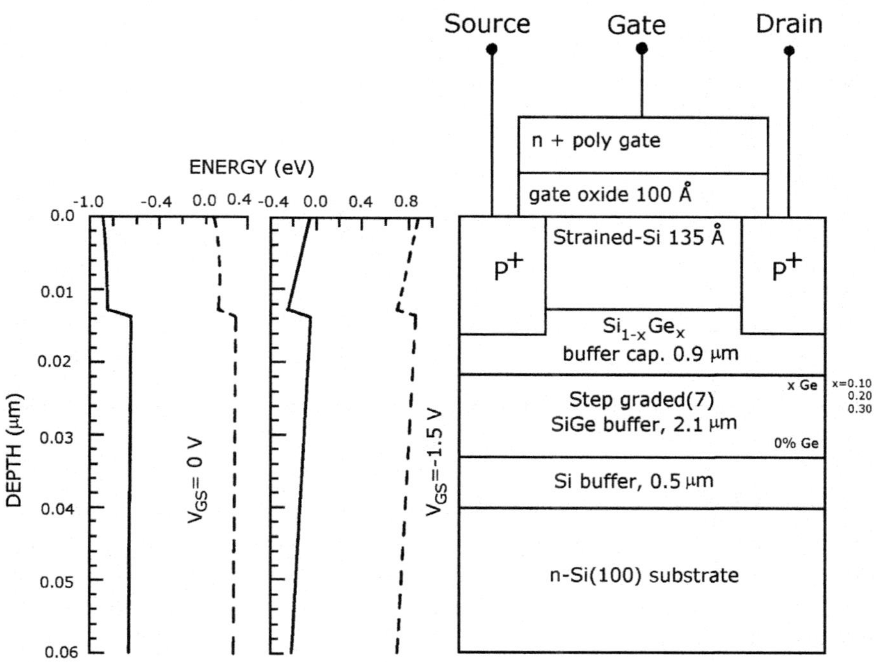

Fig. 2 Device structure and band diagram of the p-MOSFET used in simulation

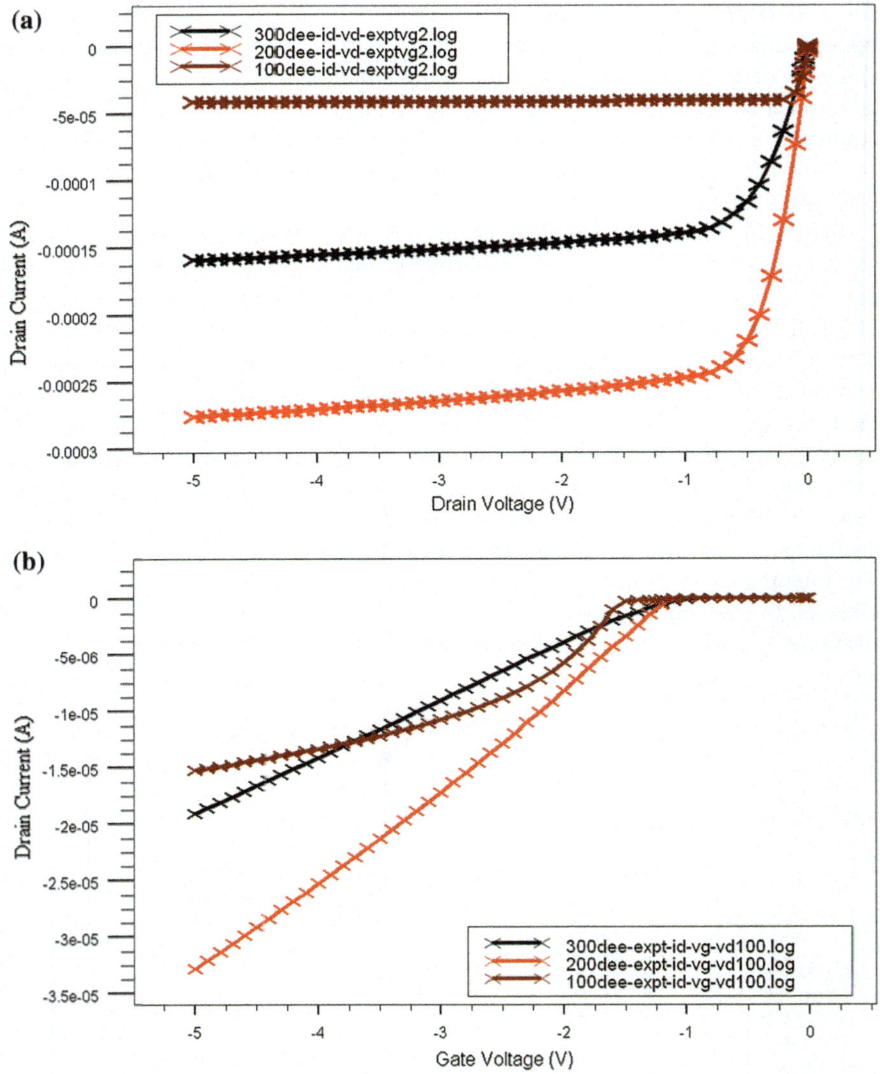

Fig. 3 **a**, **b** Output characteristics (top) and transfer characteristics (bottom) of the simulated device for room temperature (300 K) and low temperatures (200 and 100 K)

Figure 4 shows $I_d - V_d$ characteristics for four different simulation conditions for the same device structure: (a) when CVT mobility model is used, (b) bulk-Si device, (c) with Ge content 0.18, and (d) with Ge content of 0.38. It is obvious that as Ge content is increased, drain current increases and bulk-Si device shows the minimum drain current.

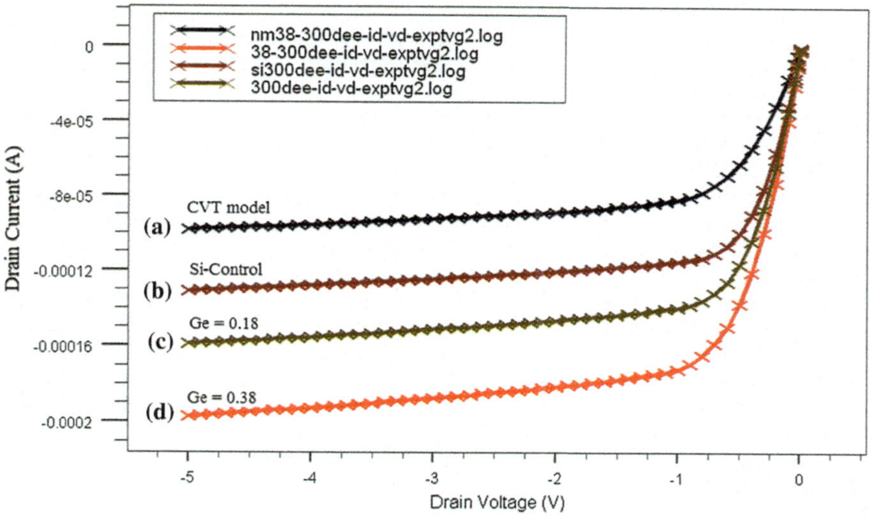

Fig. 4 Comparison of drain current for bulk-Si and strained-Si PMOS for different simulation conditions

Short channel effects were also investigated. The decrease in threshold voltage with decrease in gate length is a common short channel effect. Another short channel effect is the drain induced barrier lowering (DIBL) where threshold voltage of a transistor reduces at higher drain bias. The threshold voltage at $V_d = -0.1$ V and DIBL are found to be 1.52 V and 43.06 for the device at room temperature.

4 Conclusion

Simulation of a high mobility strained-Si channel p-MOSFET is reported. The strained-Si channel grown epitaxially on $Si_{0.82}Ge_{0.18}$ (step-graded) and layer of relaxed Si buffer (100) substrate has been used in simulation. At high vertical field channel mobility enhancement of 40% at 300 K and 200% is 200 K, for the strained-si p-channel device when compared to conventional Si device.

References

1. C.K. Maiti, G.A. Armstrong, *Applications of Silicon-Germanium Heterostructure Devices* (Institute of Physics Publishing (IOP), UK, 2001)
2. J. Franco et al., SiGe channel technology: superior reliability toward ultrathin EOT devices—Part I: NBTI. IEEE Trans. Electron Dev. **60**, 396–404 (2013)

3. N. Xu et al., Benefits of segmented Si/SiGe p-channel MOSFETs for analog/RF applications. in *VLSI Technology Symposium* (2013), pp. T142–T143
4. International Technology Roadmap for Semiconductors, 2012 (Semiconductor Industry Association)
5. I. Saad et al., Performance analysis of single and dual channel vertical strained SiGe impact ionization MOSFET (VESIMOS), in *Research and Development (SCOReD) Conference* (2013), pp. 275–280
6. H. Mertens et al., Si-cap-free SiGe p-channel FinFETs and gate-all-around transistors in a replacement metal gate process: Interface trap density reduction and performance improvement by high-pressure deuterium anneal. in *VLSI Technology Symposium* (2015), pp. T142–T143
7. P. Hashemi et al., High-mobility high-Ge-content $Si_{1-x}Ge_x$-OI PMOS FinFETs with fins formed using 3D germanium condensation with Ge fraction up to $x \sim 0.7$, scaled EOT ~ 8.5 Å and ~ 10 nm fin width. in *VLSI Circuits Symposium* (2015), T16–T17
8. L. Witters et al., Strained germanium quantum well p-FinFETs fabricated on 45 nm Fin pitch using replacement channel, replacement metal gate and germanide-free local interconnect. in *VLSI Technology Symposium* (2015), T56–T57
9. V.A. Tiwari et al., Modeling of gate-induced drain leakage mechanisms in silicon-germanium channel pFET. in *Proceedings ICEE* (2014), pp. 1–5
10. V.A. Tiwari et al., Analysis of gate-induced drain leakage mechanisms in silicon-germanium channel pFET. IEEE Trans. Electron Dev. **61**, 1270–1277 (2014)
11. C.K. Maiti, T.K. Maiti, *Strain-Engineered MOSFETs* (CRC Press (Taylor and Francis), USA, 2012)
12. Atlas Simulation Suite, User's Manual, Silvaco Inc., USA
13. D.K. Nayak et al., High-mobility strained-Si PMOSFETs. IEEE Trans. Electron Dev. **43**, 1709–1715 (1996)
14. D.K. Nayak et al., Bandedge photoluminescence of SiGe/strained-Si/SiGe type II quantum wells on Si(100). Japan. J. Appl. Phys. **32**, L1391–L1393 (1993)
15. D.K. Nayak et al., Band-edge photoluminescence of SiGe/strained-Si/SiGe type II quantum wells on Si(100). Appl. Phys. Lett. **63**, 3509–3511 (1993)
16. C.K. Maiti et al., Hole mobility enhancement in strained-Si p-MOSFETs under high vertical fields. Solid-State Electron. **41**, 1863–1869 (1997)
17. D.K. Nayak et al., High-mobility p-channel metal–oxide semiconductor field-effect transistor on strained-Si. Appl. Phys. Lett. **62**, 2853–2855 (1993)
18. D.K. Nayak, S.K. Chun, Low-field mobility of strained-Si on (100) $Si_{1-x}Ge_x$ substrate Appl. Phys. Lett. **64**, 2514–2516 (1994)

Color Image Segmentation Techniques: A Survey

Sneha Jain and Vijaya Laxmi

Abstract In today's world, where digital image processing is becoming an essential part of technology, segmentation of images poses a challenging problem. Before any complex task that has to be done on images, segmentation is a prerequisite. Segmentation ensures the simplification of a problem by changing the representation of an image from a complex one to a more analytical and easier form. Pixels of segmented regions share common characteristics. Perfect segmentation is difficult to obtain. There exist many techniques which have been applied such as edge-based segmentation, region-based segmentation, morphological operations, thresholding and clustering methods. Segmentation has a crucial role in image analysis. The accuracy of segmentation determines the success or failure of computer algorithms. Therefore, there is a need to develop efficient and less time-consuming algorithms for segmentation. This paper summarizes a number of segmentation methods.

1 Introduction

Computer vision is becoming an important part of technology. Image segmentation forms an integral part of many signal processing applications. In image processing, most of the operations need segmentation to be done. Image segmentation aims at partitioning the image into regions based on a predefined criterion. In other words, we can say that segmentation is a process of demarcating the foreground region from the background. The separate regions are homogeneous with respect to particular properties such as color, texture, brightness [1]. Extracting the region of interest from the image still remains a difficult problem. The noise present in the image along with image data ambiguity is one of the main problems in segmentation.

S. Jain (✉) · V. Laxmi
Department of Electrical & Electronics Engineering,
Birla Institute of Technology Mesra, Ranchi, India
e-mail: snehaj26@yahoo.in

V. Nath (ed.), *Proceedings of the International Conference on Microelectronics, Computing & Communication Systems*, Lecture Notes in Electrical Engineering 453,
https://doi.org/10.1007/978-981-10-5565-2_17

Selecting an appropriate technique of segmentation is a challenging issue. Assigning pixels to correct object segment is a tough task. The level of subdivisions which has to be done depends upon the problem. Medical images have incorporated algorithms based on classifiers. Image segmentation finds its applications in identifying objects based on measurements such as size and shape. Content-based image retrieval systems also deal with the process of segmentation.

This paper has been divided into four sections. Section II covers the introduction about segmentation. Section III consists of the techniques of segmentation. The last section consists of conclusion of the research work.

2 Segmentation

Segmentation is used widely to calculate the features in an image. The segmentation results help to obtain information about the various parameters in an image. Effective segmentation ensures that the objects and background do not mix with each other. Segmentation of complex image structures and backgrounds is still a problem which needs to be tackled. Segmentation methods can be contextual or non-contextual. Contextual segmentation involves grouping of pixels based on some property and the spatial relationship between the pixels. Non-contextual segmentation does not take into account the spatial relationship between the pixels.

Mathematically, segmentation can be described as:

If $f(x, y)$ is an original image, then after segmentation, disjoint subsets obtained are $f_1, f_2, \ldots, f_n$ [2]. High-level information can be extracted from the sub-regions. Information is obtained by the analysis and interpretation of the segmented regions. Research is being made to develop techniques that are fast and cost effective.

Perfect segmentation is difficult to achieve because of under-segmentation or over-segmentation. In over-segmentation, pixels that belong to the same object are classified as belonging to different segments. In under-segmentation, pixels belonging to different objects are classified as belonging to same object.

3 Techniques of Segmentation

Monochrome images are generally segmented based on two properties:

i. Discontinuity—This approach works on the concept of abrupt changes in intensity levels [2, 3].
ii. Similarity—This approach deals with partitioning of the image into similar areas in accordance with a predefined rule.

There are several segmentation methods mainly categorized as:

1. Region-based methods
   i. Region growing,
   ii. Region splitting and merging.
2. Thresholding
   i. P-tile method,
   ii. Mean value technique,
   iii. Edge maximization technique,
   iv. Optimal thresholding.
3. Clustering techniques
   i. *k*-means clustering.
4. Edge detection-based methods.
5. Morphological segmentation
   i. Watershed algorithm.
6. Matching.

3.1 *Region-Based Segmentation*

Region-based segmentation methods are broadly classified into two groups [4]:

i. Region growing,
ii. Region splitting and merging.

i. *Region growing*

As its name suggests, region growing is a process that groups pixels into larger areas based on some predefined criteria [1–3, 5, 6]. The approach starts with determining seed points or reference points. There may be one or more seed points. Same set of properties is computed at every pixel, which is used to assign pixels to growing regions. The algorithm is stated below:

a. Seed points are grouped into n clusters $C_1, C_2, \ldots, C_n$ with initial seed points labeled as $s_1, s_2, \ldots, s_n$.
b. If the difference between the pixel intensity of the seed point s_j and the neighboring pixels is less than the predefined threshold, then the pixels are assigned to the cluster C_j.
c. Reconstruct the cluster by computing the difference between neighboring pixel and seed point.

d. Repeat the steps b and c until allocation of all the pixels is completed.

Disadvantage—The determination of seed points is a difficult task. Different set of seed points yield different segmentation results. The other drawback is the time-consuming nature of the procedure.

ii. *Region splitting and merging*

The homogeneity of the image can be distinguished by this algorithm. The concept is based on quad trees. If any region is not homogeneous, it is disintegrated into square sub-regions which are represented as nodes of a quad tree. Further, if any sub-region is inhomogeneous, it is divided into four parts. If these regions have approximately the same homogeneity measure, they are merged into a single region. In this method, regions need not be contiguous (Fig. 1).

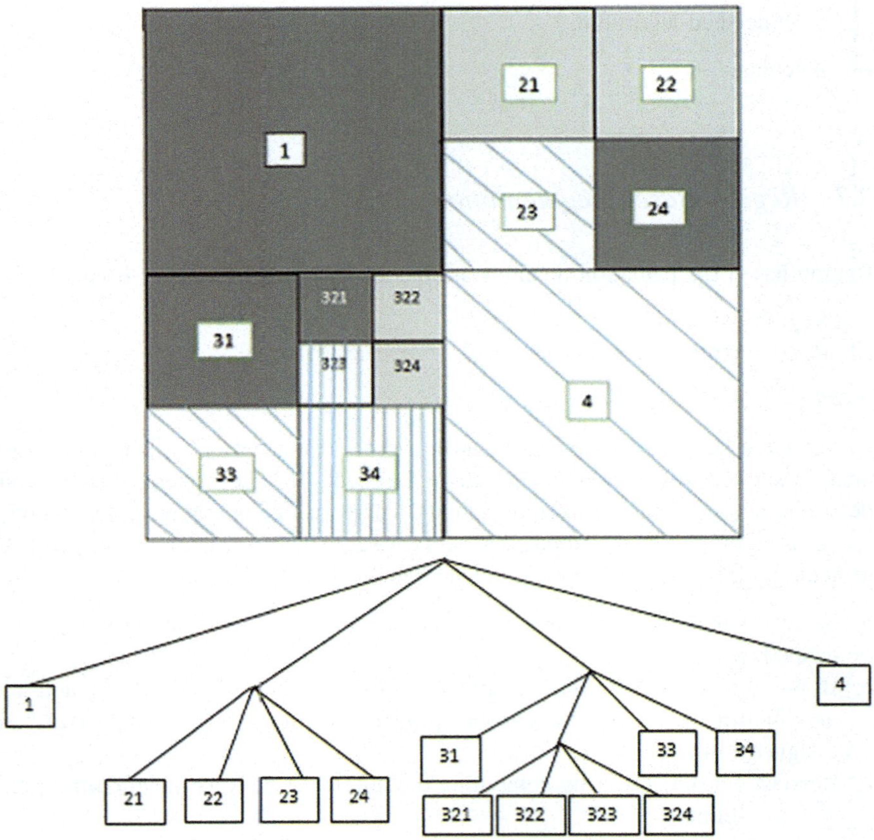

Fig. 1 Quad tree segmentation

3.2 Thresholding

Thresholding is a technique which can be used to segment objects and background. An input image *f*(*x*, *y*) is transformed into output binary image *g*(*x*, *y*) as per the following criterion [3]:

$$\begin{aligned} g(x,y) &= 1, f(x,y) \geq T \\ &= 0, f(x,y) < T. \end{aligned}$$

If $f(x, y) \geq T$, then the pixel in consideration is an object pixel, otherwise, it is a background pixel. Threshold detection methods can be local (adaptive), where there are multiple thresholds or global, where there is a single threshold for the whole image.

There are several thresholding techniques such as:

i. P-tile method,
ii. Mean value technique,
iii. Edge maximization technique (EMT),
iv. Optimal thresholding.

i. *P-tile method*

This method is based on the concept of gray level histogram. It is based on the assumption that the objects are brighter than the background and occupy a particular percentage (*P*%) of the image area. The threshold is computed as the gray level which corresponds to mapping at least *P*% of the gray level to the object.

ii. *Mean value technique*

The mean value of all the pixels is calculated which is used as the threshold. It works well in cases where approximately half of the pixels belong to the objects and the other half belong to the background.

iii. *Edge maximization technique (EMT)*

This technique is used where there are several homogeneous regions or where there is change in illuminations between background and objects. This technique has a drawback that the portions of the objects and background may be merged.

iv. *Optimal thresholding*

Optimal thresholding is based on the assumption that the histogram of an image consists of two overlapping Gaussian distribution curves [2]. The threshold is chosen as the intersection point of the two distributions which corresponds to minimum probability between the maxima of two distributions. The disadvantage of this method is that prior knowledge of object and background distributions might not be available (Fig. 2).

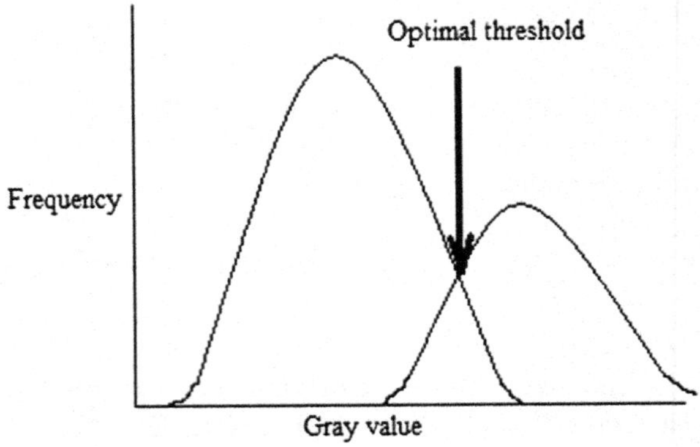

Fig. 2 Optimal thresholding

3.3 Clustering Techniques

Clustering techniques group the contents of an image into patterns that have some common attributes. *k*-means clustering is the most widely used clustering algorithm [6, 7]. The image is partitioned into *k* clusters. The following is the procedure followed for clustering:

a. Select desired number of clusters and place the centers of the clusters at locations decided arbitrarily or based on a heuristic.
b. Each pixel in the image is assigned to the cluster whose center is the closest.
c. The cluster centers are re-computed by averaging the pixels in the cluster.
d. Repeat the steps b and c until none of the pixels change their clusters.

The result of this method may not be an optimal solution.

3.4 Edge Detection-Based Methods

Edge-based segmentation methods rely on finding boundaries based on discontinuities in texture, gray levels, color, etc. [8–10]. Ambiguous boundaries where gaps occur on the edges are a common problem in edge detection. This can be rectified by using Hough transform with the help of which edges can be linked.

Various edge detectors are:

i. Sobel edge detector,
ii. Prewitt detector,
iii. Roberts cross edge detector,
iv. Log of Gaussian (LoG) detector,
v. Canny edge detector,
vi. Kiresh detector,

vii. Laplacian edge detector.

Below are the results of MATLAB simulation for various edge detectors (Fig. 3).

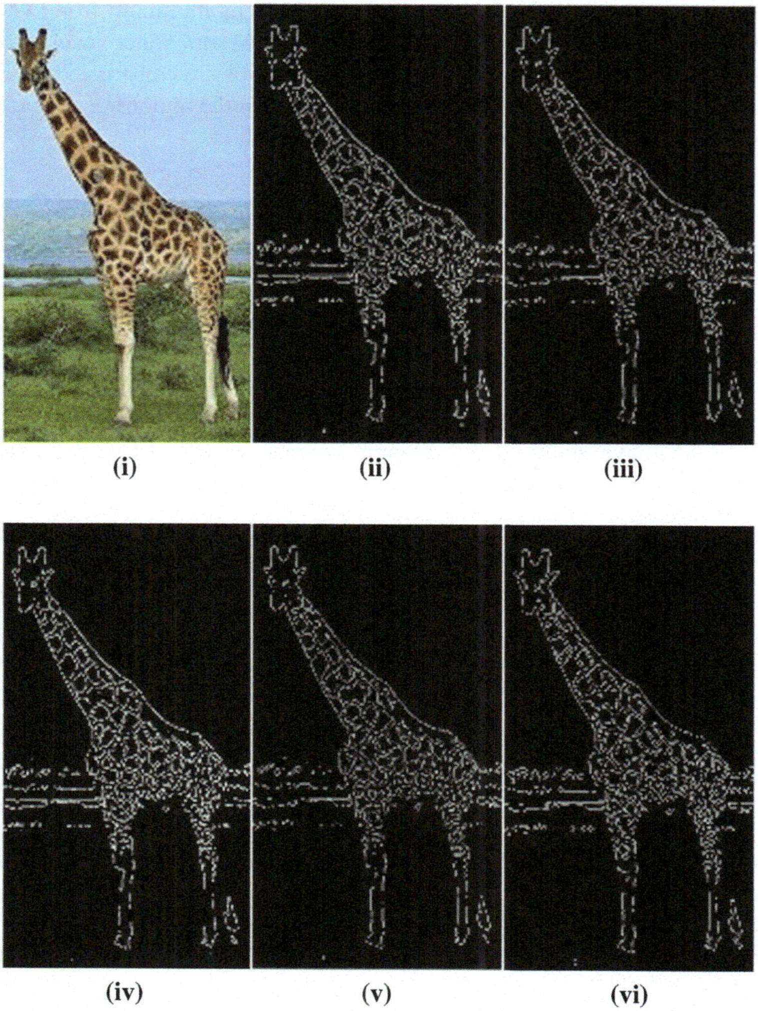

Fig. 3 MATLAB simulations: **i** Original image, **ii** Canny edge detector, **iii** Sobel edge detector, **iv** Roberts edge detector, **v** Prewitt edge detector, **vi** LoG detector

3.5 Morphological Segmentation

Watershed algorithm

The image is topographically interpreted in this method. The intensity levels correspond to the height of the terrain that represents the catchment basins and the mountains. Each basin is assumed to have a hole in its minimum from where the underground water spills and fills the catchment. As the water level rises, two adjacent basins tend to merge. To avoid merging, dams are constructed to separate the basins [5, 11]. The dams serve as the boundaries of the regions of segmentation. There is a problem of over-segmentation in this method.

3.6 Matching

In this method, a template of region of interest (object) is matched with the image locations until the object is found [2]. The template is placed over the image and the gray values of the template and the underlying image are compared. If all of the gray values match, the object is found.

The comparative study of the different segmentation techniques is shown in Table 1.

Table 1 Comparative study of segmentation methods

Parameter	Watershed algorithm	Edge based	Region based	Threshold technique	*k*-means clustering
Spatial relationship between pixels	Exists	Does not exist	Exists	Does not exist	Exists
Immunity to noise	More immunity	Less immune	More immune to noise than edge based	Less immune	Noisy data cannot be handled easily
Speed	Moderate	Moderate	Slow	Fast	Slow
Accuracy rate	Over-segmentation exists	Accurate	Accurate	Not very accurate	Moderate accuracy

4 Conclusion

In this paper, a number of image segmentation techniques have been discussed, which are encompassed by the field of computer vision. Segmentation is widely used for image database lookup, object recognition, editing of the image, optical character recognition, terrain classification in satellite images, medical images, etc. Recently, neural, fuzzy, graph cut techniques [6, 12] have also been employed for segmentation. Thus, segmentation acts as a bridge between low-level and high-level image processing.

References

1. J. Da Rugna, C. Gael, K. Hubert, *About Segmentation Step in Content-Based Image Retrieval Systems*, in Proceedings of World Congress on Engineering and Computer Science, vol. 1, 19–21, USA (2011)
2. M. Sonka, V. Hlavac, R. Boyle, *Image Processing, Analysis, and Machine Vision*, 2nd ed., Thomson
3. R. Dass, Priyanka, S. Devi, Image segmentation techniques. Intl. J. Electron. Commun. Technol. **3**(1) (2012)
4. G.S. Rawat, J. Bhattacharjee, R. Soni, Proposed method for image segmentation using similarity based region merging techniques. Intl J. Comput. Sci. Inf. Technol. **3**(5), 5128–5132 (2012)
5. S. Saini, K. Arora, A study analysis on the different image segmentation techniques. Intl. J. Inf. Comput. Technol. **4**, 1445–1452 (2014)
6. A.M. Khan, S. Ravi, Image segmentation methods: A comparative study. Intl. J. Soft Comput. Eng. **3**(4) (2013)
7. R. Kandwal, A. Kumar, S. Bhargava, Review: existing image segmentation techniques. Intl. J. Adv. Res. Comp. Sci. Sofw. Eng. **4**(4) (2014)
8. N. Senthilkumaran, R. Rajesh, Edge detection techniques for image segmentation—a survey of soft computing approaches. Int. J. Recent Trends Eng. **1**(2) (2009)
9. H.G. Kaganami, Z. Beiji, *Region-Based Segmentation Versus Edge Detection*. Fifth International Conference on Intelligent Information Hiding and Multimedia Signal Processing (2009)
10. Mr. S.S. Al-amri, Dr. N.V. Kalyankar, Dr. S.D. Khamitkar, Image segmentation by using edge detection. Intl. J. Comput. Sci. Eng. **2**(3), 804–807 (2010)
11. A.A. Aly, S.B. Deris, N. Zaki, Research review for digital image segmentation techniques. Intl. J. Comput. Sci. Inf. Technol. **3**(5) (2011)
12. P. Karch, I. Zolotova, *An Experimental Comparison of Modern Methods of Segmentation*. 8th IEEE International Symposium on Applied Machine Intelligence and Informatics, 28–30 January, 2010 (Slovakia)

Wireless Image Sensor Networks: A Review

Parivesh Pandey and Vijaya Laxmi

Abstract Wireless sensor networks (WSN) were extensively used in monitoring and observing a particular region. WSN are combination of nodes and can be sensitive to pressure, temperature, motion, sound etc. This paper represents the survey of design and implementation of Wireless Image Sensor Network, which is an integral part of monitoring and surveying the subjective region visually. Image sensor nodes are equipped with miniature visual camera and RF module for communication. The camera node provides visual information and then transmitted to another node wirelessly using ZigBee. Several challenges in sensor networks are discussed that can enhance performance and efficiency of modern day sensor networks, As it turns out FPGA can reduce computational cost through onboard image processing. We have discussed in the subsequent part about the combination of microcontrollers and FPGA which can play a major role in areas where processing capabilities such as compression, cryptography, and transmission of data are important.

Keywords ZigBee · FPGA

1 Introduction

Wireless sensor network (WSN) is a combination of sensing device, embedded processor, communication channel and power circuit. Wireless Image Sensor Network (WISN) where sensor nodes are equipped with CMOS camera which provides the application-specific information. Image can be fetched on demand or at a given specific interval. Implementing wireless image sensor with fast and efficient image readability and multi-hop transmission feature is still challenging. There are number of other groups proposed the active research of image sensor module: Panotypes—video sensor network [1], Cyclops—Image sensor daughterboard [2],

P. Pandey (✉) · V. Laxmi
Department of Electrical & Electronics Engineering,
Birla Institute of Technology, Mesra, Ranchi, India
e-mail: pariveshpandey@ymail.com

V. Nath (ed.), *Proceedings of the International Conference on Microelectronics, Computing & Communication Systems*, Lecture Notes in Electrical Engineering 453,
https://doi.org/10.1007/978-981-10-5565-2_18

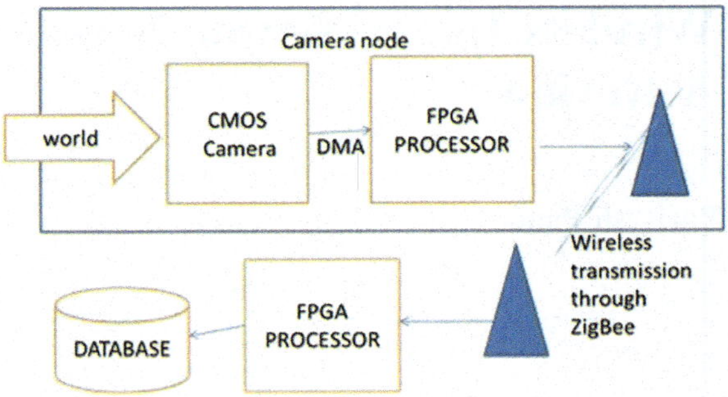

Fig. 1 Block diagram representation

EyeRis [3], SeedEyes [4], most of these platforms or modules are development of a camera board with microcontroller to perform multitask. Our motivation is to build our own image sensor platform whose response is fast and efficient. The image sensor that we propose to implement on a FPGA board which contains a wireless communication device known as ZigBee that provides information between the nodes. A simple surveillance mechanism is discussed while adding no cost in image processing. FPGA being standalone and reconfigurable is more utilizable in various scenes as it provides wide range of application and flexibility. FPGA chip consists of number of cores which is ideal for wireless image sensor to perform multiple task with ease. FPGA reconfiguration capability is necessary to improve flexibility in the field of everyday advancement in surveillance or monitoring.

A simple block diagram representation of wireless image sensor has been given in Fig. 1, in which the surveillance node consists of CMOS camera which is connected to FPGA processor. The wireless transmission of data is done with the ZigBee module.

2 Working Principle of ZigBee

ZigBee module is enhancement to IEEE 802.15.4 (low-rate wireless personal area network) that develops more sophisticated protocols for advanced application which includes encryption, authentication with valid nodes and a data routing and forwarding capability that enables mesh networking [5] and secure nodes [6] to prevent unauthorized attempts to access the node or to change message (Fig. 2).

With a rapid growth in wireless communication, wireless image sensor is becoming an important aspect of monitoring and surveying purpose. ZigBee physical layer defines three frequency operations 868 MHz, 916 MHz, and 2.4 GHz bands.

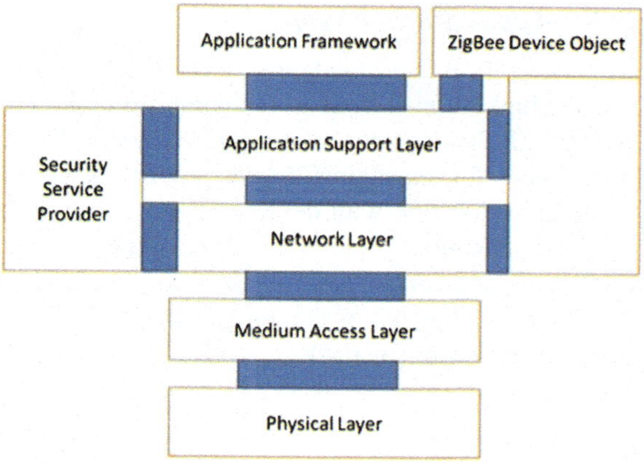

Fig. 2 Protocol stack of ZigBee

2.1 *Physical Layer*

The frequency range of the traditional lines varies from 300 to 3400 Hz. Transmission of voice takes place through this bandwidth where distortion and interference take place. 2400 Hz is the effective bandwidth that is used over the telephone lines. Digital data transmission over the analog telephone lines takes place with the help of a modem. Each modem standard has its combination of amplitude and phase called constellation pattern. Various modem standards used are V.32, V.32bis, V.34bis, V.90, and V.92.

2.2 *Network Layer*

Network layer protocol provides efficient working of network which enables the correct use of the MAC sub-layer and provides a suitable interface with application layer. ITU-specified network layer protocol is X.25. Earlier X.25 supported both connection-oriented (CO) service and connectionless (CL) service, but it gradually became more CO oriented. X.25 is a continuous protocol which allows:

1. User–DCE (Data Circuit terminating Equipment) communication
2. DCE–User communication.

Standard DCE provides particular terminal compatible with any network. If a network has several subnets interconnection between those can be possible through routing. Router routes data not between connecting paths but between networks.

2.3 Application Layer

It is the highest level layer which comprise of advanced security components added by the ZigBee specification. ZDO (ZigBee Device Object) protocol which is responsible for managing device security keys, policies and establish a secure communication link with devices.

The application support sub-layer (APS) is another component; it provides interfacing and control services. It is an association between the network layer and the other components of the application layer and also routes messages across the layers of the protocol stack.

3 Alternatives of Communication

This section describes some other communication method that can be an alternative of ZigBee which is as follows:

1. *nRF24L01/RFM12B/ RFM22B(SI4432)*—These transceiver connected via serial port interface (SPI). Transceivers possess a lot of features such as low power modes, multiple channels, channel hopping, frequency calibration. The nRF24L01 operates in the 2.4 GHz band, and others use the ISM band 433/470/868/915 MHz. These are state-of-the-art highly integrated and low-cost hardware for reliable communication application. The range varies from 10 to 150 m where SI4432/RFM22B has advance range of about 1.5 km.
2. *Bluetooth*—Bluetooth has medium data rate and medium power consumption. IEEE standard has different device profiles to enable interoperability between devices. This is the reason why it is not useful for sensor networks, but it does provide better interconnection for controlling devices using a laptop or phone which usually has Bluetooth on board.
3. Arduino can also be used for communication via serial RX and TX pins for small range purpose. Similarly, other development boards such as Raspberry pie, BeagleBone, and many others have different types of communication features.

3.1 Application of Wireless Image Sensor (WIS)

The design of WIS intended for use with enhanced processing and memory constraints gives additional consideration over generic sensors.

1. Less human intervention—once installed can easily be accessed.

2. Safety and security application—Several cryptographic techniques have been proposed for sensor nodes.
3. Environmental monitoring—sensor node based on object identification and tracking in environment, pest control, fire detection, etc.

3.2 Potentials of FPGA

FPGA's consists of interconnecting configurable logic blocks in two-dimensional array that can be reconfiguring and gives us the added advantage to implement any combinational and logical circuit. The complexity of designing in FPGA system is reduced by using hardware description language (HDL) such as VHDL and Verilog-HDL [7]. Previously, FPGAs have been mainly used for signal processing and network packet analysis [8]. DSP slices, Fast memories and high speed embedded processors can be used to enhance the utilization of FPGA system for better output and configuration. Recent development in FPGA's results in area reduction and energy consumption.

We can enhance the processing speed and reduce the computational cost just by implementing onboard image processing technique on FPGA.

3.3 VHDL and MATLAB

VHDL is the designing language used to implement programs on FPGA and sometimes it can be troublesome as we are bounded by the constraints and values of integers and bit/byte array. Xilinx collaborated with MATLAB and created an inbuilt tool to convert MATLAB code to HDL file which can be implemented on FPGA board. Thus, the complexity is reduced and providing us more freedom over the range of constraints that can't be possible through VHDL. The latest advancement in image processing through MATLAB can be processed in wireless image sensor nodes which provide the greater flexibility in recognition and motion tracking.

3.4 Related Sensor Networks

Some related sensor networks have been explained in this section, which are as follows:

1. Cyclops [2]—It's a basic image sensor, a combination of microcontroller, CPLD (complex programmable logic device) and Cyclops containing SRAM which enable us to access image data on demand. The MCU and CPLD access same address and data bus which allow us to transfer data with ease. Power

consumption reduction leads to CPLD usage instead of FPGA. CMOS static random access memory (SRAM) is used for image buffering and local inference.

2. SensEye [9]—A camera sensor which comprises three task object detection, object tracking, and object recognition. It is a multi-tier performance system and superior to single tier system as far as power consumption and object detection are concerned.
3. Wireless Multimedia Sensor Network (WMSN) [10]—WMSN gives the multimedia data, i.e., network contains camera node with microphone which enable us to create audio–video data.
4. Low-cost Wireless Image Sensor Network (WISN) [11]—WISN is for visual surveillance; it's a low-cost device implemented on Arduino Due board for processing of input data and transmitting it wirelessly with the help of XBee (IEEE 802.15.4) module. It's an example of alternative method; we cannot implement additional image processing techniques via microcontroller as we can do with FPGA, but it does produce an impressive execution time of 2.3 s. Since, microcontroller is enough for simple processing of image through a wireless network, FPGA usage can be overkill considering an external transceiver although FPGA are those of ultra low power.

The wireless multimedia sensor network will be the key research in near future. Low-cost WISN does provide us an added advantage for use in intrusion detection. Similarly the other mention platforms or applications are different as they were based on different platform and application set which gives us a brief history on wireless sensor network. We can thus create an interactive and more efficient wireless image sensor if we consider previous research on this field and design better prospects for the implementation of wireless image sensor.

4 Development of Wireless Image Sensor

The overview of the steps involved in the development of wireless image sensor has been discussed as follows:

1. IEEE 802.15.4 protocol is replaced by ZigBee module to enhance the security of transmission.
2. FPGA processor provides flexibility and complies the increasing demand in sensor technology.
3. Multi-hop transmission of image sensor nodes with combination of audio circuitry provides better understanding of subjective region.
4. Increased processing of FPGA leads to developing of the sensor nodes for advance application.
5. Intervention of MATLAB in designing sector provides the necessary cutting-edge algorithm that can be implemented on FPGA board for better result.

Table 1 Comparative study of image sensors

Image sensor	Communication medium	RF module	Bluetooth module	ZigBee module
Microprocessor	Speed	High	Moderate	High
	Encryption	Low	Moderate	High
	Range	High	Low	High
Microcontroller	Speed	High	Moderate	High
	Encryption	Low	Moderate	High
	Range	High	Low	High

4.1 Comparative Study

The comparative study of image sensors based on microcontroller and microprocessor has been shown below.

1. Multi-hopping—The image sensor based on either microcontroller or microprocessors are capable to do multi-hopping image transmission.
2. Power consumption—Microcontrollers consume less power as compared to microprocessor.
3. The advantage of microprocessor over microcontroller is on-site reconfiguration (Table 1).

5 Conclusion

In this paper, a reconfigurable image sensor node has been presented for maximum flexibility and availability to research community. Necessary characteristics of wireless image sensor have been investigated. These characteristics show much sophisticated implementation of sensor node can be developed to perform encrypted transmission. Applications such as motion tracking and facility monitoring can also be explored using the proposed method. Implementing onboard image processing through FPGA will offer significant flexibility and overcomes the bandwidth bottleneck.

References

1. W.-C. Feng, E. Kaiser, W.C. Feng, M.L. Baillif, Panoptes: scalable low-power video sensor networking technologies. ACM Trans. Multimedia Comput. Commun. Appl. **1**(2), (2005)
2. M. Rahimi, R. Baer, O.I. Iroezi, J.C. Garcia, J. Warrior, D. Estrin, M. Srivastava, *Cyclops: in situ image sensing and interpretation in wireless sensor networks*. in ACM SenSys (2005)

3. Á. Rodríguez-Vázquez, R. Domínguez Castro, F. JiménezGarrido, S. Morillas, J. Listán, L. Alba, C. Utrera, S. Espejo, R. Romay, *The Eye-Ris Cmos Vision System*, in Analog Circuit Design (2008)
4. Evidence Embedding Technology, *Seed-eye board, a multimedia WSN device*. http://rtn.sssup.it/index.php/hardware/seed-eye
5. J. Foster, *Xbee Cookbook Issue 1.4 for Series 1 with 802.15.4 Firmware*, www.jsjf.demon.co.uk/xbee/xbee.pdf. Accessed 4/12/2013
6. G. Pekhteryev, Z. Sahinoglu, P. Orlik, G. Bhatti, *Image Transmission over IEEE 802.15.4 and ZigBee Networks* in IEEE ISCAS May 2005, Kobe, Japan
7. A. de la Piedra, A. Braeken, A. Touhafi, Sensors systems based on FPGAs and their applications—a survey. www.mpdi.com/journal/sensors. ISSN 1424-8220 (2012)
8. M. Sood, M. Wagh, M. Cheema, Implementation of a wireless communication system—a review Int. J. Comput. Appl. (0975–8887) **63**(15) (2013)
9. R. Kulkarni, D. Ganesan, P. Shenoy, *The Case for Multi-Tier Camera Sensor Networks*, in *NOSSDAV'05*, Washington, USA, 13–14 June 2005
10. S. Paniga, L. Borsani, A. Redondi, M. Tagliasacchi, M. Cesana, *Experimental Evaluation of a Video Streaming System for Wireless Multimedia Sensor Networks*, in 10th IEEE/IFIP Med-Hoc-Net (2011)
11. C. Pham, *Low Cost Wireless Image Sensor Networks for Visual Surveillance and Intrusion Detection Applications*, in Proceedings of 2015 IEEE 12th International Conference on Networking, Sensing and Control, Pg. 376–381, 9–11 April 2015

Design of a Low-Cost Heart Rate Monitoring System

Suprojit Nandy and Soma Barman

Abstract With the development of health consciousness and growing of aging population, home-based health monitoring has become a key research area for information and communication technology. The objective of the paper is to monitor heart rate of a person in low cost and in reliable way. The system is implemented in LabVIEW environment.

Keywords Piezoelectric sensors · Heart rate · Microcontroller · ECG · Filters

1 Introduction

Health monitoring system is gaining importance in recent times as demands of care taking increases for fast-growing global elderly people [1, 2]. Frequent monitoring of health parameters like blood pressure, heart rate, and body temperature for elderly people in a cost-effective way is demanding. The authors in this paper described a method to measure heart rate sitting at home in low cost and in non-invasive way. The various process involved in designing a "low-cost" noninvasive heart rate monitoring system is illustrated by block representation in Fig. 1. Linear circuit components are used to design the signal conditioning and the signal processing circuits. The electrocardiogram (ECG) is widely used for diagnosis of various heart ailments. Here, a very low-cost piezoelectric sensor is placed on left wrist to capture electrical activity of heart. The ECG which is usually obtained is often contaminated with noise. The unwanted noise mainly occurs due to patient-sensor movement, baseline wander error, etc. In order to suppress such

S. Nandy (✉)
Jalpaiguri Government Engineering College, Jalpaiguri, West Bengal, India
e-mail: suprojitnandy@gmail.com

S. Barman
Institute of Radio Physics and Electronics, University of Calcutta, Kolkata, West Bengal, India
e-mail: barmanmandal@gmail.com

V. Nath (ed.), *Proceedings of the International Conference on Microelectronics, Computing & Communication Systems*, Lecture Notes in Electrical Engineering 453,
https://doi.org/10.1007/978-981-10-5565-2_19

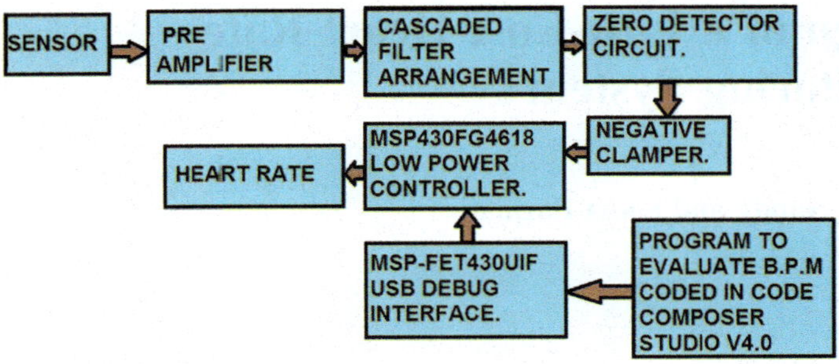

Fig. 1 Block representation of the method

noise from the ECG signal which is extracted and conditioned from the piezoelectric sensor, is processed by filter (cascade of active broad band-pass filter and active narrow band-pass filter). Initially, filters specification and component values of different circuits are checked in LabVIEW [3] and MULTISIM [4] environment prior to practical circuit realization. The signal is then processed and fed into the Texas Instruments MSP-430-FG-4618 microcontroller to process the signal further and evaluate the heart rate which is expressed in beats per minute (BPM). The LED on microcontroller blinks whenever a pulse beat is sensed by the sensor placed on the radial artery of the wrist. Code Composer Studio V4.0 core edition is used for processor programming (Fig. 1). Each block is described in detail in the next sections.

2 Signal Conditioning Circuit and the Signal Processing Circuit

The signal conditioning circuit has been designed and the following components and apparatus are used: a disk-shaped piezoelectric sensor, the NI Elvis II prototyping board, NI Elvis instrumentation CRO from LabVIEW, Texas Instruments U741A operational amplifiers, various components such as resistors, capacitors, BJTs (SL 100). The signal conditioning circuit is further divided into three stages described as follows:

2.1 Signal Conditioning Circuit Stage I

The initial phase depicts the piezoelectric sensor [5, 6] which senses the pulse from the artery on the wrist of our hand. It's shunted by a resistance of the order 100 kΩ. It is connected to the noninverting interval of the operational amplifier. The sensor which senses the pulses on our wrist develops extremely small pulse amplitude.

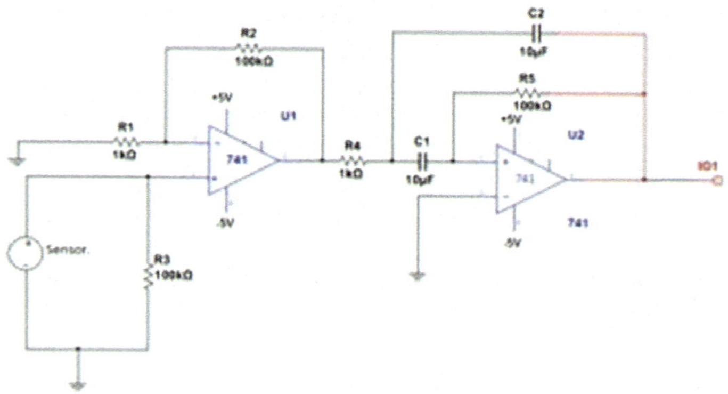

Fig. 2 Signal conditioning circuit

Therefore, there's the need to amplify the voltage levels developed across its' surface, and inverting amplifier with a gain in the range of 100 is cascaded with this stage (Fig. 2).

2.2 *Signal Conditioning Circuit Stage II*

We design a broad pass-band filter having a frequency cutoff range of 1.59–16 Hz.

Relation used for the lower and upper cutoff frequency determination of the filter is shown as below:

$$f_{c1} = \frac{1}{2\pi R_1 C_1}, \tag{1a}$$

$$f_{c2} = \frac{1}{2\pi R_2 C_2}. \tag{1b}$$

The ECG signal as derived from the radial artery, by the piezoelectric sensor, is fed into the filter, which removes most noises, especially the baseline wander error. The output of this filter is then subsequently further fed into the narrow pass-band filter. The circuit is simulated in MULTISIM platform.

2.3 *Signal Conditioning Circuit Stage III*

The third phase comprises of a narrow band-pass filter, with resonance frequency centered at 1.59 Hz roughly. The filter design was that of an infinite gain multiple feedback filter to yield a design with high *Q*-factor and a substantially steep roll-off.

A compromise had to be made in the design to have optimum values of the amplification factor/gain (A_v) and that of the Q-factor.

The resonance frequency of the designed active filter can be obtained by the relation:

$$f_r = \frac{1}{\sqrt{2\pi R_1 R_2 C_1 C_2}}. \tag{2a}$$

The relation pertaining to the maximum gain (A_v) of the circuit can also be further determined as

$$A_v = -\frac{R_2}{2R_1} = 2Q^2. \tag{2b}$$

The gain was calculated to be = 50 (inverting amplification).

The Q-factor was calculated to be = 5.

3 Signal Processing Circuit of the Heart Rate Monitoring System

Similarly as initially, the signal conditioning circuit had been described in three different stages, the signal processing circuit which deals with the conditioned signal will be described in two stages:

3.1 Signal Processing Circuit Stage 1

The conditioned signal is then passed into a threshold detector (zero detector circuit), which gives the pulse train when sensing peak of the ECG pulse. The threshold detector is form of a comparator circuit. This is absolutely essential as this pulse train will later be taken into account while obtaining the BPM as the duty cycle of the pulse train when calculated will have to be inversed and be multiplied with 60, to yield the instantaneous BPM, which is later processed by the Texas Instruments MSP430FG4618 processor. The ECG wave and threshold detector we built on MULTISIM are shown in Figs. 3, 4, and 5.

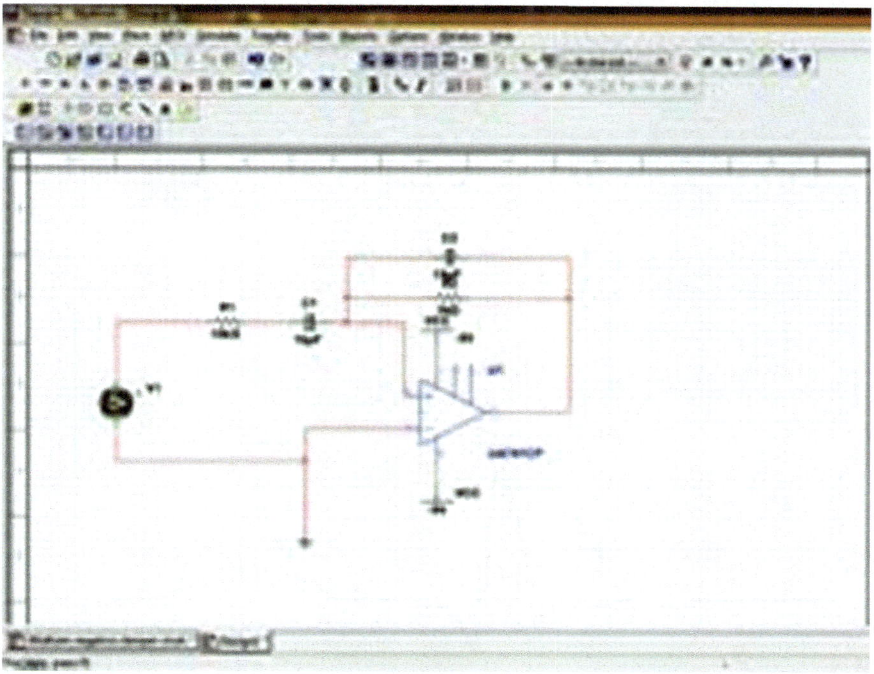

Fig. 3 MULTISIM circuit built simulation

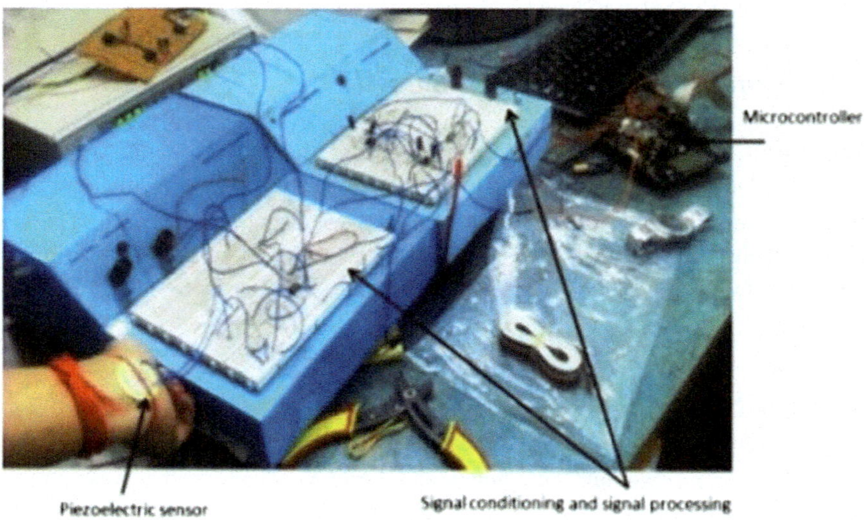

Fig. 4 The piezoelectric sensor and signal conditioning and signal processing stages as designed in our laboratory setup

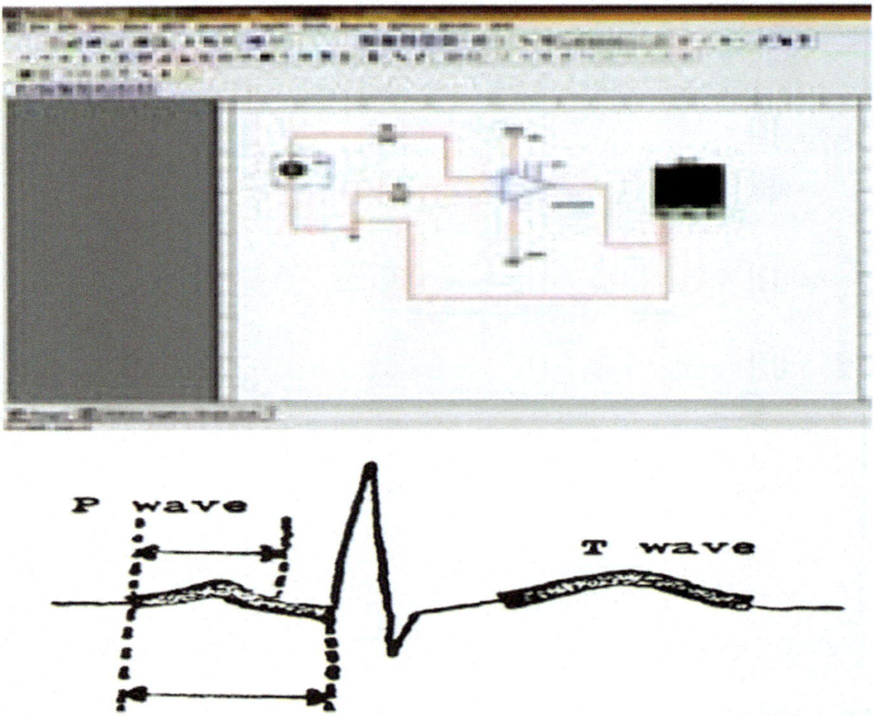

Fig. 5 LabVIEW instrumentation scope is used to display output of the Zero-Detector, and a sample example of ECG Waveform are shown

3.2 *Signal Processing Circuit Stage II*

A negative clamper circuit is constructed to clamp the negative cycle of the pulse train fed to the ADC to process. 12-bit ADC of the MSP430FG4618 Experimenter's board is used to process and measure the heart rate [7]. The simulated results of the negative clamper circuit as observed from our experimental setup are shown in Fig. 6.

4 Case Studies and Observation

The following observations were noted, and the ECG waveform that was extracted from the signal conditioning and signal processing circuit was displayed via the LabVIEW NI Elvismx instrumentation oscilloscope.

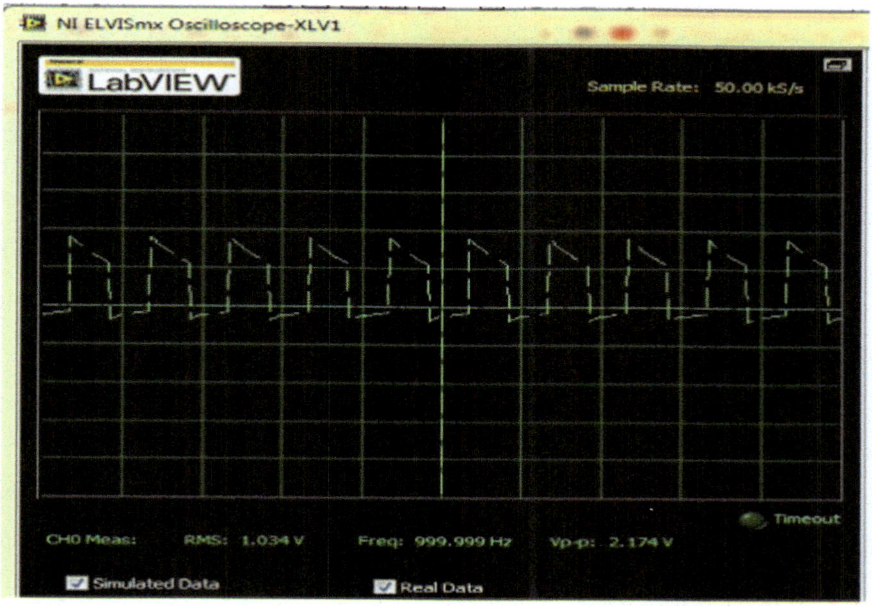

Fig. 6 Response of clamper circuit

The observations were noted in two different categories. Some of them were taken in normal condition, that is no prior physical exertions took place before the ECG reading was observed, and some were noted in an excited state, that is physical exertions were involved before the subject's reading was observed.

The BPM can be calculated by using the empirical relation as stated below:

$$\text{B.P.M} = 60 \times \text{frequency}, \tag{3a}$$

where BPM, stands for the beats per minute, used as a unit to measure the heart rate (Fig. 7).

4.1 Observation I

Patient 1 Age: 30.

Observation was noted during normal conditions, and mild physical exertions were initiated before observations.

The frequency reading = 1.313 Hz.

Heart rate (BPM) = 60 × frequency.

Heart rate = 78.78 BPM.

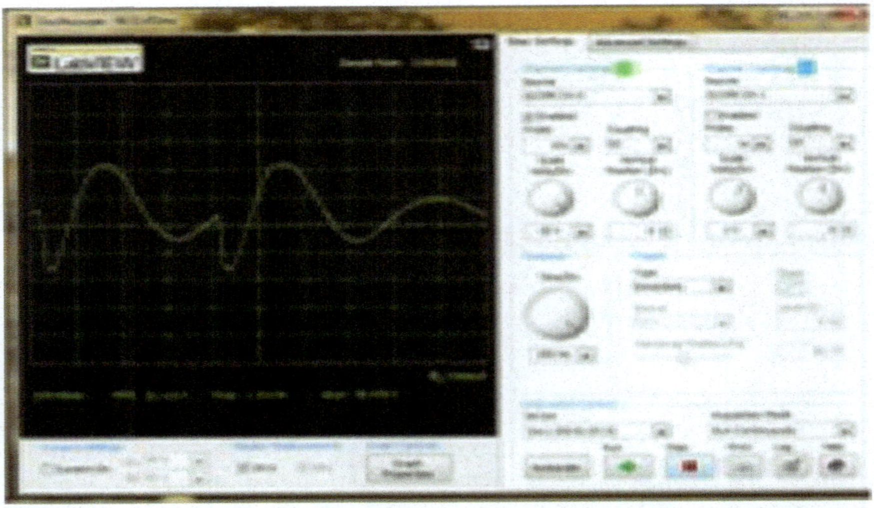

Fig. 7 ECG waveform and its frequency of Patient 1 measured using our system design setup

4.2 Observation II

Patient 2 Age: 27.

Observation was noted in normal conditions, and no physical exertions were initiated before observations.

The frequency observed = 1.163 Hz.

Heart rate (BPM) = 60 × frequency.

Heart rate = 70 BPM (Fig. 8).

4.3 Observation III

Now when the penultimate phase of the signal processing circuit is taken into account, the threshold detector would create a wave pulse train whenever a QRS complex of the ECG portion is detected, the comparator shifts up, and it reaches the lower base value as soon as the level shifts below. The following observation is depicted in Fig. 9.

Patient 3 Age: 22.

Observations were noted in normal conditions, and no physical exertions were initiated before observations.

The frequency observed = 1.181 Hz.

Heart rate (BPM) = 60 × frequency.

Heart rate = 71 BPM.

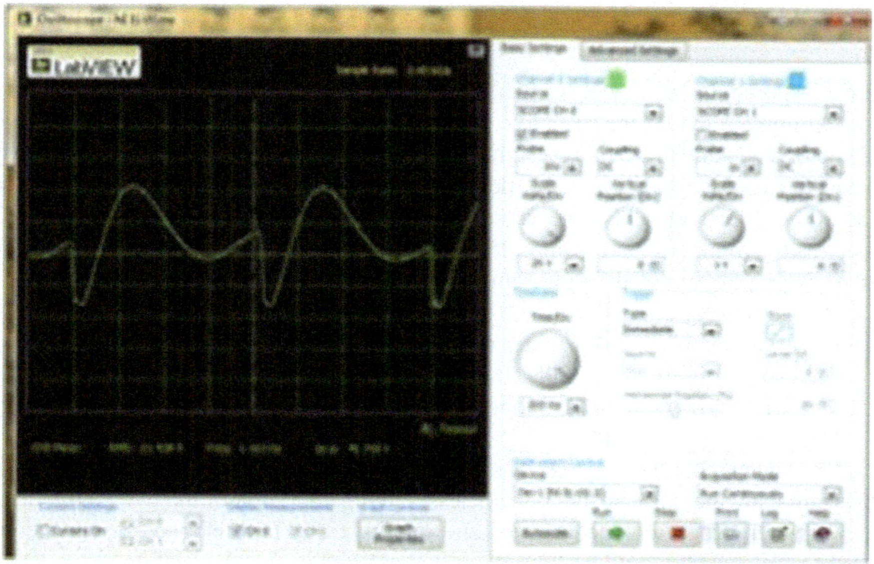

Fig. 8 Above image depicts the ECG waveform and its frequency of Patient 2 as derived from our system design setup

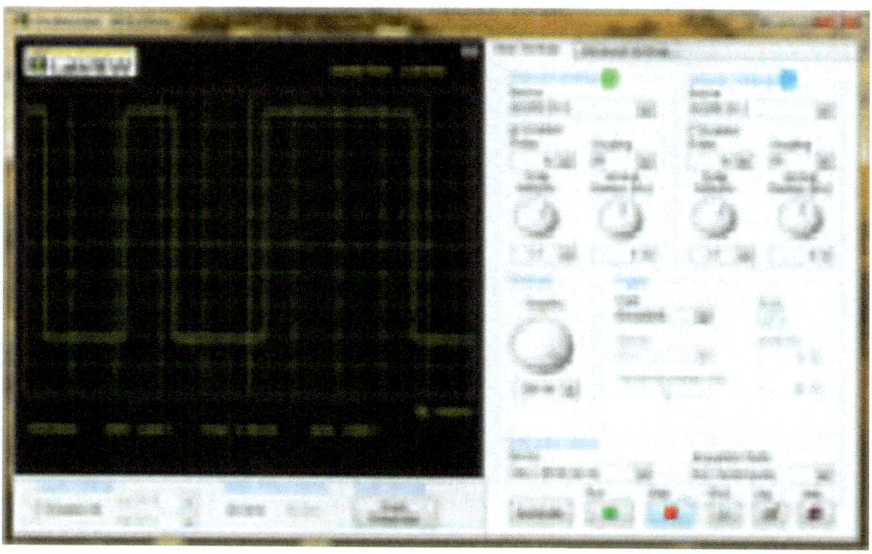

Fig. 9 Above image depicts the output from the threshold detector when fed with the ECG waveform, from Patient 2

4.4 Observation IV

In this particular observation, we need to consider the final phase of the signal conditioning circuit, the negative clamper circuit, the output of the threshold detector as deemed fit by the design is fed into the negative clamper circuit, to only keep the positive half cycle. As depicted before, the IN-4007 p-n junction diode was used. This had to be done as the ADC which will later be used, (a 12-bit ADC is already an integral part of the MSP 430 FG series from Texas Instruments). The negative half cycle of the pulse will not be processed.

Patient 4 Age: 22

Observations were noted under mild physical exertions.

Frequency observed = 1.395 Hz.

Heart rate (BPM) = 60 × frequency.

Heart rate = 83.7 BPM.

The steep charging of the capacitor and the discharge through the shunted resistor skew the pulse shape, nonetheless, the time duration of the two consecutive peaks when processed and inverted and multiplied by 60 yields the instantaneous BPM value which would be subsequently displayed in the observation/watch panel in Code Composer Studio V 4.0 core edition. The rising edge of the waveform is what will be made use of to evaluate the heart rate of the patient (Fig. 10).

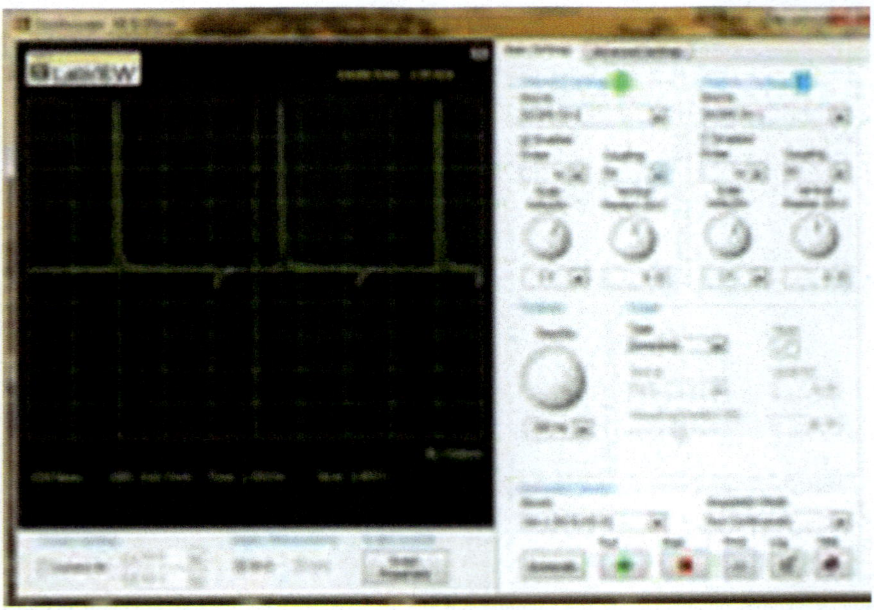

Fig. 10 Output from the negative clamper Patient 3 under physical exertion

5 A Brief Insight into Code Composer Studio and the Msp430FG4618

The Code Composer Studio V4.0 core edition was the IDE used to develop the project [8]. The program was coded onto the C project platform, and once the C project is completed in the specified workspace, we use the USB Flash emulation tool which in case of this project a MSP-FET430 UIF was used, to debug and load the program onto the MSP430FG4618. It should be further noted that MSP430 header files were linked and used to facilitate and further simplify code development.

The MSP430FG4618 experimenter's board is a comprehensive development target board, which is used for varied applications [9–11]. The device features a 16 bit RISC processor, it also boasts of 16-bit registers, dual 12-bit DAC's, three configurable op-amp, one universal serial communication link, a high performance 12-bit ADC, whose use and importance is central to this project, among several other peripherals mounted upon the experimenter's board.

The MSP430 Flash Emulation Tool is instrumental in downloading the code and debugging the MSP430FG4618. The experimenter's board is provided with two separate JTAG headers specifically to cater to this function and supporting independent debug environment.

6 The Working Overview of the Code Loaded into the Msp430FG4618

The working logic behind the code debugged and programmed is in Fig. 11. The measured BPM using Code composer studio is displayed in Fig. 12.

Table 1 shows the comparative measurement study of heart rate of a patient using standard NI Vernier hand grip sensor and using low-cost piezoelectric sensor. The BPM value measured from the ECG waveform using piezoelectric sensor is approximately same as BPM measured by costly Vernier sensor. This was a preliminary study, and further rectification is required for accurate measurement of heart rate.

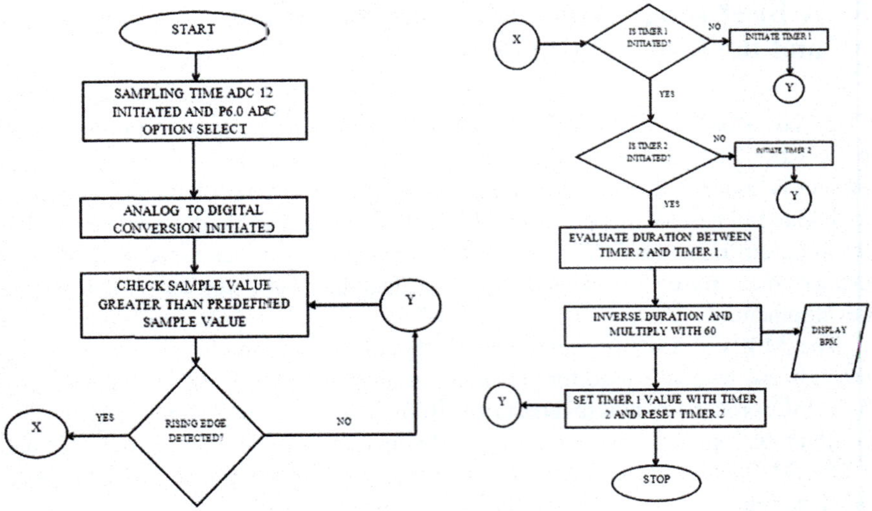

Fig. 11 Flow diagram of the code used to evaluate heart rate

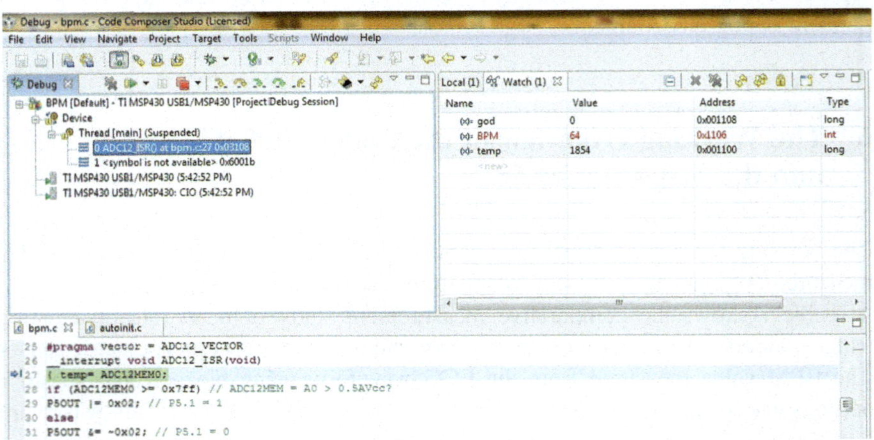

Fig. 12 The observed heart rate value of a patient in the watch section of Code Composer Studio. Patient recorded a BPM value of 64

Table 1 Comparative analysis

Sensor used	BPM calculated (heart rate)
NI Vernier sensor	82.7
Piezoelectric sensor	84

7 Conclusion

Design of a low-cost heart rate measurement system using a simple piezoelectric sensor is attempted in this paper. Signal conditioning and signal processing system is implemented with discrete components upon considering cost of the system design as an important design parameter. The system performance can be improved by using piezoresistive sensors and by using active components which may reduce the power consumption but increases the cost of the system. This work is a preliminary approach to design of health monitoring system in a cost-effective way; therefore, only one health parameter (heart rate) is considered. Other health parameters measurement may also be incorporated with this system.

References

1. Y. Shu, C. Li, Z. Wang, W. Mi, Y. Li, T.-L. Ren, A pressure sensing system for heart rate monitoring with polymer-based pressure sensors and an anti-interference post. Sensors. **15**, 3224–3235 (2015). doi:10.3390/s150203224
2. T. Lewalter, B. Lüderitz, *Historical Milestone of Electrical Signal Recording and Analysis*, vol. XVII (Springer, 2003), p. 548. ISBN:978-1-58829-069-4
3. LabVIEW getting started with LabVIEW. National Instruments, June 2013
4. National Instruments, Multisim User Guide, January 2007
5. J. Shieh, J.E. Huber, N.A. Fleck, M.F. Ashby, The selection of sensors. Prog. Mater. Sci. **46**, 461–504 (2001)
6. TI IDC 2015, TEAM ID: 444, Piezoelectric Non-Invasive Pulse Wave Analyzer, Youtube link
7. MSP430FG4618/F2013 Experimenter's Board User's Guide. Texas Instruments, October 2007
8. Code Composer Studio v5.1, User's Guide for MSP430, Texas Instruments
9. Texas Instruments Community, https://e2e.ti.com/. Accessed 17 July 2015
10. K. Quiring, MSP430 Timers In-Depth, MSP430 Advanced Technical Conference, Technology for innovators, Texas Instruments
11. J. Davies, *MSP430 Microcontroller Basics* (Elsevier). ISBN:978-0-7506-8276-3

Design of DA-Based FIR Filter Architectures Using LUT Reduction Techniques

A. Uma, P. Kalpana and T. Naveen Kumar

Abstract The multiplier-less techniques such as distributed arithmetic (DA) have gained large popularity for its high-speed processing. Architectures based on DA results in cost-efficient and area-efficient structures. This paper presents design and realization of various DA-based FIR filter architectures based on LUT reduction techniques of length $N = 4$ and also implemented using both shift accumulators and carry save shift accumulators. The larger LUT is subdivided into a number of LUTs to reduce the size of the LUT for higher order filter. FIR filter architectures designed include filter with LUT size of $2^N - 1$ words, filter with LUT size of 2^{N-1} words, filter with LUT breakup contains two $2^{N/2} - 1$ word LUTs, and also LUT-less filter but only has combinational blocks. These filter architectures have been synthesized for the target FPGA device and results are compared based on RTL area, device utilization, maximum operating frequency, and power consumption.

Keywords Distributed arithmetic (DA) · Carry save shift accumulator (CSSA) Shift accumulation (SA)

1 Introduction

Multiply and accumulate operation is very common in all digital signal processing algorithms such as finite impulse response (FIR) filter. As multipliers consume more area and power in multiply and accumulate (MAC) operation of FIR Filter, several multipliers-less schemes have emerged. Distributed arithmetic (DA) method is one of the multiplier-less techniques which uses memories (RAMs, ROMs) or LUTs to store precomputed values of coefficient operations [1]. DA is an efficient technique for performing multiply-and-add in which the multiplication is reorganized such that multiplication and addition are performed on data and single bits of the coefficients, at the same time [2]. The inner products containing many terms can

A. Uma (✉) · P. Kalpana · T. Naveen Kumar
Department of ECE, PSG College of Technology, Coimbatore, India
e-mail: umavithy22@yahoo.com

V. Nath (ed.), *Proceedings of the International Conference on Microelectronics, Computing & Communication Systems*, Lecture Notes in Electrical Engineering 453,
https://doi.org/10.1007/978-981-10-5565-2_20

be partitioned into a number of smaller inner products which can be computed and summed by using either DA or an adder. Hence, DA is widely used for the implementation of digital filters [3].

The advantages of DA are best exploited in data-path circuit designing. DA efficiently implements the MAC using basic building blocks (Look-up Tables) in FPGAs. These DA structures can be used even in adaptive filters. An adaptive filter which is commonly used in devices such as mobile phones, camcorders. This adaptive structure is a system with a linear filter that has a transfer function controlled by variable parameters and a means to adjust those parameters according to an optimization algorithm [4]. Adaptive filters are required for some applications because some parameters of the desired processing operation are not known in advance or are changing [5].

In conventional DA-based FIR filter, as filter size increases, the memory requirements of the implementation also grow exponentially. This in turn increases the look-up table (LUT) size. For example, a 128-tap DA-FIR filter will require a prohibitively large 2^{128} entries in the DA-based LUT [6]. The larger LUT is subdivided into a number of LUTs to reduce the size of the LUT for higher order filter. The FIR filter architectures are designed using four LUT reduction methods. They include filter with LUT size of $2^N - 1$ words, filter with LUT size of 2^{N-1} words, filter with LUT breakup contains two $2^{N/2} - 1$ word LUTs, and also LUT-less filter [4] but only has combinational blocks. This paper presents various DA-based FIR filter architectures with LUT reduction techniques and its implementation using both shift accumulator and carry save shift accumulator (CSSA). The results are compared with area, operating frequency, and power.

The paper is organised as follows: In Sect. 1, a brief introduction of DA-based filter structure is given. In Sect. 2, DA-based FIR filter structures and their operations are discussed. In Sect. 3, modified DA architectures and their hardware complexities derived through LUT reduction techniques are described. The synthesis and comparison results of DA-filter structures are described in Sect. 4. Summary and conclusions are given in Sect. 5.

2 DA-Based FIR Filter Structure

Many digital signal processing (DSP) applications require linear filters that can adapt to changes in the signals they process. Adaptive filters find extensive use in several DSP applications including acoustic echo cancellation, signal de-noising, sonar signal processing, clutter rejection in radars, and channel equalization for communications and networking systems [3]. DA is one method often preferred since it eliminates the need for hardware multipliers and is capable of implementing large filters with very high throughput [6].

Distributed arithmetic (DA) is an efficient technique for performing multiply-and-add where both the multiplication and addition is done at same time on coefficients which has single bit [6]. Distributed arithmetic is a bit level rearrangement of a multiply accumulate in order to eliminate the multiplications and also an efficient method for computing inner products between a fixed and a variable data vector [6]. The basic principle of the DA is the computed values are stored rather than carry out the computation. It reduces the size of parallel hardware implementation of multiply accumulate which is more preferred for FPGA. This feature can be extended to sum functions used in complex multipliers, Fourier Transforms. The basic DA technique which is bit serial has ROM look-ups an efficient technique to implement on Field Programmable Gate Arrays (FPGAs). The often encountered forms of computation in DSP are sum of product, dot product [7], and inner product which can be implemented using distributed arithmetic.

The block diagram of DA-based FIR filter is shown in Fig. 1. The DA based FIR filter has to perform an inner-product computation which produces the critical path [8]. The critical path is the longest delay free path which affects the speed of the architecture.

Let the inner product of FIR filter be given by

$$y = \sum_{k=0}^{N-1} w_k x_k \tag{1}$$

where w_k and x_k for $0 \leq k \leq N-1$ form the N-point vectors corresponding the current weights and most recent $(N-1)$ input, respectively [6]. Assume L be the bit

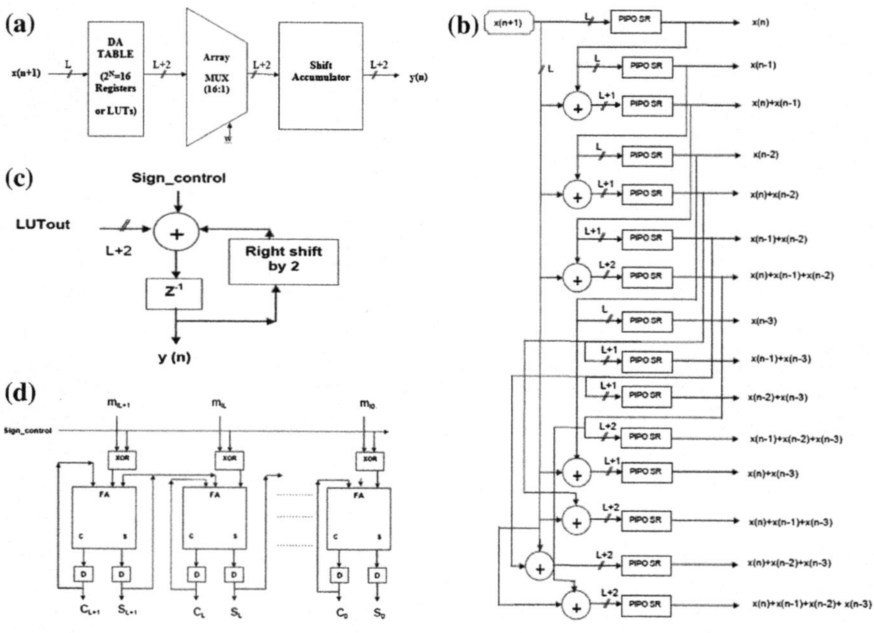

Fig. 1 **a** Block diagram of DA-based FIR filter, **b** generation of possible sum of input combinations (DA table), **c** shift accumulator, **d** carry save shift accumulator

width of the weight and w_k be a N-bits of two's complement number representation where $|w_k| < 1$ and w_k: $\{w_{k0}, w_{k1}, w_{k2}, \ldots, w_{k(N-1)}\}$ where w_{k0} is the sign bit.

$$w_k = -w_{k0} + \sum_{l=1}^{L-1} w_{kn} 2^{-l} \tag{2}$$

Substituting (2) in (1) and interchanging the order of summations over the indices k and l will give the reformulation of inner product to form distributed inputs as

$$y = -y_0 + \sum_{l=1}^{L-1} 2^{-l} \cdot y_l \quad \text{where} \quad y_l = \sum_{k=0}^{N-1} x_k \cdot w_{kl} \tag{3}$$

Any weight bit of the N-point w_k: $\{w_{k0}, w_{k1}, w_{k2}, \ldots, w_{k(N-1)}\}$ will either be one or zero, so the partial sum y_l will have 2^N possible values. DA-based computation requires storing all possible values of partial sum y_l in LUT of size 2^N words. The weight bit sequence $\{w_{kl}\}$ can be used as address vector to read out the corresponding partial sum from LUT, for computing the inner product. The hardware realization of inner product needs PIPO SR, shift accumulators, Vectored mux, etc. The calculation of inner product requires L cycles of shift accumulation, then read by LUT-read operations based on the number of bit slices $\{w_{kl}\}$ where $0 \leq 1 \leq L - 1$.

The block diagram representation of DA-FIR filter architecture is shown in Fig. 1a. It contains DA table, Array Multiplexer followed by shift accumulator. The various DA-based FIR filters are implemented and compared with both shift accumulator and carry save shift accumulator (CSSA). The matched LUT content should be selected by the weight vector in a bit serial vector. The detailed blocks of possible sum of input generation is shown in Fig. 1b.

DA Table Generation: The DA table stores ($2^N - 1$) words using Parallel In Parallel Out Shift Registers (PIPO SR), whereas RAM-based LUT stores 2^N words [6]. For example, considering $N = 4$, the number of registers required is only 15, to store the precomputed possible sum of input sequences and seven adders are required to produce the seven new values of input sums.

Parallel In Parallel Out Shift Registers (PIPO—SR): A L-bit PIPO SR constructed by L number of D flip-flops. Once the register is clocked, all the data at the D inputs appear at the corresponding Q outputs simultaneously. N is the number of taps in FIR filter. In the process of constructing DA table, $2^N - 1 = 15$ PIPO registers with data bus sizes in the range from L to $L + 2$ are used as shown in Fig. 1b.

Bus Mux: Vectored Array Multiplexer is used to transfer the selected input bit combination to the output. The weight vector which are fed in bit serial fashion acts as select lines for the 16:1 vectored mux, for $N = 4$ and $L = 8$. The selected combination of inputs has data width size of $L + 2 = 10$ bits. The output of vectored mux is passed to input of adder-based shift accumulator or carry save shift accumulator [6].

Shift Accumulator: The shift register is shifted right at every bit clock cycle to feed the weight vector inputs serially to mux [6]. The corresponding LUT entries are also shifted and accumulated using L consecutive times to generate the output using adder-based conventional shift accumulator shown in Fig. 1c. The sign bit control is used to change the addition to subtraction for the sign bits.

Carry Save Shift Accumulator: The shift accumulation has large critical path and the longest delay free path that decides the ultimate speed of the architecture. The carry save accumulation serves as an alternate block for conventional shift accumulation and is shown in Fig. 1d. The bit-sliced vector w is fed as inputs from the least significant bit (LSB) to the most significant bit (MSB). The ex-or gate is used to pass all the bits of table output, and the sign bit is one.

The byte clock and the bit clock are the two clock signals used in the design. The byte clock synchronizes with sampling period of input sequence. The carry save accumulation block uses bit clock. The byte clock is used in the remaining circuits.

3 Modified DA-Based FIR Filter Architectures

It can be noted that as the filter size increases, the memory requirements of the implementation in Fig. 1 grow exponentially. This in turn increases the look-up table (LUT) size. This problem can be rectified by altering LUT sizes and use of combinational blocks. The DA-FIR filter architectures are modified and implemented based on shift accumulator and also carry save shift accumulator (CSSA) [9]. The LUT reduction technique is incorporated by breaking up the filter into smaller base DA filtering units that require tractable memory sizes and then summing up the outputs of these units. By incorporating additional Multiplexers and use of only combinational blocks such as adders and multiplexers in realizing inner-product block of DA-FIR filter further reduces the LUT size [10].

Using the LUT reduction techniques, the different DA-based FIR filters are

(1) DA-based filter with LUT size of 2^{N-1} words.
(2) DA-FIR filter with partitioned LUTs where breakup contains two $2^{N/2} - 1$ word LUTs.
(3) DA-FIR filter with No LUTs, but only having combinational blocks.

All the above architectures can reduce the memory requirement and also modified using both CSSA and shift accumulator.

3.1 Modified DA-Based Filter with LUT Size of 2^{N-1} Words

The conventional DA-based FIR filter is implemented with LUT size of 2^{N-1} words, and it uses shift accumulator. It is shown in Fig. 2. From the DA table, it is observed that lower half of LUT (locations where $w_3 = 1$) is the same with the sum of the upper half of LUT (locations where $w_3 = 0$) and $x\,(n-3)$, i.e., Lower Half of LUT outputs = Upper Half of LUT outputs + $x\,(n-3)$.

In the DA-based filter, LUT size is reduced to half of the memory LUT size in original DA table. This architecture includes one more 2:1 multiplexer and one adder additionally, which occupy very less area. This mux selects $x\,(n-3)$ input when the MSB bit of weight vector is high ($w_{31} = 1$), else 0. This is then added to output of main LUT to be sent to conventional shift accumulator to produce the output for inner-product computation of DA-FIR filter [7]. The architecture is modified by replacing the shift accumulator with carry save shift accumulator.

For instance, $N = 4$, this architecture requires 8 memory LUTs, one 8:1 mux, one 2:1 mux and one adder. For higher order filters, example of $N = 16$, this architecture requires 256 memory LUTs, one 256:1 mux, one 2:1 mux, and one adder, whereas in conventional DA-FIR, it needs $2^{16} = 65536$ memory LUTs and 65536:1 mux.

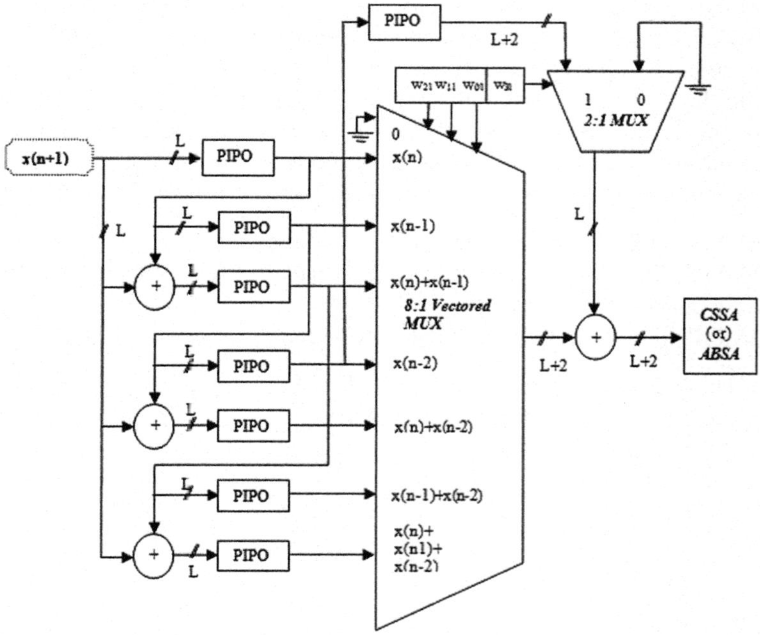

Fig. 2 DA-based filter with LUT size reduced by ½ for $N = 4$

3.2 *Modified DA-Based Filter with Partitioned LUTs*

The conventional DA-based filter is implemented with partitioned LUTs where breakup contains two $2^{N/2} - 1$ word LUTs. In this architecture, LUT is divided into LUT 1, LUT 2 so on. The N-tap filter is divided into p smaller filters each having q-tap DA base units. Here, it is assumed that N is not prime and also $N = p * q$. The total number of memory elements requirement to implement this structure is $p * 2^q$ [3]. Thus, there is a marginal decrease in throughput of this architecture.

For instance, if $N = 128$, $p = 32$, and $q = 4$ can be chosen, which would only require 512 memory elements. The number of clock cycles required for this implementation would be 21 clock cycles as compared with the single LUT implementation that would require 16 clock cycles [4]. Similarly for $N = 4$, p and q can be chosen as $p = 2$ and $q = 2$. This requires $2 * 2^2 = 8$ memory elements only for the implementation along with two 4:1 Multiplexers and one adder, as shown in Fig. 3. The partitioned LUT architecture is modified by replacing shift accumulator with CSSA and results are compared.

3.3 *Modified DA-Based Filter with No LUTs*

The conventional DA-based filter with no LUTs or LUT less is implemented using shift accumulator and then modified with carry save shift accumulator. By the same

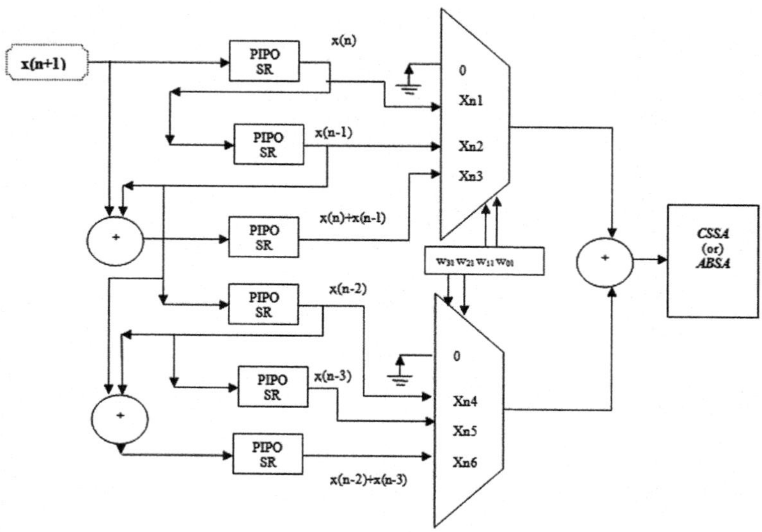

Fig. 3 DA-based filter with partitioned LUT for $N = 4$

LUT reduction procedure, the final LUT-less DA-FIR filter architectures can be realized [1]. For $N = 4$, it requires four 2:1 Multiplexers and two stages of three adders as shown in Fig. 4.

Similarly, for $N = 8$, it requires 8, 2:1 Multiplexers and 3 stages of 7 adders, where first, second, and third stage contains 4, 2, and one number of adders, respectively. So, in general, for any N-tap DA-FIR filter, it requires N, 2:1 Multiplexers, and ($N/2 + N/4 + \cdots + 1$) number of adders. For the use of combination logic circuit, the filter performance will be affected. But when the taps of the filter is a prime, 4-input LUT units with additional multiplexers and full adders can be used to get the trade-off between filter performance and small resource usage [1].

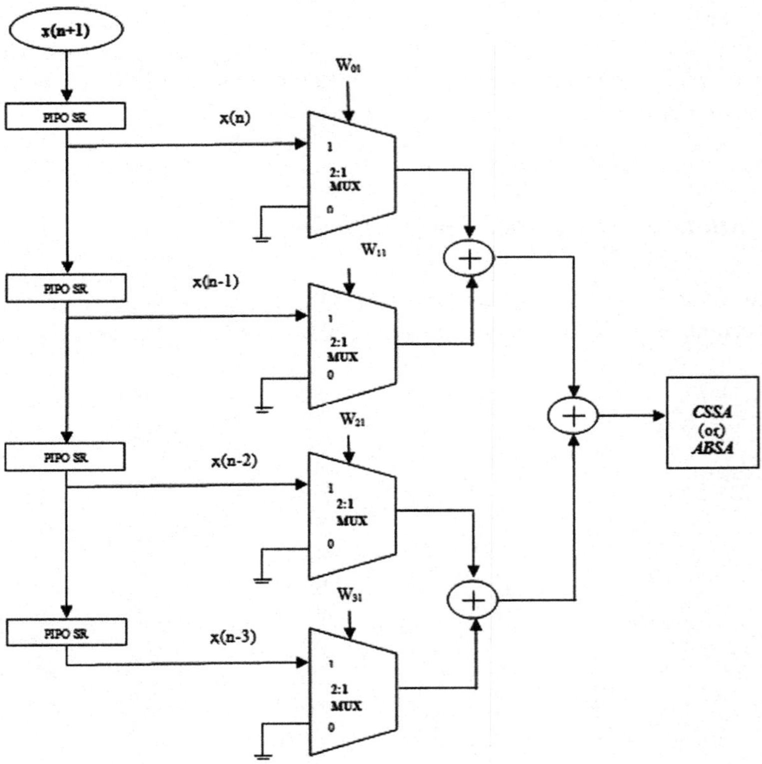

Fig. 4 DA-based filter with no LUT

4 Synthesis and Comparison Results of DA-Filter Structures

The conventional DA-based FIR filter architectures (for $N = 4$ and $L = 8$) with shift accumulator and modified architecture where shift accumulator replaced with carry save shift accumulator are implemented. The design is coded in Verilog HDL and simulated and synthesized for the target device of Xilinx's FPGA SPARTAN-3E (XC3S500E-FG320). The comparison results of inner-product block replaced by conventional DA and other architectures based on shift accumulator and CSSA are shown in Tables 1 and 2. The comparison results are tabulated based on maximum operating frequency area and power consumption for the given clock frequency = 10 MHz, bit clock frequency = 80 MHz. The default activity factor = 12.5% and capacitive load of 35 pF has been taken.

The comparison results show that LUT-less DA-filter has less power consumption and area when compared to other structures and also it can be used for high-speed applications.

Table 1 Comparison of maximum operating frequency, power, and area for DA-based filters using CSSA

Features and architectures	Power (mW)	Maximum operating frequency (MHz)	No. of slice (FFs)	No. of slices	No. of 4 input LUT
Conventional DA-FIR filter	91	252.972	146	121	147
Filter with LUT of 2^{N-1} words	87	256.805	79	67	80
Filter with partitioned LUTs	88	254.388	60	55	69
Filter with no LUTs	83	446.429	51	46	54

Table 2 Comparison of maximum operating frequency, power, and area for DA-based filters using shift accumulator

Features and architectures	Power (mW)	Maximum operating frequency (MHz)	No. of slice (FFs)	No. of slices	No. of 4 input LUT
Conventional DA-FIR filter	95	252.972	155	130	167
Filter with LUT of 2^{N-1} words	87	256.805	88	76	99
Filter with partitioned LUTs	88	254.388	69	64	88
Filter with no LUTs	86	255.820	42	37	35

5 Conclusion

The conditional signed carry save accumulation which replaces the conventional adder-based shift accumulation for DA-based inner-product computation, in order to reduce the sampling period and to decrease the total area occupied, has been realized. By LUT reduction techniques, modified DA-FIR filter architectures such as filter with partitioned LUTs, LUT size of 2^{N-1} words, and also LUT-less filters have been realized. The comparison results show that LUT-less DA has low power and even it can be used for high-speed applications. For higher order filters, the LUT reduction based DA-FIR filter architectures can be adopted to reduce the resource utilization. Hence, it is shown that modified DA architectures are hardware efficient for FPGA implementation.

References

1. J. Xie, J. He, G. Tan, FPGA realization of FIR filters for high-speed and medium-speed by using modified distributed arithmetic architectures. Micro Electron. J. 365–370 (2010)
2. Y.-H. Chen, J.-N. Chen, T.-Y. Chang, C-W. Lu, High-throughput multistandard transform core supporting MPEG/H.264/VC-1 using common sharing distributed arithmetic. IEEE Trans. Very Large Scale Integration (VLSI) Syst. **22**(3), Mar 2014
3. J. Xie, P.K. Meher, J. He, Hardware-efficient realization of prime-length DCT based on distributed arithmetic. IEEE Trans. Comput. **62**(6), June 2013
4. D.J. Allred, H. Yoo, V. Krishnan, W. Huang, D.V. Anderson, LMS adaptive filters using distributed arithmetic for high throughput. IEEE Trans. Circ. Syst. I Reg. Pap. **52**(7), 1327–1337 (2005)
5. M.S. Prakash, R.A. Shaik, Low-area and high-throughput architecture for an adaptive filter using distributed arithmetic. IEEE Trans. Circ. Syst. Ii Express Briefs **60**(11), Nov 2013
6. S.Y. Park, P.K. Meher, Low-power, high-throughput, and low-area adaptive FIR filter based on distributed arithmetic. IEEE Trans. Circ. Syst. II Express Briefs **60**(6), June 2013
7. R. Guo, L.S. DeBrunner, Two high-performance adaptive filter implementation schemes using distributed arithmetic. IEEE Trans. Circ. Syst. Ii Express Briefs **58**(9), Sept 2011
8. E. Özalevli, W. Huang, P.E. Hasler, D.V. Anderson, A reconfigurable mixed-signal VLSI implementation of distributed arithmetic used for finite-impulse response filtering. IEEE Trans. Circ. Syst. I Regul. Pap. **55**(2), Mar 2008
9. B.K. Mohanty, P.K. Meher, A high-performance energy-efficient architecture for FIR adaptive filter based on new distributed arithmetic formulation of block LMS algorithm. IEEE Trans. Sig. Process. **61**(4), 15 Feb 2013
10. S.Y. Park, P.K. Meher, Efficient FPGA and ASIC realizations of DA-based reconfigurable FIR digital filter. IEEE Trans. Circ. Syst. Ii Express Briefs doi:10.1109/TCSII.2014.2324418

VLSI Implementation of Smith–Waterman Algorithm for Biological Sequence Scanning

K. Rajalakshmi and R. Nivedita

Abstract This paper presents the design and implementation of Smith–Waterman algorithm. The aim of this work is to improve the speed of the algorithm by applying optimization concepts of VLSI signal processing such as retiming and parallelism. This facilitates the reduction of critical path and computational time of the algorithm. The algorithm is implemented in Simulink-MATLAB 2013, and the corresponding Verilog codes are written and synthesized in Xilinx ISE Design Suite 14.7.

Keywords Smith–Waterman algorithm · Biosequence · VLSI signal processing

1 Introduction

Deoxyribonucleic Acid (DNA) sequences are made up of exons, and protein sequences are made up of amino acids. These two biological sequences determine the characteristics of any organism. So, if the genomic markers and characteristics of a new biological sequence are to be explored, it can be compared to an existing sequence and regions of similarity between the two sequences can be identified. Also the percentage of biological connectivity between any two organisms can be established by finding out the extent of similarity between the two sequences.

The most important algorithms used in biological sequence scanning are Needleman–Wunsch algorithm and Smith–Waterman algorithm. The former applies the principle of global alignment, whereas the latter follows local alignment of sequences. Global alignment gives the similarity score of the sequences by matching as many characters as possible from end to end. But the local alignment focuses more on finding regions of matching between them. So local alignment gives a better idea of where the two sequences are related. Therefore, the local alignment method of Smith–Waterman algorithm is widely used in the field of bioinformatics.

K. Rajalakshmi (✉) · R. Nivedita
Department of Electronics & Communication Engineering,
PSG College of Technology, Coimbatore, India
e-mail: krl@ece.psgtech.ac.in

V. Nath (ed.), *Proceedings of the International Conference on Microelectronics, Computing & Communication Systems*, Lecture Notes in Electrical Engineering 453,
https://doi.org/10.1007/978-981-10-5565-2_21

The paper is organized as follows. Section 2 gives an overview of the Smith–Waterman algorithm. Section 3 traces the evolution of the implementation of the algorithm. The proposed implementation in VLSI is given in Sect. 4. Results are discussed in Sect. 5. Conclusion is provided in Sect. 6.

2 Smith–Waterman Algorithm

The algorithm compares an unknown sequence with a known sequence. The unknown sequence is the "query sequence" denoted by $a1, a2, \ldots, am$. The known sequence is called as the "subject sequence" denoted by $b1, b2, \ldots, bn$. m and n are the lengths of query sequence and subject sequence, respectively. Equation (1) shows the mathematical function of the Smith–Waterman algorithm.

$$H(i,j) = \max\{0, H(i,j-1) - \alpha, H(i-1,j) - \alpha, H(i-1,j-1) + \mathrm{sbt}(i,j)\}, \quad (1)$$

where "i" varies from $1 \leq i \leq m$ and "j" varies from $1 \leq j \leq n$. Here, $H(i, j)$ is the output matrix and its dimensions are determined by the lengths of the two sequences. sbt(i, j) represents the substitution matrix which stores the result after comparison of the two elements of the sequences as shown by Eq. (2).

$$\left.\begin{array}{ll} \mathrm{sbt}(i,j) = x, & \text{if } i = j \\ \mathrm{sbt}(i,j) = y, & \text{if } i \neq j \end{array}\right\}, \quad (2)$$

where x and y are constants which can be chosen by the user. To achieve maximum scores, insertions or deletions of null space can be performed to the query sequence. This null space is called as gap penalty. The first gap penalty in the algorithm is denoted by α. The subsequent gap penalties are denoted by β. These gap penalties can be classified as linear gap penalties and affine gap penalties. If $\alpha = \beta$, then the case is known as linear gap penalty as used by Eq. (1). If $\alpha \neq \beta$, then it is called as affine gap penalty. Equation (3) shows the implementation with affine gap penalty (Fig. 1).

$$\left.\begin{array}{l} H(i,j) = \max\{0, E(i,j), F(i,j), H(i-1,j-1) + \mathrm{sbt}(i,j)\} \\ E(i,j) = \max\{H(i,j-1) - \alpha, E(i,j-1) - \beta\} \\ F(i,j) = \max\{H(i-1,j) - \alpha, F(i-1,j) - \beta\} \end{array}\right\}, \quad (3)$$

3 Evolution of Smith–Waterman Algorithm Implementations

Smith–Waterman algorithm was proposed and is explained in [1]. In [2], the algorithm is mapped into fine-grained processing elements (PE) where each PE consists of 14 components for linear gap penalty and 23 components for affine gap

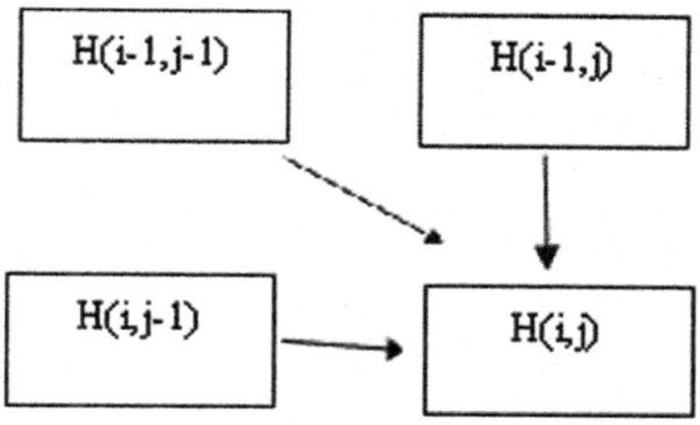

Fig. 1 Data dependencies of a cell of the score matrix on the previous cells

penalty. The original computational time for the Smith–Waterman algorithm is $O(mn)$ which is reduced to $O(m + n)$ in [3]. Four different stages have been introduced in [4] for the implementation which leads to an extra auxiliary array where the dimensions are determined by the two sequences. A linear systolic array was used in [5, 6] to transfer the biosequence database to the implementation. Another algorithm named CAST (complexity analysis of sequence tracts) was invented by reformulating the Smith–Waterman algorithm in [7].

In [8], the algorithm was implemented in a Von Neuman architecture involving concurrent computation of the arithmetic and logic operations. In [9], the concept of Network on Chip (NoC) was applied to facilitate large-scale integration. The logic is realized by bit-serial systolic PE in [10]. The algorithm is coded specifically to the CPU, and the sequence database is preprocessed extensively in [11]. Graphics Processing Unit (GPU) and Compute Unified Device Architecture (CUDA) are used for the implementation in [13]. Single Instruction Multiple Data (SIMD) instructions are used at instruction level to parallelize the implementation in [14]. The algorithm is approached in [12] by VLSI signal processing concepts such as retiming and look-ahead pipelining. The computational time is reduced from $O(m + n)$ to $O((m + n)/j)$ where "j" represents the level of pipelining. In [15], an Electronic System Level (ESL)-based development is adopted, and the design space for Smith–waterman implementation is explored using ESL method which led to increase in speed up to 2.5 times with roughly 1.6 more gate count.

4 Proposed VLSI Implementation of Algorithm

Critical path in VLSI signal processing can be defined as the longest execution path without any delays from the source node to sink node. It can be considered as the sequence that connects critical events and critical jobs from the source to destination node. The significance of the critical path is that its computation time determines the minimum feasible clock period of the network. So reducing the critical path results in minimizing the least execution time of the network thus resulting in the increase of total computational speed of the architecture.

Equation (1) can be realized by the architecture shown in Fig. 2 where T_{m} represents "maximum" function. The critical path is $2T_{\mathrm{m}} + T_{\mathrm{add}}$. A cut-set is introduced in order to reduce the critical path to $2T_{\mathrm{m}}$ as shown in Fig. 3.

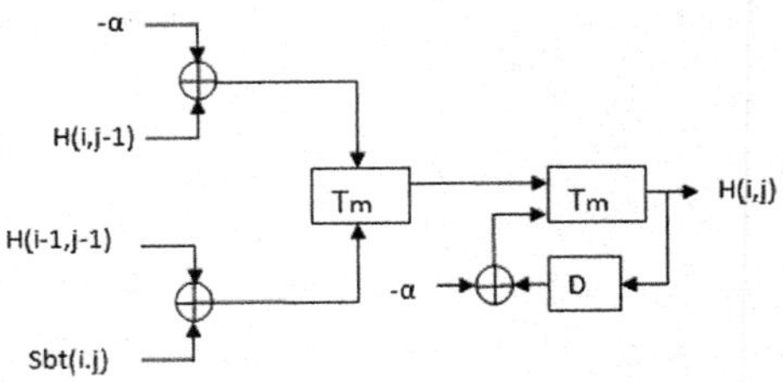

Fig. 2 Architectural realization with linear gap penalties

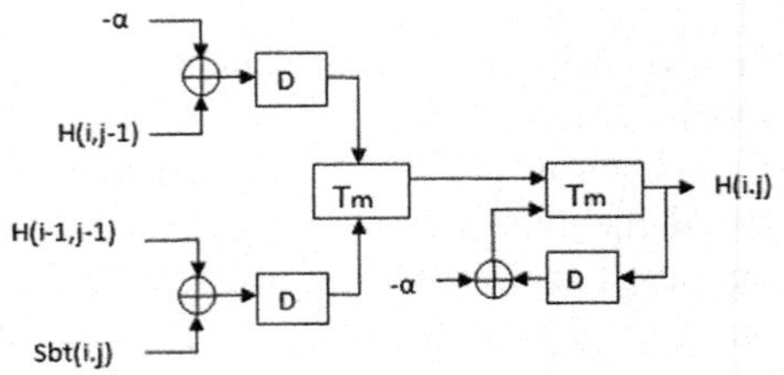

Fig. 3 Pipelined architecture of Fig. 2

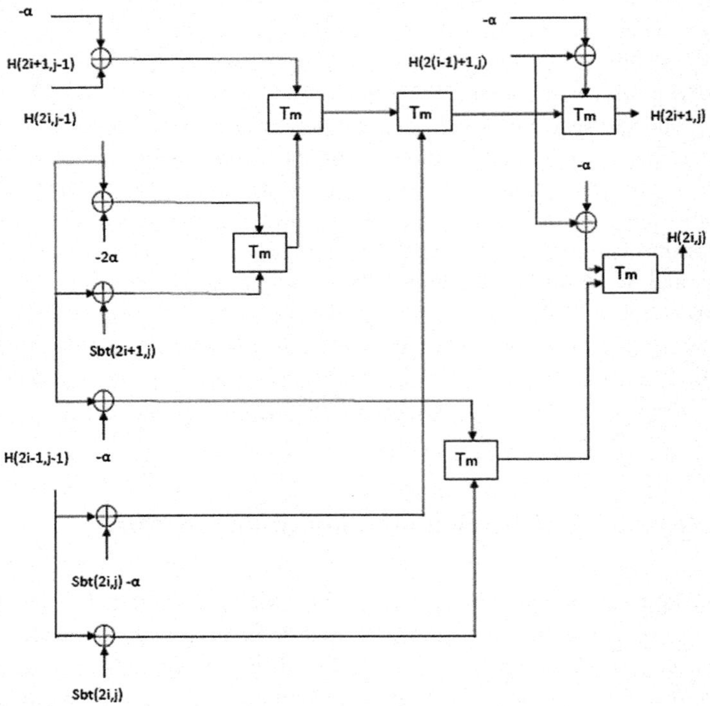

Fig. 4 2-parallel architecture with linear gap penalties

To increase the throughput, a two-parallel implementation is done by replacing $i = 2i$ to get the architecture in Fig. 4.

The critical path for parallel architecture is $4T_{\text{m}} + T_{\text{add}}$. Pipelining is performed, and the critical path is reduced to $3T_{\text{m}}$ as shown in Fig. 5.

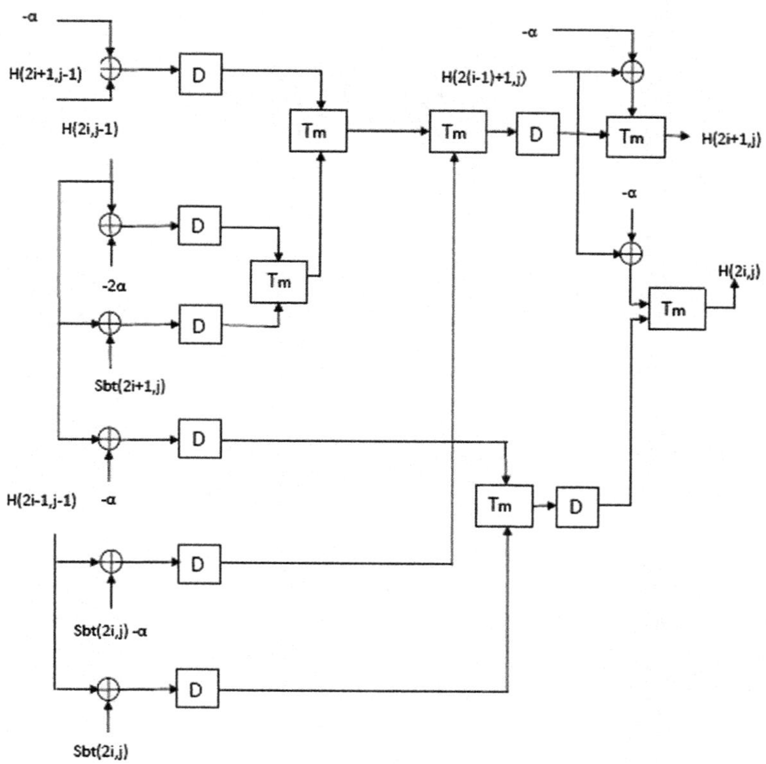

Fig. 5 Pipelined architecture of 2-parallel implementation

To reduce the critical path further, retiming is performed to the pipelined architecture. As shown in Fig. 6, the critical path is reduced to $T_m + T_{add}$.

To increase the throughput further, 3 level-parallel implementation of Fig. 2 is obtained by substituting $i = 3i$. The architecture obtained is shown in Fig. 7 whose critical path is $3T_m + T_{add}$.

Pipelining is now introduced in order to reduce the critical path to $2T_m$ as shown in Fig. 8.

Retiming is performed to the above circuit in order to reduce three delay elements as shown in Fig. 9 but the critical path remains at $2T_m$.

Smith–Waterman algorithm is implemented with affine gap penalties, given by Eq. (3) as shown in Fig. 10. The critical path is $3T_m + T_{add}$.

Pipelining is performed in order to reduce the critical path to $3T_m$ as shown in Fig. 11.

Retiming is performed to the pipelined architecture in order to reduce the critical path to $T_m + T_{add}$ as shown in Fig. 12.

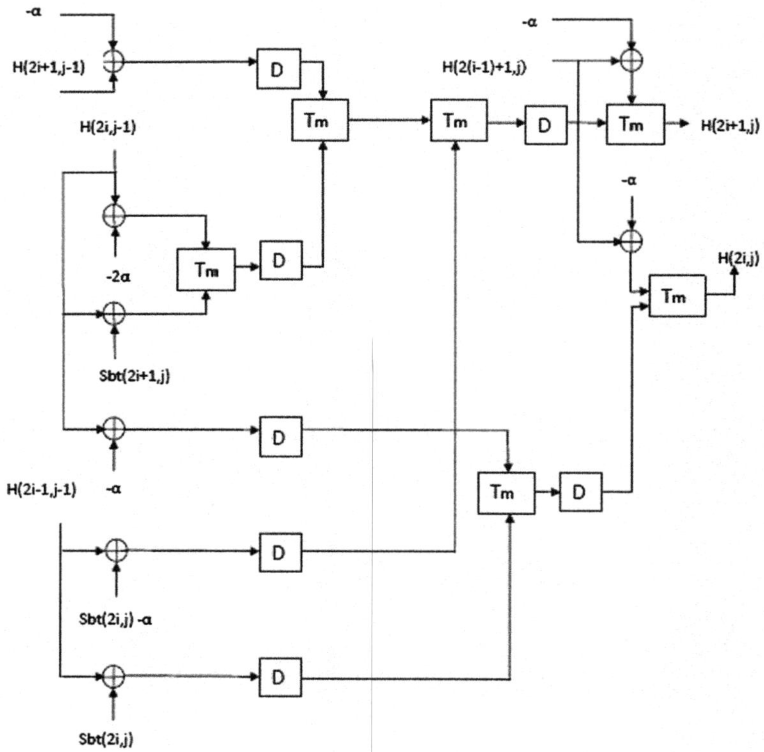

Fig. 6 Retimed architecture of pipelined architecture of 2-parallel implementation

To increase the throughput, 2-parallel implementation is obtained by substituting i with $2i$. The 2-parallel implementation has a critical path of $5T_m + 2T_{add}$ as shown in Fig. 13.

Pipelining is performed to reduce the critical path to $3T_m$ as shown in Fig. 14.

Retiming is performed in order to further reduce the critical path to $T_m + T_{add}$ as shown in Fig. 15.

5 Results and Discussion

5.1 Need for Hardware Implementation

The biological sequence scanning is usually carried out by software. But as software executes sequentially and hardware executes in parallel, the execution time of the algorithm is decreased when implemented in hardware due to the high processing speed on hardware. The timing constraints are generally low, and hardware

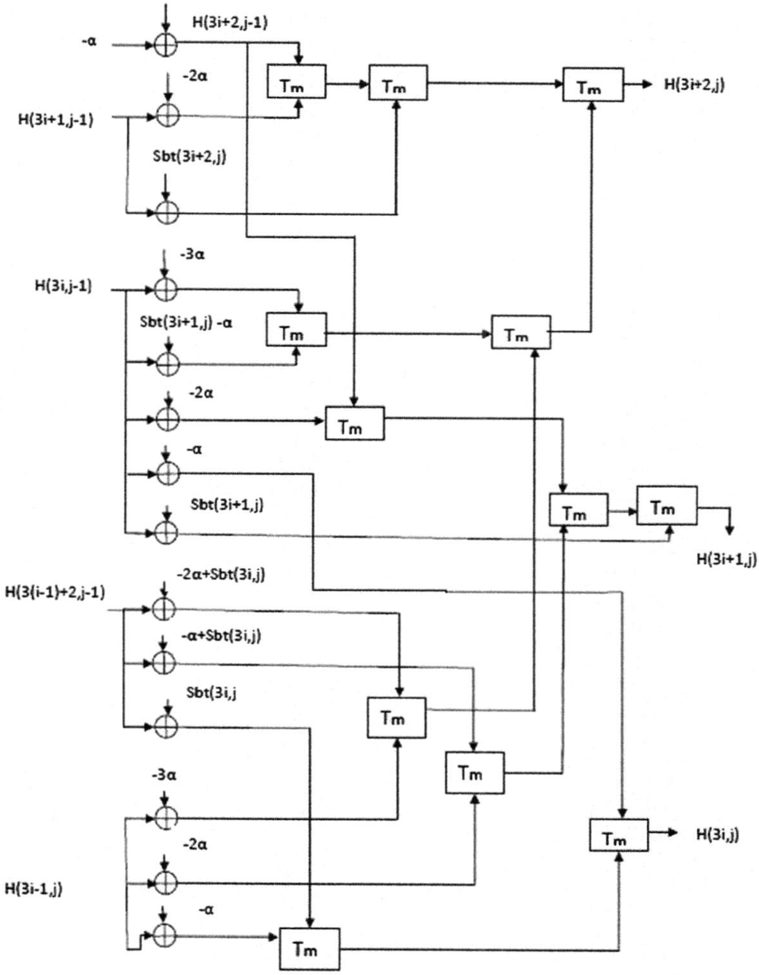

Fig. 7 3 level-Parallel implementation of Fig. 2

facilitates real-time validation. Verilog codes for the above implementations are written, simulated, and synthesized in Xilinx ISE Design Suite 14.7. The output score matrix of the algorithm is shown in Fig. 16.

Table 1 summarizes the reduction of critical paths by various VLSI signal processing techniques.

Table 2 shows the timing summary of the proposed implementation of Smith–Waterman algorithm.

Table 3 shows the comparisons of critical path of proposed implementation and the existing implementation.

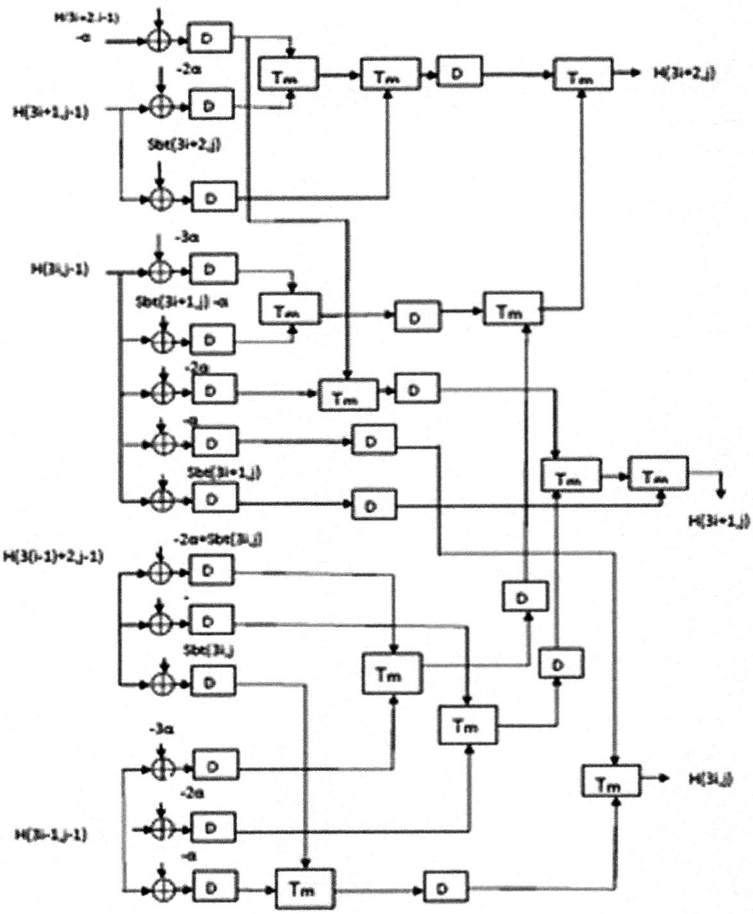

Fig. 8 Pipelined architecture of 3 level-parallel implementation

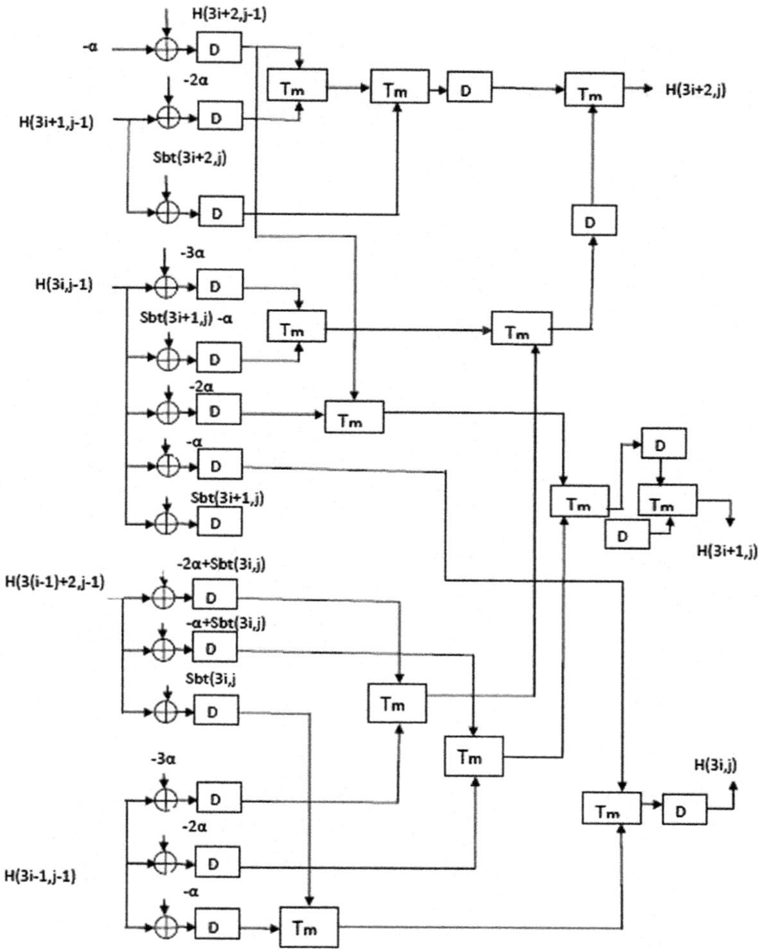

Fig. 9 Retimed architecture of the pipelined implementation 3 level-parallel implementation

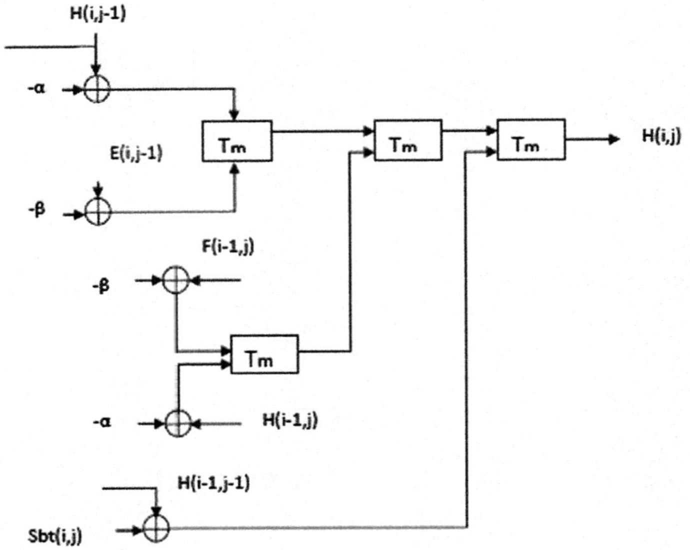

Fig. 10 Smith–Waterman algorithm with affine gap penalties

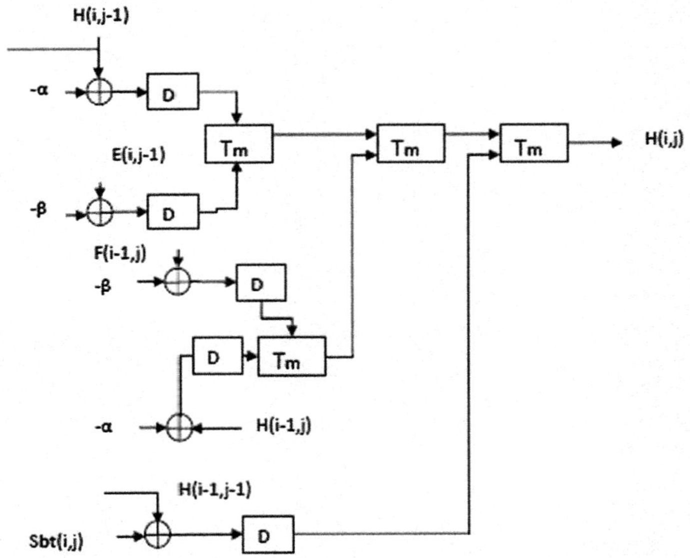

Fig. 11 Pipelined architecture of Fig. 10

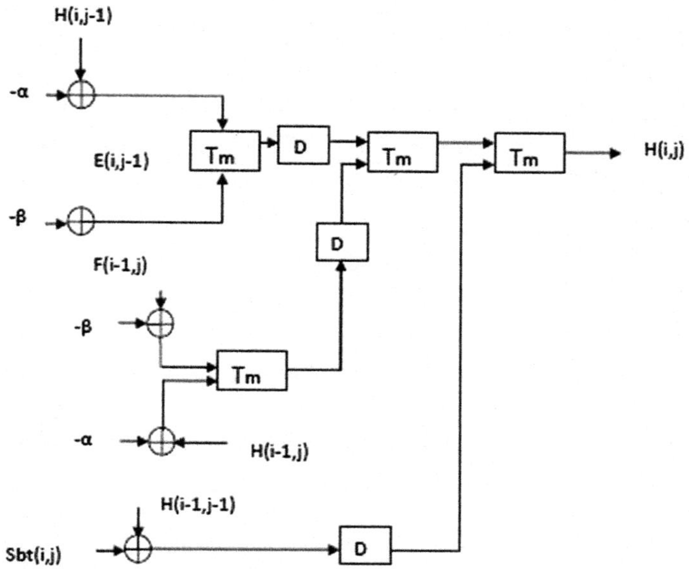

Fig. 12 Retimed implementation of pipelined architecture

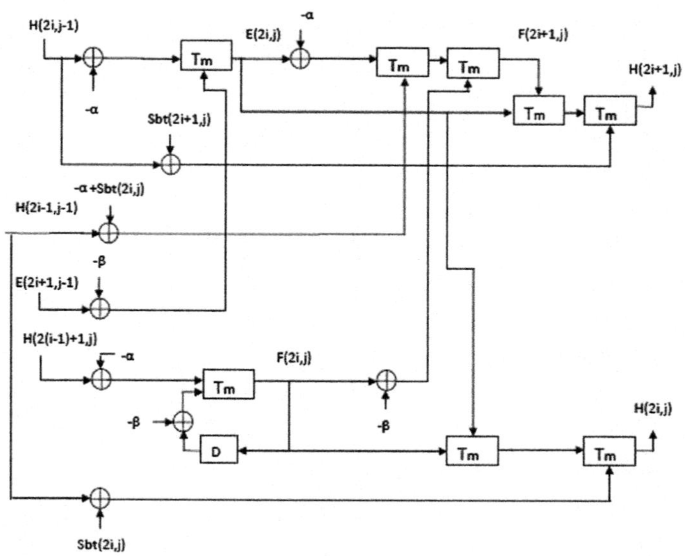

Fig. 13 2 level-parallel implementation of Fig. 10

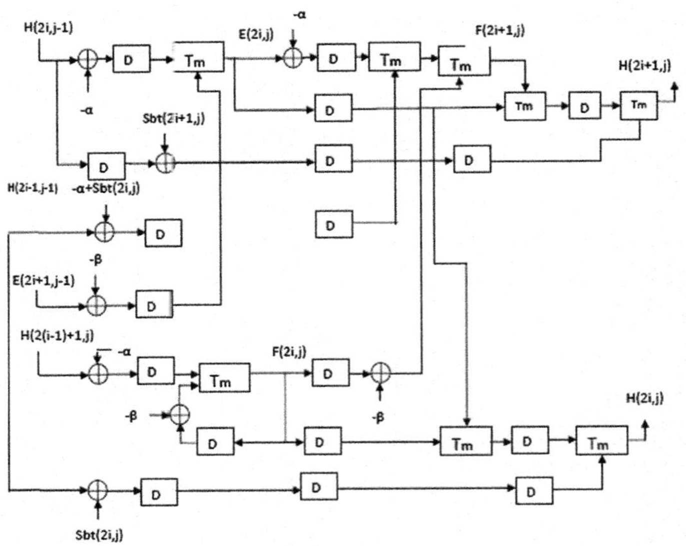

Fig. 14 Pipelined architecture of 2-parallel implementation

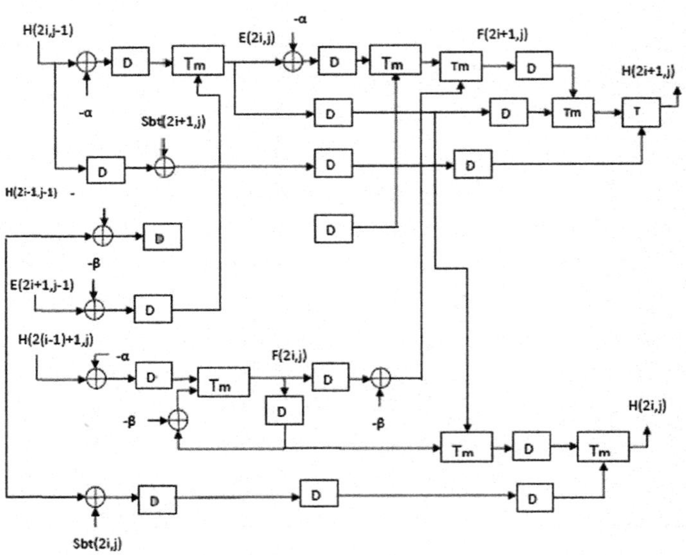

Fig. 15 Retimed implementation of pipelined architecture of 2-parallel implementation

0	0	0	0	0	0	0	0	0	0	0	0
0	0	3	3	2	1	0	0	3	2	1	3
0	0	3	6	5	4	3	2	3	2	1	4
0	0	3	6	5	4	3	2	5	4	3	4
0	0	2	5	9	8	7	6	5	4	7	6
0	0	1	4	8	8	11	10	9	8	7	6
0	0	1	3	7	6	11	14	13	12	11	10
0	0	0	2	6	6	10	14	13	12	11	10
0	3	2	1	5	9	9	13	13	16	15	14
0	3	2	1	4	3	2	12	11	16	15	14
0	2	2	1	4	3	2	11	11	15	19	18
0	1	5	5	4	3	2	10	14	14	18	22
0	3	4	3	4	7	6	9	13	17	17	21

Fig. 16 Output score matrix of Smith–Waterman algorithm

Table 1 Summary of reduction of critical path for the proposed architecture of Smith–Waterman algorithm

Critical path		
Linear gap penalty	**Normal**	$2T_m + T_{add}$
	After pipelining	$2T_m$
	2-parallel architecture	$4T_m + T_{add}$
	After pipelining	$3T_m$
	After retiming	$T_m + T_{add}$
	3-parallel architecture	$3T_m + T_{add}$
	After pipelining	$2T_m$
	After retiming	$2T_m$ (3 delay elements reduced)
Affine gap penalty	**Normal**	$3T_m + T_{add}$
	After pipelining	$3T_m$
	After retiming	$T_m + T_{add}$
	2-parallel architecture	$5T_m + 2T_{add}$
	After pipelining	$3T_m$
	After retiming	$T_m + T_{add}$

Table 2 Minimum computational time for various proposed implementations of Smith–Waterman algorithm

Minimum computational time (in ns)		
Linear gap penalty	Normal	6.42
	2-parallel architecture	3.283
	3-parallel architecture	2.231
Affine gap penalty	Normal	6.487
	2-parallel architecture	3.425
	3-parallel architecture	2.307

Table 3 Comparison of critical paths between existing and proposed implementation of Smith–Waterman algorithm

Critical path		Cheng [12]	Proposed
Linear gap penalty	Normal	$2T_m + T_{add}$	$2T_m$
	2-parallel architecture	$2T_m + T_{add}$	$T_m + T_{add}$
	3-parallel architecture	$2T_m + T_{add}$	$2T_m$
Affine gap penalty	Normal	$2T_m + T_{add}$	$T_m + T_{add}$
	2-parallel architecture	$2T_m + T_{add}$	$T_m + T_{add}$

6 Conclusion

Thus, the Smith–Waterman algorithm is implemented with reduced critical path and increased speed by applying the VLSI signal processing techniques of pipelining, retiming, and parallelism.

References

1. T.F. Smith, M.S. Waterman, Identification of common molecular subsequences. J. Mol. Biol. **147**, 195–197 (1981)
2. T.F. Oliver, B. Schmidt, D.L. Maskell, Reconfigurable architectures for bio-sequence circuits and systems—II: express briefs, in *IEEE Transactions on Database Scanning on FPGAs*, vol. 52, 12 Dec 2005
3. H.-Y. Liao, M.-L. Yin, Y. Cheng, A parallel implementation of the Smith-Waterman algorithm for massive sequences searching. IEEE Eng. Med. Biol. Soc., 1–5 (2004)
4. E. Rucci, A. De Giusti, M. Naiouf, Smith Waterman algorithm on heterogenous systems: case study, in *IEEE International Conference on Cluster Computing* (2014)
5. T.T. Ngoc, S. Kittitornkun, Y.H. Hu, Mass-similarity search of biological sequences using FPGA. Adv. Parallel Comput. **15** (2008)
6. C.W. Yu, K.H. Kwong, K.H. Lee, P.H.W. Leong, A Smith–Waterman systolic cell, in *Proceedings of the 13th International on Workshop Field Programmable Logic and Applications*, pp. 375–384 (2003)
7. A. Papadopoulos, I. Kirmitzoglou, V.J. Promponas, T. Theocharides, FPGA-based hardware acceleration for local complexity analysis of massive genomic data. Integr. VLSI J. 0167-9260 (2012)
8. D. Lavenier, Dedicated hardware for biological sequence comparison. J. Univ. Comput. Sci. **12** (2006)
9. S. Sarkar, G.R. Kulkarni, P.P. Pande, A. Kalyanaraman, Network-on-chip hardware accelerators for biological sequence alignment. IEEE Trans. Comput. **59**(1) (2010)
10. G. Pfeiffer, H. Kreft, M. Schimmler, Hardware enhanced biosequence alignment, in *International Conference on Mathematics and Engineering Techniques in Medicine and Biological Sciences* (2005)
11. G. Ivan, D. Banky, V. Grolmusz, Fast and exact sequence alignment with the Smith–Waterman algorithm: the SwissAlign webserver. PIT Bioinform. Group, Sep 2013
12. C. Cheng, K.K. Parhi, High-speed implementation of Smith-Waterman algorithm for DNA sequence scanning in VLSI, in *IEEE Conference on Signals, Systems and Computers* (2008)

13. Y. Munekawa, F. Ino, K. Hagihara, Design and implementation of the Smith–Waterman algorithm on the CUDA-compatible GPU. IEEE Int. Conf. Bioinform. Bioeng., Jan 2009
14. M. Farrar, Striped Smith–Waterman speeds database searches six times over other SIMD implementations. Int. J. Bioinform. **23**, 156–161 (2007)
15. Y. Li, Z. Jia, S. Xie, ESL based Smith-Waterman engine. Future Wirel. Netw. Inf. Syst., LNEE **144**, 65–70 (2012)

A Clusterhead Selection Technique for a Heterogeneous WSN and Its Lifetime Enhancement Using HeteroLeach Protocol

Yogesh Kumar Sharma and Sanjeet Kumar

Abstract WSN consists of hundreds or even thousands of nodes, which increases the reliability of the data but at the same time it also increases the redundancy of the collected data. So, the role of cluster head is important to reduce the redundancy generated in a sensor network, since early die out of cluster head may result in network breakdown or lifetime reduction of a WSN. This paper proposes modified LEACH algorithm in heterogeneous network named as HETEROLEACH. It increases the lifetime of a WSN by properly choosing a cluster head in a cluster, based on energy and predefined range. This reduces the energy consumption of nodes especially cluster heads in such a manner that redundancy is reduced and no overload takes place at CH.

Keywords WSN · LEACH · Clustering · Routing protocol · Modified HETEROLEACH

1 Introduction

Wireless sensor networks consisting of hundreds or thousands of low-power, low-cost nodes deployed to monitor physical or environmental conditions, such as temperature, sound, vibration, pressure, motion, and to cooperatively pass their data through the network. Heterogeneous sensor network consists of sensor nodes with different capabilities, such as different energy level, sensing range, and computational power. Among the total number of nodes in WSN, 10% of the nodes have double the energy than other nodes.

Y. K. Sharma (✉) · S. Kumar
Department of Electronics and Communication Engineering,
Birla Institute of Technology, Mesra, Ranchi, India
e-mail: yogeshzsharmaz@gmail.com

S. Kumar
e-mail: sanjeet@bitmesra.ac.in

V. Nath (ed.), *Proceedings of the International Conference on Microelectronics, Computing & Communication Systems*, Lecture Notes in Electrical Engineering 453,
https://doi.org/10.1007/978-981-10-5565-2_22

A sensor node is made up of four basic components: a sensing unit, a processing unit, a transceiver unit, and a power unit [1]. Sensing units are usually composed of two subunits: sensors and analog-to-digital converters (ADCs). The analog signals produced by the sensors based on the observed phenomenon are converted to digital signals by the ADC and then fed into the processing unit. The processing unit, which is generally associated with a small storage unit, manages the procedures that make the sensor node collaborate with the other nodes to carry out the assigned sensing tasks. A transceiver unit connects the node to the network. One of the most important components of a sensor node is the power unit. Power units may be supported by power scavenging units such as solar cells and rechargeable battery. Some design challenges in WSN are physical resource constraints, adhoc deployment, scalability, quality of service, unattended operation, security, fault tolerance. A major benefit of WSN systems is that they perform in-network processing to reduce large streams of raw data into useful aggregated information. Some applications of WSN are military, medical, telematics, buildings, environment, precision agriculture, machine surveillance, and preventative maintenance. In hostile environment, charging and recharging battery of sensor nodes becomes quite difficult. Although replacement of battery option is available, it will obstruct the continuous operation of WSNs.

Section 2 presents an analysis of the related work which has been done on lifetime of WSN. Section 3 describes the proposed protocol with flowchart. Simulation results performed on MATLAB are reported in Sect. 4. Finally, conclusions are drawn in Sect. 5 (Fig. 1).

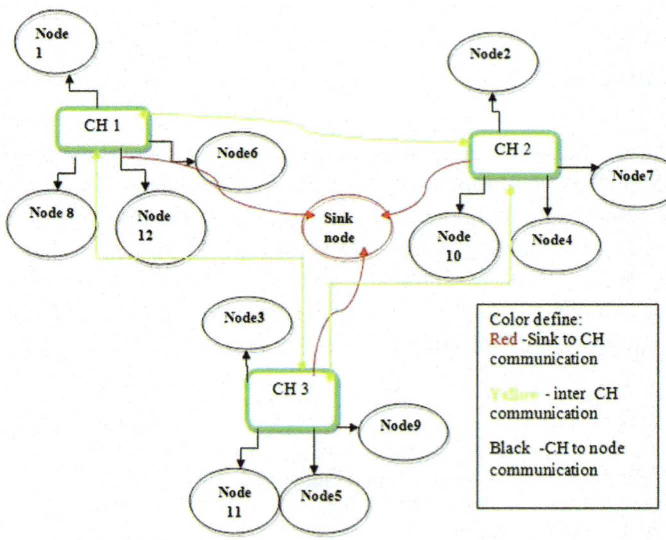

Fig. 1 WSN architecture with CH formation

2 Related Work

Wireless sensor nodes mainly depend on batteries, which get depleted at a faster rate as they have to perform computation and communication operations. There are various ways of enhancing the lifetime of a WSN like energy efficient routing, node scheduling, clustering, and also like the introduction of virtual coordinator in a WSN [1]. This paper mainly focuses on routing protocols and clustering methods for enhancing the lifetime of WSN. Two routing protocols called Low Energy Adaptive Clustering Hierarchy (LEACH) and Energy Aware Multi-hop Multipath Hierarchical Protocol (EAMMH) have been modified to provide solutions for low-power consumption. These modifications have exploited the heterogeneity of the nodes in both the protocols. LEACH is one of the famous clustering algorithms used for minimizing energy dissipation. It allows rotation around all sensor nodes to randomly select cluster heads for uniformly distributing energy among all sensor nodes [2]. The operation of LEACH is achieved by rounds. LEACH provides a balancing of energy usage by random rotation of cluster heads. The algorithm is also organized in such a manner that data fusion can be used to reduce the amount of data transmission. In LEACH, a randomized rotation technique of cluster head position is used such that the energy is equally distributed among all sensor nodes in the network. The cluster head selection is random and is based on the probability of the node to become cluster head. The operation of LEACH is achieved by rounds. Hence, the network is reclustered periodically in order to select energy-abundant nodes to serve as CHs, thus distributing the load uniformly on all the nodes. Hence, lifetime improvements can be achieved if data aggregation is exploited and the network is reclustered periodically [3]. Various techniques were proposed to improve the lifetime of a wireless sensor network by introducing some modifications in LEACH. Then came the concept of applying LEACH in heterogeneous networks, which gave better result than homogeneous networks [4]. Other improvements in LEACH include LEACH-C, LEACH-F, LEACH-V, TL-LEACH. While there are advantages of using LEACH's distributed cluster formation algorithm, this protocol offers no guarantee about the placement and number of cluster head nodes. Since the clusters are adaptive, obtaining a poor clustering setup during a given round will not greatly affect overall performance [5].

3 Proposed Protocol

An efficient heterogeneous WSN based on Range v/s Distance relationship between cluster heads and nodes was successfully deployed using the proposed algorithm as described in progress. It is an improved version of HETEROGENEOUS LEACH which is highly efficient in order to minimize energy expenditure and also responsible to enhance the node lifetime of WSN. As WSN consists of hundreds or even thousands of nodes with limited energy, computation, and communication capabilities, hence the data sensed by a node and other nodes which are in its sensing range leads to redundancy. To minimize the redundancy and enhancing the lifetime, hence a new method is proposed in which first selection of the cluster heads takes place based on energy of nodes and then nodes are associated with the cluster head depending on the predefined user range of cluster head. Only the nodes which are in the range of cluster heads will associate with it, and the nodes will not communicate with each other or other cluster heads. The data is aggregated by the cluster head, and later data is send to the base station or sink by them. Also the cluster head will communicate among themselves for security purpose timely. Performing the regular interval checking will not only help in maintaining the security but also help in determining whether the network is performing well without any node die out or knowing the nodes which are died out already for managing the network efficiently.

This new method for enhancement of lifetime is MODIFIED HETEROLEACH which is purposed and successfully implemented increases the lifetime of WSN by reducing the energy consumption of nodes especially cluster heads [6]. Earlier cluster heads die out before the other nodes resulting in network breakdown or the half of the nodes of network die out very soon reducing the efficiency of network. Clustering technique can also be used to perform data aggregation. Data aggregation is to combine the data from source nodes into a small set of meaningful information, and hence the fewer messages are transmitted thus saving communication energy.

3.1 Flowchart of Proposed Protocol

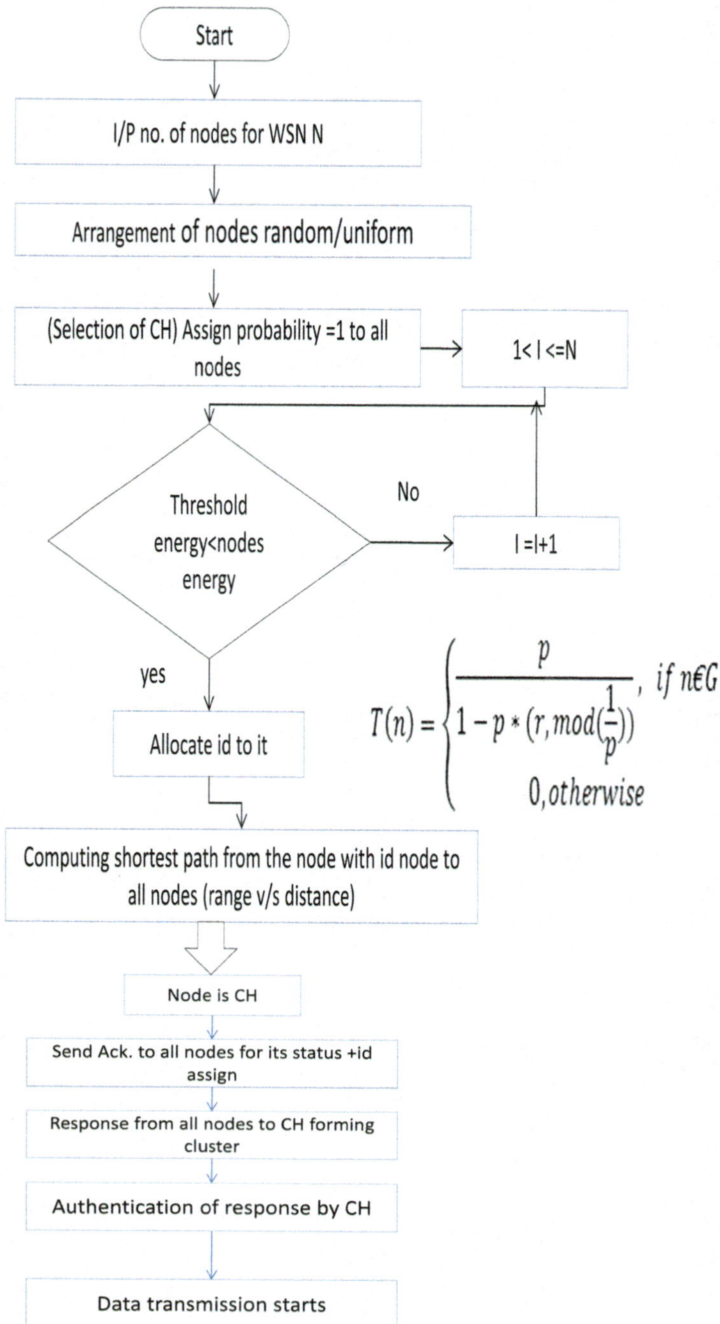

3.2 Algorithm of Proposed Protocol

- *// n is number of nodes*
- //M IS FRACTION OF NODES WHICH HAVE ENERGY GREATER THAN REST OF NODES
- //ENOR IS ENERGY OF NORMAL NODES
- //EADV IS ENERGY OF ADVANCED NODES
- //ETH IS THRESHOLD ENERGY USER DEFINED
- //DMIN IS THE DISTANCE BETWEEN CLUSTER HEAD AND NODES
- //R IS THE RANGE PREDEFINED BY USER
- //ENODE IS ENERGY OF NODES
- //ACK IS ACKNOWLEDGEMENT

➢ *Initial condition: a heterogeneous network N formation with (n, m ,Enor, Eadv, Eth, dmin, R) attributes.*

➢ *If arrangement =='random'*
Else arrangement =='uniform'

a) FOR SELECTION OF CH:

➢ ***Input:*** *A heterogeneous network N(n, m, Enor, Eadv)*

➢ *Initial condition: Eadv = Enor*

➢ ***Output:*** *Cluster head selection and cluster formation for a WSN*

➢ ***Steps:***

1. *Initialization of heterogeneous network*
2. *Assignment of energy to nodes(advanced nodes and normal nodes):*
   for each node i∈(1.....n)

➢ *Assign probability p==1 to all nodes.(comparing each node energy with the cluster head[CH])*

```
if (Enode >= Enor && Enode>Eth)
      set E(i) = "cluster head"
else
set node i = 'Normal'
      end if
                end for
for each round r∈(1.........RND)
```

➢ *Now allocate id to all the CH*

b) FOR CLUSTER FORMATION:

➢ *for each node i∈(1......n)*

```
    if dmin <=R
 set node n(i) == 'member' (formation of one link at a time)
    end if
```

➢ Again repeat selection of CH until all advanced nodes with Eadv are selected as CH

c) AUTHENTICATION OF ALL CH:
For each CH

➢ *Send ACK to all nodes by all CH*

```
 If(ACK(id) ==allocated id)
 Nodes type ="CH"
Else nodes == 'normal nodes'
  end if
          end for
```

➢ *Authentication of response by CH successful*

➢ *Data transmission can now begin*

4 Simulation Results and Discussion

We have performed simulation in order to evaluate the proposed algorithm. We used MATLAB as simulator to evaluate the performance of Modified heteroleach using different scenarios such as random 90-node network, different area dimensions, or rounds. The first scenario as in Fig. 2 describes as sink is located at (80, 80) in a 160 m × 160 m field with 90 nodes along with the different parameters defined in Table 1. Similarly for second scenario also we will refer the same parameters from Table 1, but now the sink is located at (15, 15) in a 30 m × 30 m field with 24 nodes as shown in Fig. 3.

The focus of simulation is on energy efficiency and number of nodes alive as they are important factor for lifetime enhancement of WSN.

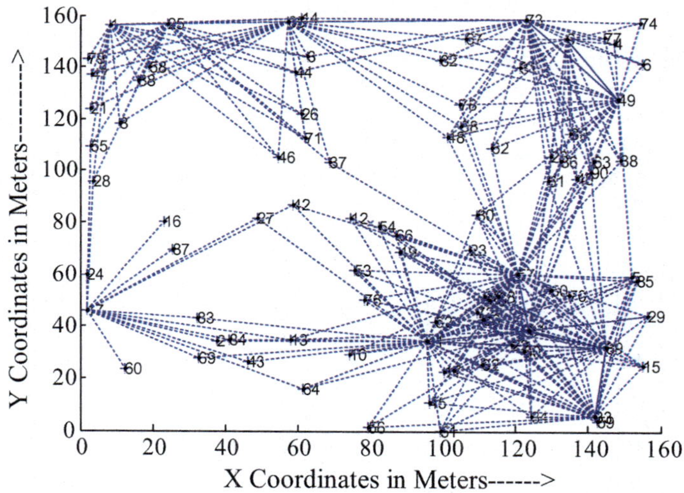

Fig. 2 Network (160 × 160) m with 90 nodes

Table 1 Performance comparison of Mhl and hl [1]

Sl. no.	Simulation parameters and their values		
	Parameters	Scenario 1	Scenario 2
1.	Number of nodes	90	24
2.	Location of sink	Center of the network	Center of the network
3.	Election probability value of CHs (p)	0.1	0.6
4.	Number of rounds initial	2000	2000
5.	Initial energy of normal nodes (E_{nor})	0.05 J	0.05 J
6.	Initial energy of normal nodes (E_{nor})	0.1 J	0.1 J
7.	Sensing range of nodes	12 m, 15 m, 18 m	12 m, 15 m, 18 m

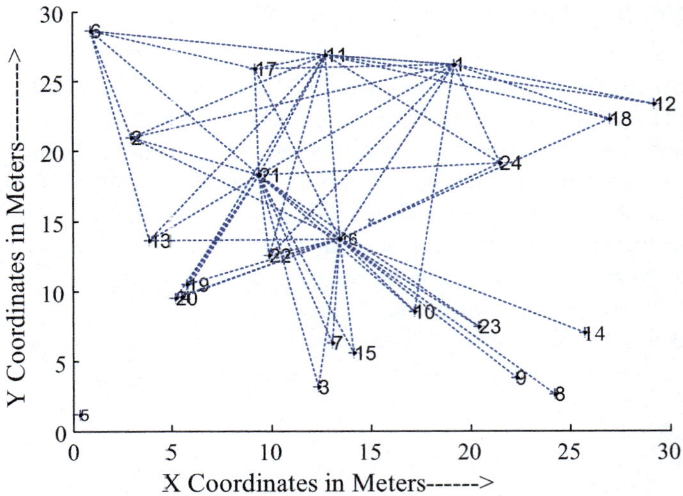

Fig. 3 Network (30 × 30) m with 24 nodes

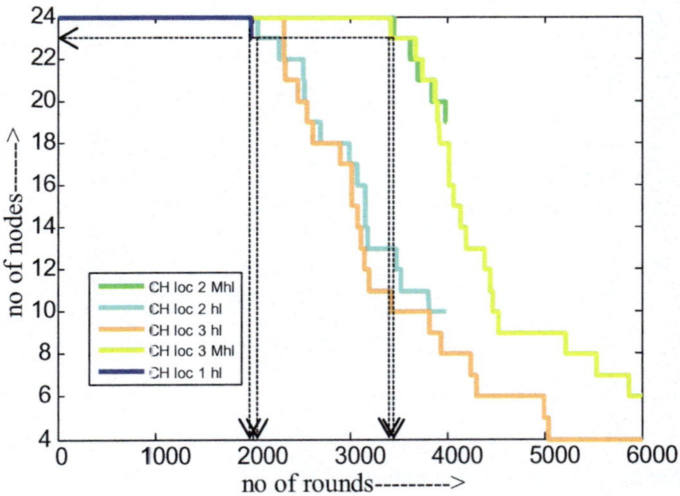

Fig. 4 Case 1 rounds increment

Figure 4 describes first node die out and total number of nodes alive as the rounds been increased. It can be clearly observed that the number of nodes alive in Modified heteroleach (Mhl) is larger than the one in heteroleach (hl) after each round ends, which shows Mhl makes the energy consumption more efficient.

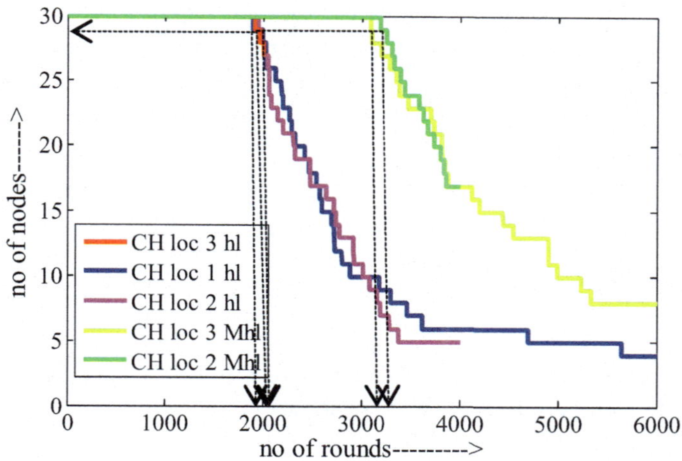

Fig. 5 Case 2 nodes increments

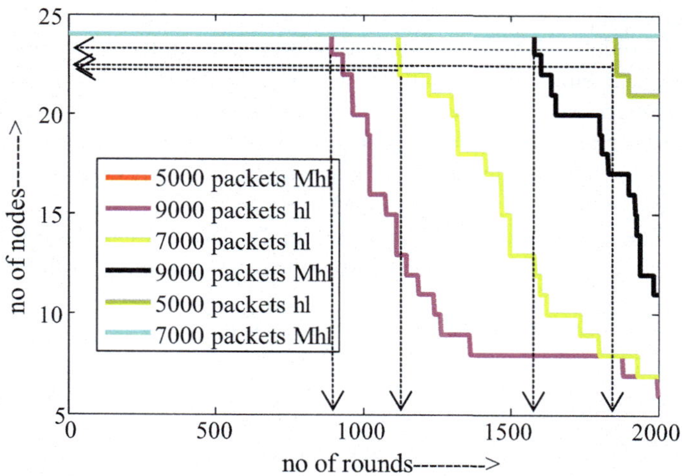

Fig. 6 Case 3 packets increment

Figure 5 describes first node die out and total number of nodes alive as the nodes been increased, and it can be seen that the number of nodes alive in Mhl is larger than the one in hl after each round ends.

Figure 6 describes first node die out and total number of nodes alive as the packets size been varied, it can be clearly observed that keeping the number of nodes and rounds constant while increasing size of packets the proposed protocol Mhl makes the energy consumption more balanced which effectively prolong the network lifetime compared with hl.

Table 2 Numerical analysis of first node die out round with respect to LEACH [1]

Network area	First node die out round w.r.t. HETEROGENEOUS LEACH	
Fig. 1.	Modified heterogeneous LEACH (Mhl)	Heterogeneous LEACH (hl)
100 × 100	3700 rounds	1400 rounds
200 × 100	3100 rounds	1350 rounds
300 × 100	2600 rounds	1100 rounds
500 × 100	900 rounds	400 rounds
700 × 100	250 rounds	125 rounds
1000 × 100	70 rounds	25 rounds

The heteroleach (hl) and modified heteroleach (Mhl) have been compared according to their performances and are shown in Table 2.

5 Conclusion

This paper is proposed on the basis of comparison of Range v/s Distance network which was initially deployed. Using this proposed protocol, we minimize the redundancy and enhance the lifetime of WSN. The first selection of the cluster heads takes place based on energy of nodes and then nodes are associated with the other cluster heads depending on the predefined user range of cluster head. Hence, it provides a way to avoid the redundant data transmission and save the energy about 66% in compare with the HETEROLEACH protocol.

In the future, we wish to use mobile sink in the network and security in order to prolong the network life expectancy and security enhancement of the WSN.

References

1. N. Swarup, S. Kumar, Lifetime enhancement of a wireless sensor network based on modifications in leach and EAMMH protocols. ME thesis. BIT, Mesra, May 2015
2. N. Swarup, S. Kumar, Lifetime enhancement of heterogeneous wireless sensor network using modified leach protocol, in *2nd International Conference on Knowledge Collaboration in Engineering*. Coimbatore, March 2015
3. M.Z. Hussain, M.P. Singh, R.K. Singh, Analysis of lifetime of wireless sensor network. Int. J. Adv. Sci. Technol. **53**, April, 2013, India
4. V. Katiyar, N. Chand, S. Soni, Improving lifetime of large-scale wireless sensor networks through heterogeneity, in *Proceedings of the International Conference on Emerging Trends in Electrical and Computer Technology (ICETECT)*, pp. 1032–1036, March 2011

5. N. Zivic, C. Ruland, O.U. Rehman, Error correction over wireless channels using symmetric cryptography, in *Proceedings of the International Conference on Wireless VITAE'09*, Feb 2009
6. F. Bajaber, I. Awan, Energy efficient clustering protocol to enhance lifetime of wireless sensor network. J. Ambient Intell Human Comput **1**, 239–248 (2010)
7. W. Heinzelman, A. Chandrakasan, H. Balakrishnan, Energy-efficient routing protocols for wireless microsensor networks, in *Proceedings of the 33rd Hawaii International Conference on System Sciences* (*HICSS*). Maui, HI, Jan 2000

Investigation of Microgripper Using Thermal Actuator

N. Chattoraj, Abhijeet Pasumarthy, Rajeev Agarwal and Asifa Imam

Abstract In the recent years, MEMS technology, because of its micro-size, has matured as a field of research. Micro-gripper is one of the applications of MEMS technology. This paper describes the design, simulation and analysis of micro-gripper based on electrothermal actuator. An electrothermally actuated micro-gripper has been designed, optimized and simulated using COMSOL Multiphysics simulation tool. The simulation of the gripper design is done by using copper as a structural material. Different parametric studies have been carried out such as displacement, stress and deformation by varying the driving voltage and temperature.

Keywords Micro-gripper · Finite element method (FEM) · Meshing
Electrothermal actuators · MEMS · Joule heating principle · Thermal expansion

1 Introduction

The new methods of micro-fabrication techniques are the major cause which is responsible for the advancement in the miniaturization of semiconductor devices [1, 2]. These contraptions have been among the catalyst giving travail to the micro-electromechanical systems (MEMS) technology.

MEMS technology implements the fabrication of a vast variety of miniaturized sensing and actuating devices [1, 3]. Micro-grippers are important emerging tools which could be used as end effectors for systems that can handle and manipulate

N. Chattoraj (✉) · A. Pasumarthy · R. Agarwal · A. Imam
Department of Electronics and Communication Engineering, Department of Production Engineering, Birla Institute of Technology, Mesra, Ranchi 835215, Jharkhand, India
e-mail: nchattoraj@bitmesra.ac.in

A. Pasumarthy
e-mail: pabhijeet@gmail.com

R. Agarwal
e-mail: rajeevagarwal@bitmesra.ac.in

V. Nath (ed.), *Proceedings of the International Conference on Microelectronics, Computing & Communication Systems*, Lecture Notes in Electrical Engineering 453,
https://doi.org/10.1007/978-981-10-5565-2_23

micro-scaled objects with applications in various fields such as biomedical science and industries [3]. For example, in the case of micro-robots (very small size about in mm), micro-grippers are used for an assembly where the components are in the order of micrometres which are produced using micro-machining fabrication process. Several types of micro-grippers have been designed and developed, based on different actuation schemes such as mechanical, electrostatic, piezoelectric, electromagnetic, vacuum and electrothermal actuator. Mechanical micro-grippers are not of much use because using such actuators makes micro-objects more vulnerable to fatigue. Electrostatic micro-gripper cannot be used in biomedical applications, as biological medium has electrolytic property. Since these grippers operate due to electrostatic force attraction [4], it may damage biological cells as operating voltage required is very high in the order of 100 V. Electromagnetic micro-gripper works on the principle of electromagnetic field induced due to the flow of current. It has less accuracy because of the fact that the magnitude of current is difficult to control. Vacuum micro-gripper needs negative and positive pressure for picking and placing the micro-object. In case of piezoelectric micro-gripper, large gripping stroke is required in the range of (0–70 V) for small displacements [5, 6]. Electrothermal micro-gripper works on the principle of Joule's law and expansion law. It is widely used due to its capability for generating large displacement, force and due to its structural and functional robustness when compared with standard IC fabrication process and materials. The main advantage of this micro-gripper is the requirement of less driving voltage which is in the range of 0–5 V. But given a proper design, the driving voltage may well be in the order of few hundreds of millivolts. Many micro-grippers with different materials have been developed based on nickel [7, 8], poly-silicon [9, 10], gold [10–12], PMMA [12] and copper [13] which are some of the common materials used in the fabrication of micro-grippers.

In this paper, three models of an electrothermally actuated micro-gripper have been designed using copper as the structural material. The design and analysis is done using COMSOL Multiphysics software. The designed model consists of mainly three components, i.e. hot arm (flexure), cold arm (flexure) and central arm (anchor). All the three components together form a beam. The main objective of the proposed work aims at large displacement at a low driving voltage for manipulation of micro-objects. An optimization has been tried by varying the parameters of the above-mentioned components. Mostly, the lengths of the components have been varied, which is done in the case of the central arm. The length is determined keeping the anchor as the reference point. A total of three structures shown in Fig. 2a–c have been worked out for further analysis. In the first structure, the length of the central arm is fixed at 20 μm from the anchor. Subsequently, it is varied from 50 to 110 μm. The range of driving voltage was varied from 0.1 to 0.5 V. Investigation has been carried out by using different materials with the same structure, and different parameters are studied and analysed. Results are compared to obtain the design optimization parameters for developing an application-specific design.

2 Design of Micro-gripper

2.1 Design Configurations

The design of the electrothermal actuator is done keeping in mind the principle requirement of hot and cold beams which are the major actuation components. Figure 1 shows the basic conceptual design of electrothermally actuated micro-gripper. The micro-gripper structures consist of two anchors (A1 and A2), two hot and cold arms (H1, H2, C1 and C2), central arm, supporting arm, connecting arm (connects the hot and cold arm), gripper arm and arm tip. The overall dimensions of the micro-gripper are 350 μm × 220 μm × 10 μm. The design specifications are mentioned in Table 1. Initial gap between tips of the end effector is 50 μm. The length and width of the hot and cold arms (H1 and C1) are more than the H2 and C2. In further sections for optimization purpose, three models of micro-gripper are designed based on the base design shown in Fig. 1. All the structures and dimensions are unaltered except the dimension of the central and supporting arm.

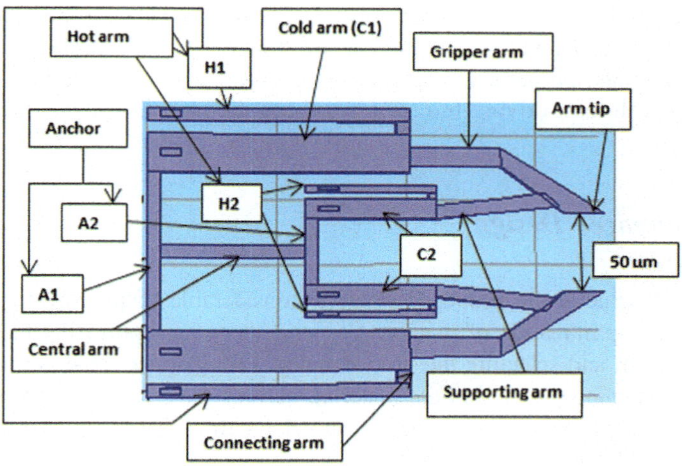

Fig. 1 Topology of the micro-gripper

Table 1 Structural dimensions

Parameters	L × W (μm^2)
Cold arms (C1)	200 × 30
Cold arms (C2)	120 × 17
Hot arms (H1)	200 × 10
Hot arms (H2)	120 × 7
Central arm	110 × 10
Supporting arm	70 × 15
Anchor (A1)	10 × 120
Anchor (A2)	10 × 36

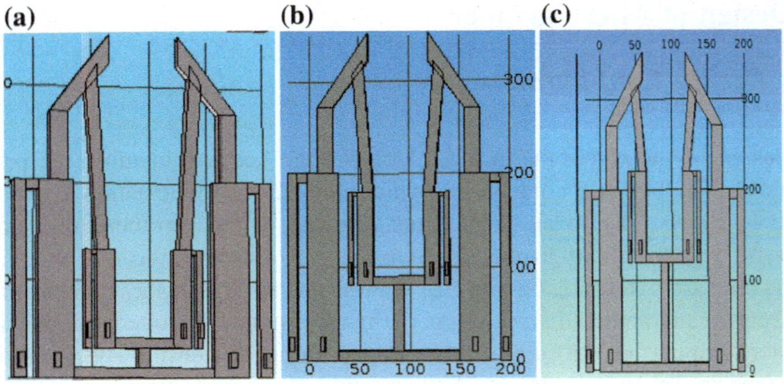

Fig. 2 Comparison between three structures of micro-gripper for different lengths of the central arm. **a** length of central arm 20 μm. **b** length of central arm 70 μm. **c** length of central arm 110 μm

The design is later altered based on the requirement for optimum performance which is discussed in the later sections. Later, residual stress and total displacement analysis is done for the design iterations as well as gripper with different material properties. The different design iterations are shown in Fig. 2. Further, the study of various materials' properties is carried out to investigate and obtain optimum displacement with least driving force for satisfying the objective mentioned above.

2.2 *Optimized Design*

In order to achieve the proposed goal of considerable displacement, this work proposes the optimization of mainly two parameters, i.e. optimizing the length of the central arm and selecting the proper material for gripper construction.

For determining an appropriate material, various materials are compared based on the electrical properties and mechanical behaviour. A list of suitable materials is given in Table 2.

For determining the parameter suitable for design optimization, mechanical behaviour was studied for the gripper structure. It is observed that, by varying the length of the central beam, its displacement also varies. So, three structures are designed with different length of the central arm shown in Fig. 2. In the first design iteration Fig. 2a of the micro-gripper, the length of the central arm is determined at 20 μm, and subsequently in second (Fig. 2b) and third (Fig. 2c) micro-gripper design, the length has been increased to 70 and 110 μm, respectively.

Table 2 Properties of materials

Properties	Unit	Copper (Cu)	Gold (Au)	Nickel (Ni)	Titanium (Ti)	Poly-Si
Coefficient of thermal expansion	1/K	16.5e−6	14.2e−6	13.4e−6	8.60e−6	2.6e−6
Heat capacity	J/Kg K	384	129	445	522	678
Electrical conductivity	s/m	58.1e6	45.6e6	13.8e6	2.6e6	5e4
Thermal conductivity	W/m K	401	317	90.7	21.9	34
Relative permittivity	–	8960	19,300	8900	4506	2320
Young's modulus	Pa	120e9	70e9	219e9	115.7e9	160e9
Poisson's ratio	–	0.34	0.44	0.31	0.321	0.22
Relative permittivity	kg/m^3	–	–	–	–	4.5

3 Working Principle and Methodology

MEMS electrothermal actuation is principally based on the conversion of electric energy into heat energy which structurally deforms the material hence giving some noteworthy displacement. Since there are several stages involved right from applying potential to getting an actuation. It is based on the governing equations which help us to understand the behaviour of the model in a better way. In order to observe the mechanical behaviour of the structures, an actuation scheme of the micro-gripper is required.

Basic governing equation for electrothermal actuation scheme consists of Joule's law (2) and thermal expansion law (4). Joule's effect interrelates the electric potential Eq. (1) and the heat flow Eq. (4). The heat causes the mechanical expansion (5) and manipulation of arm. Equations for the thermal energy (4) and heat transfer in solids are given below (3).

Electric potential,

$$-\nabla \cdot (\sigma \nabla V_e) = 0. \tag{1}$$

Equation (1) determines the amount of voltage that needs to be supplied initially. This eventually creates an inductive heating field which heats the material excluding the potential terminals. The effect of heating is explained by Eq. (2), and the property of heat transfer in a given material is determined by the Eq (4).

Equation of Joule's law,

$$Q = \sigma\left|\nabla V_{\mathrm{e}}^2\right|. \tag{2}$$

Heat transfer in solids,

$$\rho C_{\mathrm{p}} u \cdot \nabla T = \nabla(K \nabla T) + Q, \tag{3}$$

where c_{p} is the heat capacity at constant pressure, ρ is the density of the material, Q is the heat flux, ∇T is the change in temperature and K is the thermal conductivity.

Transfer of heat within the cross section due to the flow characteristics in the gripper gives us the energy thus generated. The energy equation is stated below.

Equation for thermal energy,

$$-\nabla(K(T)\nabla T) = \sigma(T)\left|\nabla V^2\right|, \tag{4}$$

where σ is the electrical conductivity, V is the potential (voltage), T is the temperature and k is the thermal conductivity.

Based on the thermal energy generated, there is a change in the orientation of the underlying particles which ultimately results in displacement. The expansion is directly proportional to the heat generated by the material.

Thermal expansion law,

$$\varepsilon_{\mathrm{th}} = \alpha \Delta T. \tag{5}$$

Based on the above equation, it is certain that not only the material but the design is also crucial in determining the obtaining displacement.

4 FEM Simulation and Analysis

The working equations are taken into consideration and tried to be implemented on the designed model. But before improvising the theoretical equations in a simulation environment, it is to be noted that the input and the output are both of different genre of physics. Thus, a virtual environment is required which can combine the space of electricity, heat transfer and structural mechanics under one platform. To cater such cases, COMSOL Multiphysics has been used for studying the behaviour of an electrothermal gripper. To give a clear understanding of the results obtained, an example where the effect of potential on displacement has been demonstrated in the form of a 3D surface plot in Fig. 3.

With the application of the driving voltage, the gripping movement occurs and the jaw tips come closer and hold the micro-object. With the withdrawal of the driving voltage, the jaw tips release the micro-object.

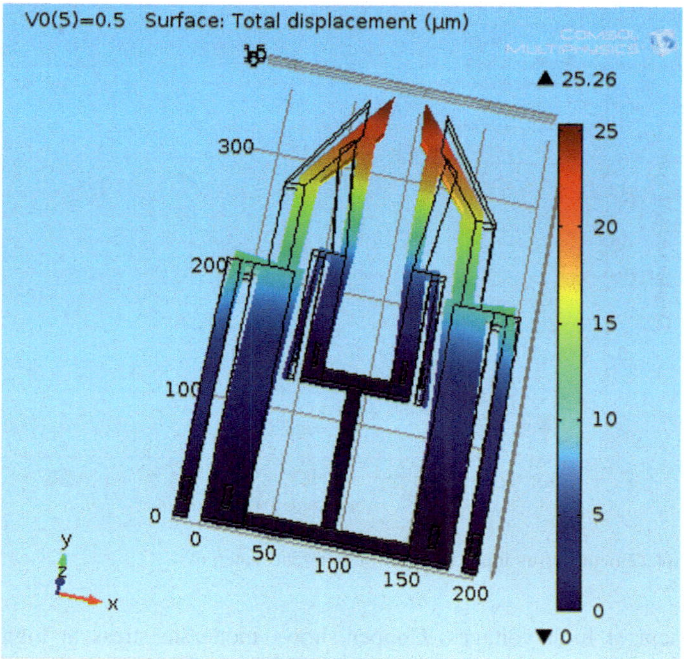

Fig. 3 3D-FEA model of the optimized micro-gripper after actuation

Finite element methods have been used to carry out the simulations of the micro-gripper for all the designs for analysis of displacement, stress and strain. The software used for FEM simulations were based on COMSOL Multiphysics. Here, temperature was kept constant at 293 K. Thermal expansions and mechanical deformation was obtained.

A comparative study wherein different materials as mentioned in Table 2 have been applied to the design shown in Fig. 1 and are subjected to variable voltages ranging from 0.1 to 0.5 V is graphically represented in Fig. 4. Due to different Young's modulus and the conductivity coefficient, there is a stark behaviour noticed in the deformation characteristics. Poly-silicon is observed to be the least reactive due to its insulating properties. Materials such as gold, copper and nickel show fairly good deflection for small amounts of voltage. Here, copper tops with the highest amount of deflection. The behaviour is almost linear in nature. Titanium has moderate deflection as compared to copper.

The above-mentioned model comprising of various materials one at a time is subjected to stress analysis while encountering displacement at variable voltages ranging from 0.1 to 0.5 V and is graphically represented in Fig. 4. It is noticed that poly-silicon due to its least displacement experiences the least stress. On the other extreme, nickel happens to be the most strenuous material when subjected to high

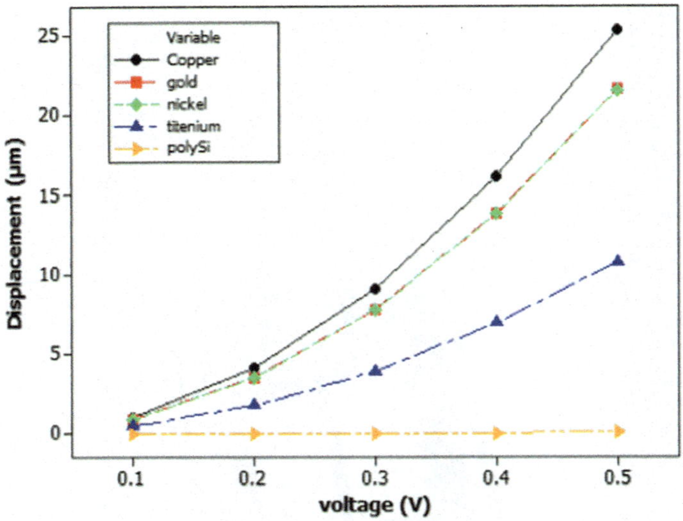

Fig. 4 Displacement versus input voltage for different materials

displacement at high voltages. Copper shows moderate stress at high voltages, which is a good sign considering the massive displacement at the same voltages.

A comparative study has been done for suggesting the best structural design by subjecting the model to voltages ranging between 0.1 and 0.5 V. Displacement characteristics is studied for designs suggested in Fig. 2a–c. The material of the gripper is chosen based on the analysis in Fig. 4. The graphical representation in Fig. 5 suggests the relation that the length of the central arm component is directly proportional to the displacement achieved. Maximum displacement is noticed at 25.26 µm. A trade-off could be observed as the displacement of the central arm with length 70 µm has good enough displacement when compared to the displacement offered by the longest central arm.

The temperature range is varied from 313 to 473 K at an interval of 80 K. The model showcasing the maximum displacement in Fig. 6 was chosen for this analysis. The response of the gripper is almost unaffected by the change in temperature.

5 Result

The designed micro-gripper shows total displacement of 25.26 µm at the driving voltage 0.5 V. This displacement is suitable for the purpose of manipulation of micro-object between the ranges of 20–45 µm. The overall dimension of the micro-gripper is found to be 350 µm × 220 µm × 10 µm.

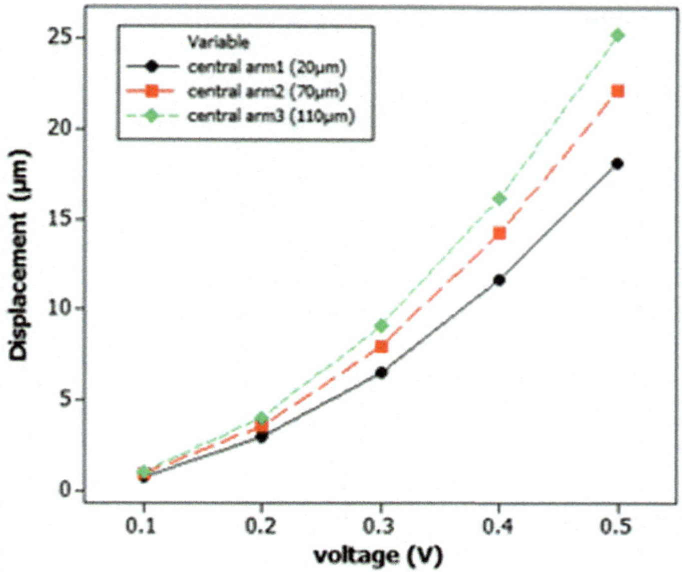

Fig. 5 Displacement versus input voltage for different structures

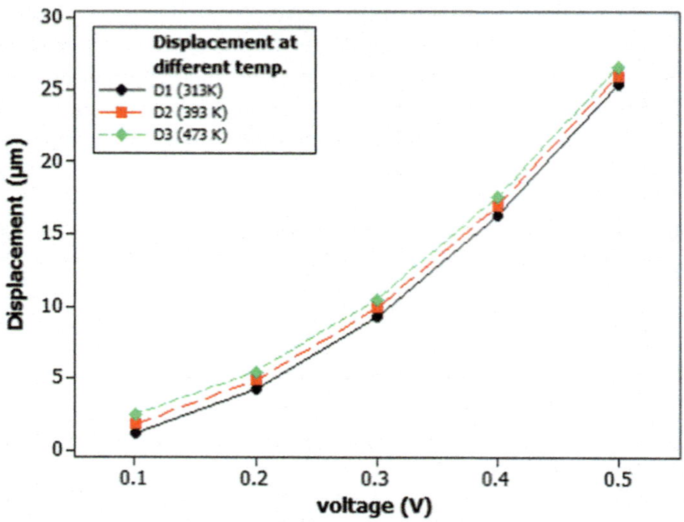

Fig. 6 Displacement versus input voltage for different temperature (K)

6 Conclusion

A novel design of micro-gripper presented in this paper is based on electrothermal actuation for achieving large displacements at low driving voltages. The electrical and mechanical properties of five different materials copper, gold, nickel, titanium and poly-Si were compared and analysed. After optimization, copper material was selected as the structural material for the modelling of the micro-gripper. Three different structures are represented by varying length of central arm represented in Fig. 2(a–c) and compared. A finite element-based COMSOL Multiphysics was used for the modelling, analysis and simulation of the thermal, electrical and mechanical properties of the micro-gripper. After optimization, third model shows the best result among the proposed three models. Total displacement of 25.26 μm is obtained at driving voltage of 0.5 V.

7 Future Scope

The work proposed has numerous possibilities which need further investigation into areas such as better optimization parameters and realistic analysis. Apart from the central arm, there are many components such as the anchor length and the hot/cold arm length which might be optimized. Moreover, optimizing the points where the potential could be supplied other than anchor for achieving large displacement.

Realistic analysis includes the cost estimation for fabrication of the gripper. This suggests that we could select design parameters or materials which could provide substantial results at low cost. Besides, combining more than one material could also be an interesting area to dig into.

Acknowledgements The authors would like to thank Hemant Kumar and COMSOL India for their software support, P. Banerjee, R Agrawal for discussions and BIT Mesra for resources.

References

1. M. Bao, W. Wang, Future of microelectromechanical systems (MEMS). Sens. Actuators **56**, 135–141 (1996)
2. H. Fujita, A decade of MEMS and its future, in *Proceedings of the IEEE, 10th Annual International Workshop on Micro Electro Mechanical Systems*, pp. 1–8 (1997)
3. H. Fujita, Future of actuators and microsystems. Sens. Actuators A **56**, 105–11 (1996)
4. C.J. Kim, A.P. Pisano, R.S. Muller, Silicon-processed overhanging microgripper. J. Microelectromech. Syst. **1**(1), 31–36 (1992)
5. H. Xinhan, C. Jianhua, W. Min, L. Xiadong, A piezoelectric bimorph micro-gripper with micro-force sensing. In: *IEEE International Conference on Information Acquisition*, p. 5 (2005)

6. S.K. Nah, Z.W. Zhong, A microgripper using piezoelectric actuation for micro-object manipulation. Sens. Actuators A **133**(1), 218–224 (2007)
7. N. Ali, M.M. Hassan, R.I. Shakoor, Design, modeling and simulation of electrothermally actuated microgripper with integrated capacitive contact sensor. IEEE, 978-1-4577-0657-8/11 (2011)
8. J.J. Khazaai, H. Qu, M. Shillor, L. Smith, Design and fabrication of electro-thermally activated micro gripper with large tip opening and holding force. IEEE, 978-1-4244-9289-3 (2011)
9. A.A. Geisberger, N. Sarkar, M. Ellis, G.D. Skidmore, Electrothermal properties and modeling of polysilicon microthermal actuator. J. Microelectromech. Syst. **12**(4) (2003)
10. G. Adilfo, A. Rodriguez, C. Rossi, Multiphysic modeling os a microactuator based on the decomposition of an energetic material: application to fluidisk, Hal-00150263, version 1 (2007)
11. S.M. Karbasi, M. Shamshirsaz, M. Naragh, M. Maroufi, Optimal design analysis of electrothermally driven microactuators. Microsyst Technol **16**,1065–1071 (2010)
12. V. Vidyaa, G. Arumaikkannu, Hybrid design of a polymeric electrothermal. IJMIE **1**, ISSN No. 2231-6477 (2011)
13. R. Zhang, J. Chu, H. Wang, Z. Chen, A multipurpose electrothermal microgripper for biological micromanipulation. Microsyst. Technol. (2012) doi:10.1007/s00542-012-1567-0

An Ultra-Low-Power Internet-Controlled Home Automation System

Pooshkar Rajiv, Rohit Raj, Ramakant Singh, Rishabh Nagarkar, Anurag Kumar Chaurasia, Sushant Agarwal and Vijay Nath

Abstract In this paper, an ultra-light low-power and unique smart automation system has been implemented which interfaces the internet communication protocols like Hyper Text Transfer Protocol (HTTP) through an embedded Linux platform to obtain home automation. It has improved upon the preexisting work by removing MOM type middleware, providing efficient M2M communication by implementing a direct virtual link between the transport layer of the two communicating devices. This smart automation system is ultra-low powered and improves the efficiency and throughput involved in communication. An embedded Linux platform, connected to the Internet through its Ethernet port, is used for the demonstration of this smart automation system. A Web portal interface is used to give commands and receive updates about the result of automation.

Keywords Internet of Things (IoT) · Embedded Linux board · SSH tunneling Home automation system · Message-oriented middleware architecture

P. Rajiv (✉) · R. Nagarkar · S. Agarwal · V. Nath
Department of ECE, Birla Institute of Technology Mesra, Ranchi, India
e-mail: pooshkar.01@gmail.com

R. Nagarkar
e-mail: nagarkarrishabh@gmail.com

S. Agarwal
e-mail: sushantagarwal1412@gmail.com

V. Nath
e-mail: vijaynath@bitmesra.ac.in

R. Raj · R. Singh · A. K. Chaurasia
Department of CSE, National Institute of Technology, Patna, India
e-mail: rohitshubham@gmail.com

R. Singh
e-mail: singh.ramakant8@gmail.com

A. K. Chaurasia
e-mail: anuragchaurasia.93@gmail.com

V. Nath (ed.), *Proceedings of the International Conference on Microelectronics, Computing & Communication Systems*, Lecture Notes in Electrical Engineering 453,
https://doi.org/10.1007/978-981-10-5565-2_24

1 Introduction

There has been a tremendous growth in the segment of Internet of Things (IoT) in the recent years. This growth has been fueled by the consumer's requirement for device automation. Automation is now not only limited to sophisticated laboratories or robotic organizations rather it has percolated in the realms of everyday lifestyle of people. The demand for IoT in everyday life mostly stems from the requirement of ingenious home automated solutions. The tech giants of the industry are increasingly concentrating resources to drive innovations and produce solutions in this field.

IoT produces solutions by interfacing end-devices by Internet. These devices can be subdivided broadly into two subcategories i.e., resource constrained and resource rich. Resource-constrained devices are those devices that do not have support of TCP/IP suite on the board, whereas resource-rich devices are those who have native support for TCP/IP suite. The former type of boards exchanges data by protocols such as ZigBee or Bluetooth which are lightweight in nature, and they need to communicate with at least one resource-rich device to interact with the Internet. Resource-constrained devices generally operate on batteries and rely on local wireless sensor.

Many protocols are available for resource-rich devices which work on application layer. These application layer protocols follow different architectural paradigms.

For example, service-oriented architecture (SOA) follows the protocols design and mechanism of [1], whereas DDS [2] follows a broker-less Publisher-Subscriber model (Pub/Sub). Software-defined networking (SDN) is yet another IoT interfacing mechanism. MQTT [3] is another famous protocol which is based on broker-based Pub/Sub architectural middleware oriented messaging (MOM). However, these architectural paradigms along with the associated protocols are often difficult to setup and even more difficult to manage.

This paper proposes a novel mechanism which eliminates the need of a MOM type middleware between machines for M2M communication thereby increasing direct communication between machines and severely decreasing latency for interaction. As opposed to data-centric communication, our work uses a more reliable network-centric approach. In network-centric approach, there is direct referencing of the communicating devices which increases the communication reliability. Moreover, our work also decreases the throughput of data exchanged for M2M communication between devices. This communication architecture establishes a direct virtual link between the client and the end-device. This virtual link exists above the transport layers of two devices. Since, transmission control protocol (TCP) is being used for data exchange, guaranteed delivery is provided inherently in the proposed architecture. Furthermore, this work is characterized by an increase in control flexibility of the resource-rich board as there is direct link between the controller and the controlled device.

This paper is divided into five sections. Section 2 does a critical analysis of the related work in this field. Section 3 gives a detailed insight on the proposed work and explains the architecture. Section 4 presents the favorable results and output of simulation of the work on board to demonstrate real-life scenario. This paper concludes with remarks of authors and future scope of the work in Sect. 5.

2 Related Work

The adoption of home automation has been relatively slower than the demand for it. Complex network topologies and inefficient middleware design are partly responsible for it. Work done by Tu et al. [4] focuses on smart home middleware. The experimental results in the paper pointed out that the use of context manager in the communication led to error rate of 42.5% while it increased the runtime to 3.4 s for a single operation on the board.

Similarly [5] talks about AwareHome, which was a project dealing with culmination of technologies for deploying ubiquitous computing in the house. This project used plethora of sensors and dealt with multiple challenges in the domain of home automation. It talked about content-based automation in the houses which would be very costly and range of sensors used would make the design impractical.

Kim et al. [6] used an RFID middleware framework for providing IoT-related services to clients. They used ucode system for differentiating the requests of the system from multiple sources and used a central managing server for coordinating requests. This architecture is highly prone to single point of failure (SPOF). The work used highly scalable architecture to distribute the workload into event managers, legacy applications, and workload services.

Atukorala et al. [7] proposed home automation and monitoring using SmartEyes, which again uses a central server. The flexibility available in providing security levels are inherently increased in such subsystems where decoupling of middleware takes place. This work moreover provided flexibility as it divided the services into three subsystems namely, mobile subsystem, Web config subsystem, and home subsystem. It allowed a Rabbitcore to use TCP stack for communicating with other nodes.

Several such works either use middleware or use direct Hypertext Transfer Protocol (HTTP) to communicate with the board. Stateless protocols are characterized by lack of their capability to maintain session. As HTTP is a stateless protocol, this scheme makes it increasingly difficult to track the users using the system. Moreover, HTTP uses high amount of data transfer to communicate with the other end and generally demands higher resources with the communicating nodes.

3 Proposed Architecture

The application domains for our proposed architecture are smart home environments. In the proposed architecture, we have a client (mobile) connected to the embedded Linux board. Clients are lightweight and easily connected over Internet. When they are connected over Internet they possess a number of advantages, including the reduced costs of not having to run virtual private networks (VPNs) and being able to deploy software updates in a timelier manner (Fig. 1).

The work uses direct communication with the embedded Linux platform using the Secure Shell (SSH) protocol which is a popular application layer protocol. This ensures higher reliability and secure exchange of data between the communicating nodes.

Because of the higher security requirements of managing client on a network, Internet-based client management requires that the site is using certificates. This ensures that connections to the management point and distribution points are authenticated by an independent authority and that data to and from these site systems is encrypted using Secure Shell protocol (SSH).

SSH is a secure way for you to remotely access your hosting account. By using SSH, you are able to log into a command prompt and execute commands just as if you were sitting at the server itself. SSH works on port no. 22. SSH allows for the encryption of data so that those malicious would-be attackers cannot access your

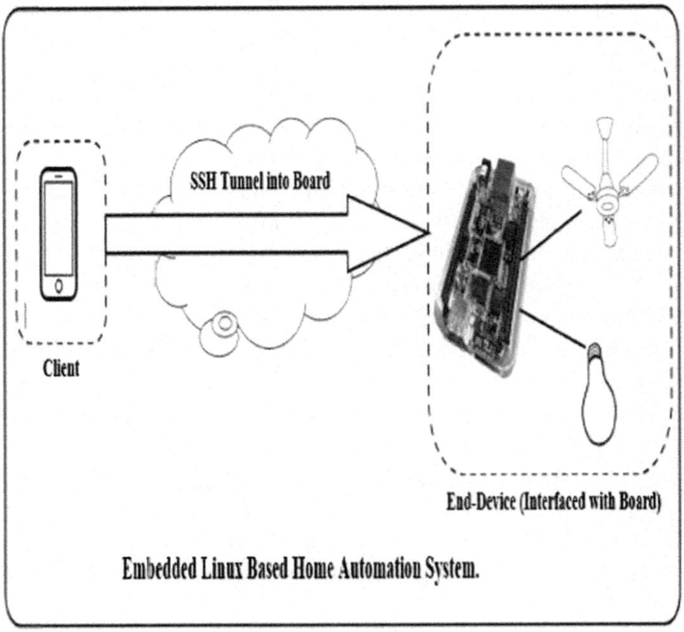

Fig. 1 Proposed architecture

user information and passwords. SSH also allows for the tunneling of other protocols such as FTP. SSH protects clients from IP source routing, DNS spoofing, data manipulation, eavesdropping or sniffing of the transmitted data, and IP address spoofing. SSH authenticates user include password and public key authentication methods but others (e.g., Kerberos, NTLM, and keyboard-interactive) are also available. SSH flexibility allows new authentication methods to be incorporated into the system as they become available. Passwords, in combination with a username, are a popular way to tell another client that you are who you claim to be. If the username and password given at authentication match the username and password stored on a remote system, you are authenticated and allowed access. When connection is establishment and authentication of client is done, then we enlighten the LED bulb using the GPIO pin of embedded Linux board.

The application layer provides the user interface via a network virtual terminal and supports network-based services such as access to file, file transfer, mail services through its protocols. To provide the user interface and services, the application layer identifies communication clients, determines resource availability, and establishes cooperation between clients (end-devices). The application layer defines the languages and its rules that programs need to use to establish communication between client and server (using protocols such as Association Control Service Element, ASCE).

4 Automation Demonstration and Results

Internet-controlled low-power automation as proposed in this paper is achieved by using a resource-rich embedded Linux device such as BeagleBone Black which is controlled through a Web page. Previously, such implementations required the use of large computers as middleware which could not be automated and also consumed a lot of energy.

The home automation system introduced in this paper is demonstrated by controlling the blinking of an LED from a Web page. The corresponding measure of response time, latency, power consumption, and system performances shows that the introduction of the virtual link layer significantly improved the performance.

The BBB is a microcomputer on a chip, which has embedded Linux distribution and it uses Ubuntu, Debian, Angstrom, or even Android as an OS. The end devices that have to be controlled have been connected to the BBB, and the communication and control is implemented through the common communication protocols. The embedded Linux device is put into its low-power deep sleep mode, in which the operation frequency decreases from 1000 to 275 MHz, and this deep sleep mode of the device can be awaken only through interrupts that are triggered by the pressing of the buttons on the Web page. The fact that the above-mentioned board is resource-rich with 4x LAN Ethernet for Internet connectivity and has 512 MB DDR3 RAM making it ideal for usage.

The user accesses a Web portal, and after authenticating, he is provided with a personalized list of devices which are under his domain of control. The device control page has a menu-driven approach with each device having its own control parameters that can be varied by the user. When the parameters are changed, asynchronous requests are sent by the client's browser to the server. The client's browser now waits for the status of the request. The validations of few parameters are done on the client side while few validations are left for server scripts. The scripts residing on the HTTP server are then executed. The scripts then forward the request to the embedded Linux platform. Thus, the BBB (embedded Linux platform) is accessed through an HTTP virtual link above the TCP layer; thus this implementation requires no MOM.

Figure 2 shows the menu that is presented to the user. Now, the direct GPIO control over the LED is implemented as shown in Fig. 3.

The embedded Linux device is connected to the Internet through its Ethernet port and is connected to other devices for performing various control operations as required by the distant user. The mobile phone as shown is used to access the Web portal for control purposes.

It can be noted that as soon as the button was slid into the on position, the LED connected to the GPIO pin of the Beaglebone Black is turned on; the status of the LED was received and displayed through a pop-up box.

Figure 4 shows the variation in the current consumed by the embedded Linux board (for the automation purpose) with time. The embedded Linux device is used to control a number of other devices such as TIVA and MSP430 for the purpose of performing various operations like temperature sensing or data logging. The figure

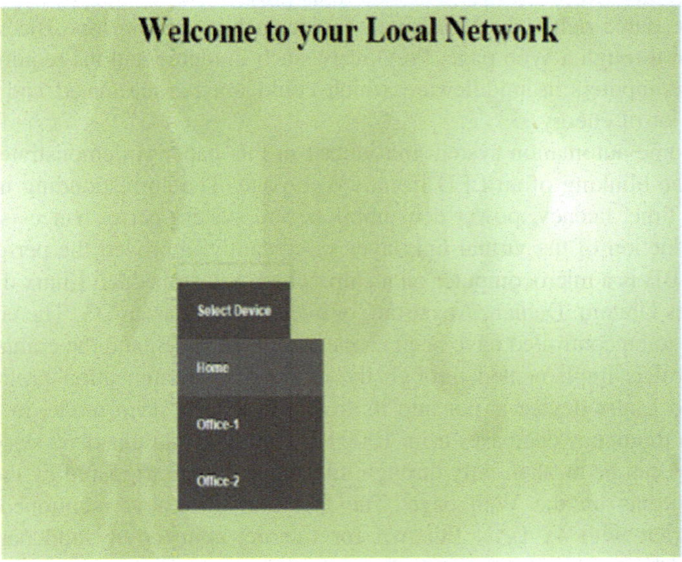

Fig. 2 Home Web page

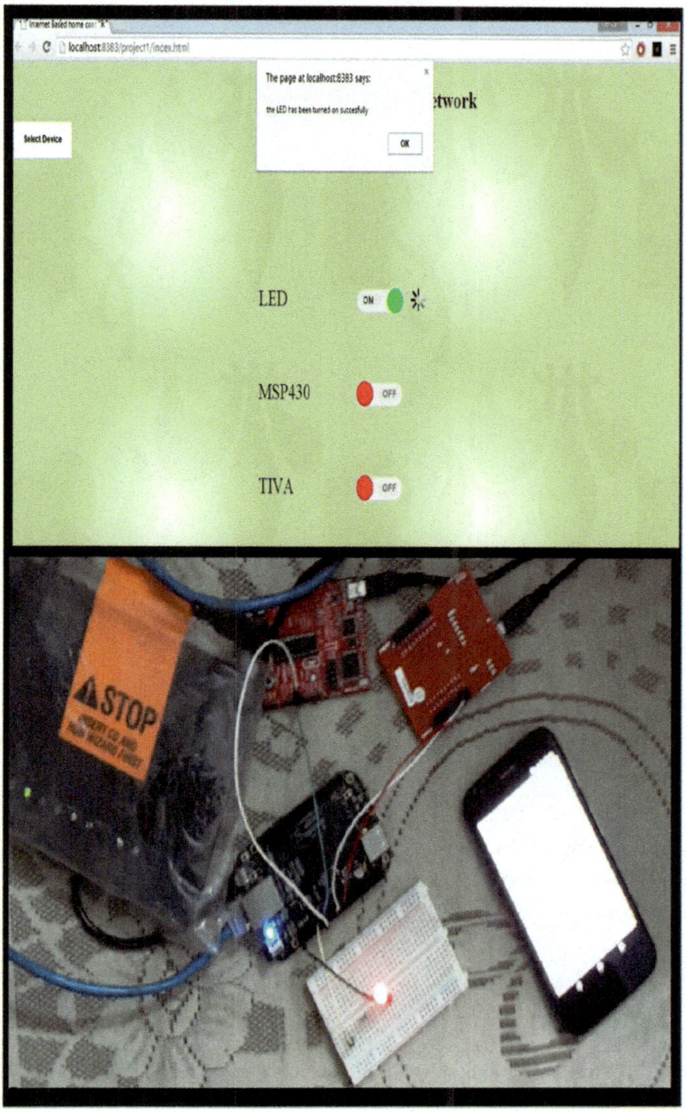

Fig. 3 Implementation of direct GPIO control over the LED

clearly demonstrates the efficiency of the deep sleep low-power mode which facilitates the ultra-low-power automation of the system. The system remains in its deep sleep mode consuming very little current with time until it is awakened by an interrupt due to the pressing of the button of the Web page which in turn causes it to execute the desired automation for an extended period before returning to its deep sleep.

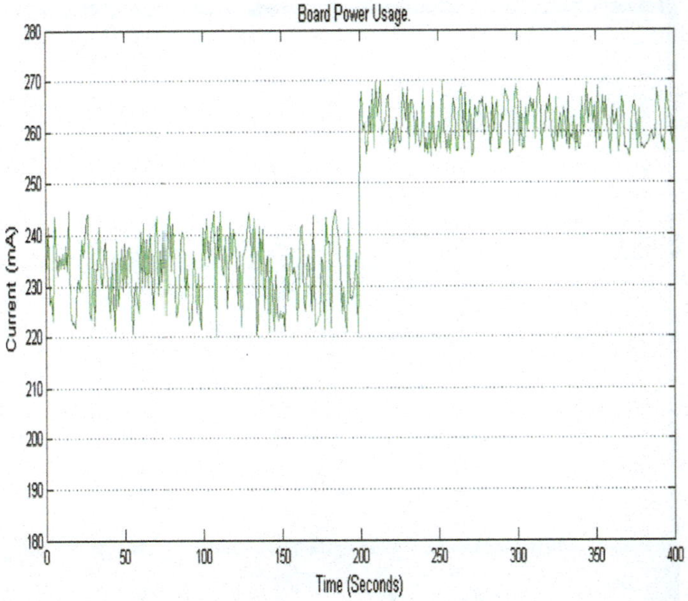

Fig. 4 Power usage of the board

Table 1 Comparison with middleware-based approaches

	UBISOAP [8]	SERVILLA [9]	Proposed architecture
Resource discovery	DD, ND	SD	ND, SD
Resource management	RA, RM, RCA, RCL	RA, RM, RCA	RA, RM
Code management	NS	CA, CM	CA, CM
Event management	SS	SS	SS
Security	NS	NS	Very high
Reliability	NS	NI	Very high
Scalability	AL, NL	AL, NL	AL

Legends Device Discovery (DD), Network Discovery (ND), Service Discovery (SD), Resource Allocation (RA), Resource Monitor (RM), Resource Conflict (RCL), Resource Composition Adaptive (RCA), Small Scale (SS), Not Supported (NS), No Information (NI), Code Allocation (CA), Code Migration (CM), Application Level (AL), Network level (NL)

Table 1 showcases the detailed comparison of the proposed architecture with other middleware-based systems as proposed by Capurisco et al. [8] and Fok et al. [9]. It was observed that the proposed system has better security, reliability, and scalability as compared to previously existing models.

5 Conclusion

The emerging Internet of Things (IoT) market place [10] aims to integrate smart objects and smart things while satisfying all the service requirements [11, 12], facing the challenges involved [13] to develop smart automated systems like smart home, smart City. Most of the IoT solutions make use of specially customized hardware and software and commonly make use of Wi-Fi or Bluetooth for all kinds of systems that are present. This paper proposed an Internet-controlled embedded Linux-based automation system which is ultra-low power, low on latency, and high on efficiency. The elimination of the middle layer (MOM) used by other techniques is an important improvement.

The server to BBB communication makes use of SSH (RSA 256), making access to BBB by an intruder very difficult. The proposed smart automation system works on low power, as low as 3.3 V. High performance is guaranteed even if disk space is low as the script storage is done on the server.

Thus, the system proposed in this paper can become the key to automation in the future.

Acknowledgements The authors would like to thank the Department of Electronics and Communication, BIT Mesra, for providing all the facilities for the implementation of this project.

References

1. I.M. Delamer, J.L.M. Lastra, Service-oriented architecture for distributed publish/subscribe middleware in electronics production, in *IEEE Transactions on Industrial Informatics*, vol. 2, no. 4, pp. 281–294, Nov 2006
2. A. Corradi, L. Foschini, A DDS-compliant P2P infrastructure for reliable and QoS-enabled data dissemination, in IEEE International Symposium on Parallel & Distributed Processing (IPDPS 2009), pp. 1–8, 23–29 May 2009
3. International Business Machines Corporation (IBM), MQTT v3.1 Protocol Specification, Eurotech-2013. http://docs.oasis-open.org/mqtt/mqtt/v3.1.1/os/mqtt-v3.1.1-os.html
4. M.C. Tu, D. Shin, D. Shin, J. Choi, Fundamentals and design of smart home middleware, in International Joint Conference on Computational Sciences and Optimization (CSO 2009), vol. 1, pp. 647–650, 24–26 April 2009
5. C.Y. Yang, Y.J. Chen, H.Y. Kung, C.H. Lin, S.D. Li, A service-aware home networking system based on OSGi, in 2009 Fourth International Conference on Innovative Computing, Information and Control (ICICIC), pp. 1268–1272, 7–9 Dec 2009
6. Y.i. Kim, J.S. Park, T.S. Cheong, Study of RFID middleware framework for ubiquitous computing environment, in The 7th International Conference on Advanced Communication Technology (ICACT 2005), vol. 2, pp. 825–830
7. K. Atukorala, D. Wijekoon, M. Tharugasini, I. Perera, C. Silva, SmartEye integrated solution to home automation, security and monitoring through mobile phones, in Third International Conference on Next Generation Mobile Applications, Services and Technologies, (NGMAST '09 2009, pp. 64–69, 15–18 Sept 2009
8. M. Caporuscio, P.G. Raverdy, V. Issarny, UbiSOAP: A service oriented middleware for ubiquitous networking. IEEE Trans. Serv. Comput. **5**(1), 86–98 (2012)

9. C.L. Fok, G.C. Roman, C. Lu, Servilla: a flexible service provisioning middleware for heterogeneous sensor networks. Sci. Comput. Program **77**(6), 663–684 (2012)
10. C. Perera, C. H. Liu, S. Jayawarde, The emerging internet of things marketplace from an industrial perspective: a survey. Emerg. Top. Comput. IEEE Trans. 99,1
11. 3GPP TS 22.368, Service requirements for machine-type communications (MTC) stage 1. (2013)
12. ETSI TS 102 689, Machine-to-machine communications (M2M); M2M service requirements (2010)
13. M. Starsinic, System architecture challenges in the home M2M network. in 2010 Long Island Systems, Applications and Technology Conference (LISAT), 2010, pp. 1–7

Depth-Averaged Velocity Distribution for Symmetric and Asymmetric Compound Channels

Kamalini Devi, Jnana Ranjan Khuntia and Kishanjit K. Khatua

Abstract Movement of water during flood creates a compound channel appearance which consists of a main channel and its adjoining floodplains are very important for environmental, ecological and design issues. The structures of the flow in such channels are rigorous. The principal reason of this flow structure is due to the momentum transfer mechanism between the main channel and the floodplain. Flow mechanism in an asymmetric compound channel is different than that of a symmetric compound channel. There is a stronger interaction exists between the main channel and floodplain in asymmetric compound channel as compared to symmetric compound channel where the interaction is distributed to the both sides of the floodplain. Analysis of depth-averaged velocity distribution in both compound channels is strongly influenced by width ratio, aspect ratio and relative flow depth. The variation of depth-averaged velocity distribution in such channels for different geometry and flow conditions has been analysed. Proper prediction of depth-averaged velocity distribution in a compound channel is depending upon the magnitude of shear layer for which the advanced software CES is not providing accurate prediction especially for asymmetric compound channel. Suggestions and improvements to predict depth-averaged velocity distribution in both symmetric and asymmetric compound channels have been made.

Keywords Symmetric compound channel · Asymmetric compound channel
Momentum transfer · Interface · Floodplain · Main channel · Regression analysis
Width ratio · Relative depth

K. Devi (✉) · J. R. Khuntia · K. K. Khatua
Department of Civil Engineering, N.I.T., Rourkela, India
e-mail: kamalinidevi1@gmail.com

J. R. Khuntia
e-mail: jnanaranjan444@gmail.com

K. K. Khatua
e-mail: kkkhatua@yahoo.com

V. Nath (ed.), *Proceedings of the International Conference on Microelectronics, Computing & Communication Systems*, Lecture Notes in Electrical Engineering 453,
https://doi.org/10.1007/978-981-10-5565-2_25

1 Introduction

Mostly, compound channels are consisting of a main river channel which is flanked by two or one side floodplains. The first case is known as symmetric compound channel where the both floodplains are equal; otherwise, the channel is said to be asymmetric whether it is flanked by either side [1–3]. The design of such channel is very important for environmental, ecological and design issues. Generally, in dry seasons of low water levels, normally, the main channel conveys the entire flows. When flood occurs, the flow rate for a particular river may change drastically that the overbank flow condition is breached and inundates the surrounding floodplain area [4–6]. This causes a threat to the environment and population. Here some experimental investigations in symmetric and asymmetric compound channels of trapezoidal configuration have been extensively carried out in laboratory of the Fluid Mechanics and Hydraulics Laboratory, NIT, Rourkela, Odisha [7]. The structures of the flow in such channels are rigorous because of the momentum transfer mechanism between the main channel and the floodplain. Flow mechanism in an asymmetric compound channel is different than that of a symmetric compound channel. There is a stronger interaction between the main channel and floodplain in an asymmetric compound channel as compared to symmetric compound channel. In symmetric compound channel, the interaction is distributed from the both sides of the floodplain. Analysis of depth-averaged velocity distribution in a compound channel is strongly influenced by both geometric and hydraulic parameters such as width ratio, aspect ratio and relative depth. Conveyance estimation system (CES) is a software tool, used for simulating the flow variables in both types of channels and the applicability of this CES is discussed. Here the numerical results of depth-averaged velocity from CES are compared with their corresponding experimental values for both types of channels. The variation of depth-averaged velocity distribution in such channels for different geometry and flow conditions has been analysed. The aim of this study is to evaluate the CES software for depth-averaged velocity computation in symmetric and asymmetric trapezoidal compound channels. The results are compared satisfactory.

2 Theory of Application

There is a large difference in velocity exists between river channel and its floodplain so that the mean velocities are different in both the subsections. This leads to the exchange of momentum between the main channel water and that of the floodplain making complex flow structure. During inundation of the compound channel, the mean velocity on the floodplain is lower than that in the main channel as the depth of flow on the floodplain is lower than the main channel and also having high-hydraulic roughness. The velocity difference results in turbulence at the transition zone forming a bank of vertices.

Previously, the complex mechanism of momentum transfer at the junction of a compound channel has been noticed by many researchers and analysed the horizontal coherent structure in both laboratory and field study. The experimental evidence of occurrence of such structures in the transition region is shown in Fig. 1b, c. Figures 2 and 3 show the cross-sectional details of asymmetric and symmetric compound channels.

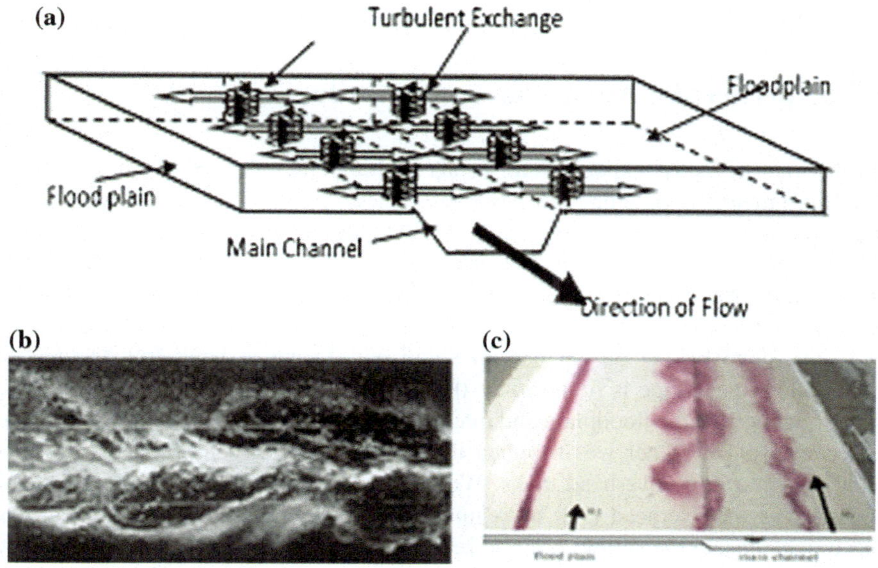

Fig. 1 **a** Schematic view of momentum transfer between main channel and floodplain of a symmetric compound channel, **b** macro-vortices in transition region, **c** large coherent structures in mixing layer made visible by dye injection [8]

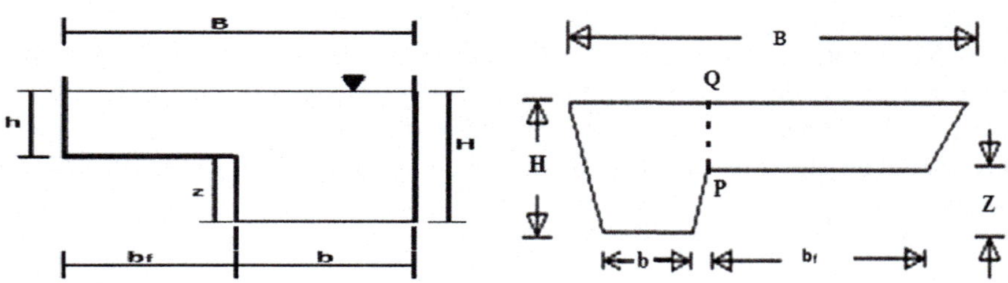

Fig. 2 Asymmetric compound channels

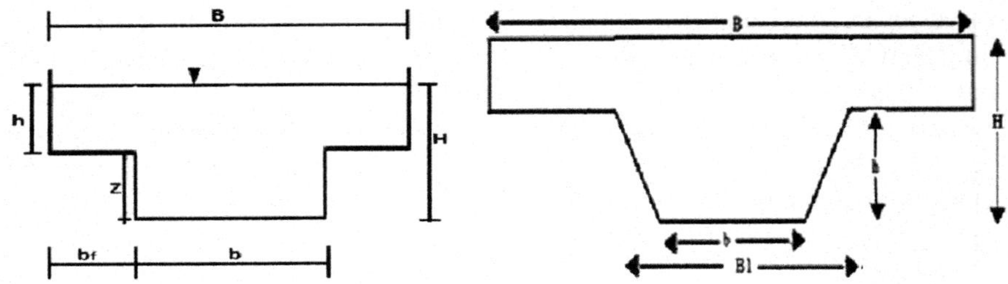

Fig. 3 Symmetric compound channels

3 Experimental Apparatus and Procedure

In this part of study, experimental results of compound channels with symmetric and asymmetric floodplains are described. These channels are constructed using neat finished plain cement concrete at the hydraulic engineering laboratory of the Civil Engineering Department, National Institute of Technology, Rourkela, India. For asymmetric channels, floodplains are at right side making the total width of the compound section 168 cm. The main channels are trapezoidal in cross section with 1.1 side slope having 33 cm bottom width and 11 cm at bank full depth. The longitudinal bed slope is found to be 0.001 satisfying subcritical flow conditions. The roughness of the floodplain and main channel is kept same having Manning's n is equal to 0.01. Water was supplied through numbers of centrifugal pumps discharging into large overhead tanks. Water is made to flow to the stilling tank of flume from the overhead tank by regulating valves. Baffle walls arrangement has been made inside the stilling basin to reduce the turbulence. Water is made to flow under gravity from the head gate end to the tail gate of the flume under uniform and steady flow conditions. At the downstream end, there is a measuring volumetric tank followed by a large underground sump which feeds the water to the overhead tank through pumping. This is an arrangement of complete recirculation system of water for the experimental channels.

For analysis of symmetric compound channels, 1 number of experimental data set from NITR, India and 3 numbers of large channel facility channel, UK data sets have been considered (Fig. 4). The depth-averaged velocity data from point-to-point along the wetted perimeter of both symmetric and asymmetric compound channels have been acquired experimentally to study the variation of shear layer width with different geometrical and hydraulic conditions.

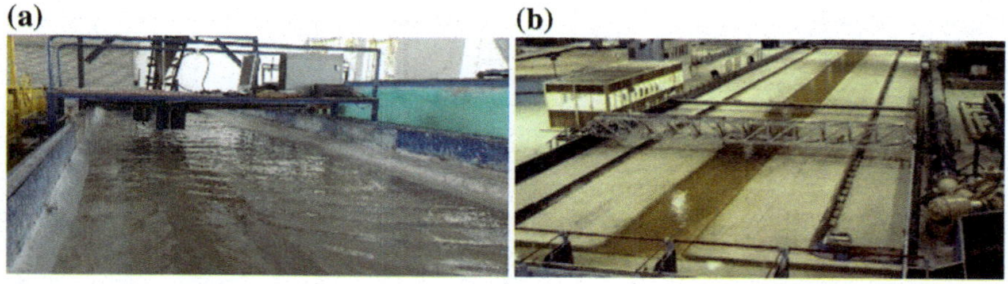

Fig. 4 Photographs of compound channels fitted with instruments: **a** NITR-5.1, **b** Flood channel facility at HR Wallingford [9]

3.1 Depth-Averaged Velocity

The depth-averaged velocity U_d *for* the entire overbank cases was calculated using the equation

$$U_d = \frac{1}{H}\int_0^H u dz \tag{1}$$

where u is the local point streamwise velocity at a vertical line. U_d is calculated by integrating point velocities (u) over a flow depth H for each vertical line. Then by joining all the values of depth-averaged velocities across the lateral cross section of the channel, the distribution of depth-averaged velocity plots can be obtained.

After finding out the depth-averaged velocity, the total discharge in symmetric and asymmetric compound channels is evaluated by multiplying it with the corresponding sectional area [10–12]. Experimental depth-averaged velocity distribution results of two typical flow depths of asymmetric compound channel are provided in Figs. 5 and 6. Likewise, the graphical presentation of depth-averaged velocities distribution of symmetric compound channel is also demonstrated in Figs. 7 and 8.

Fig. 5 For flow depth 12.5 cm

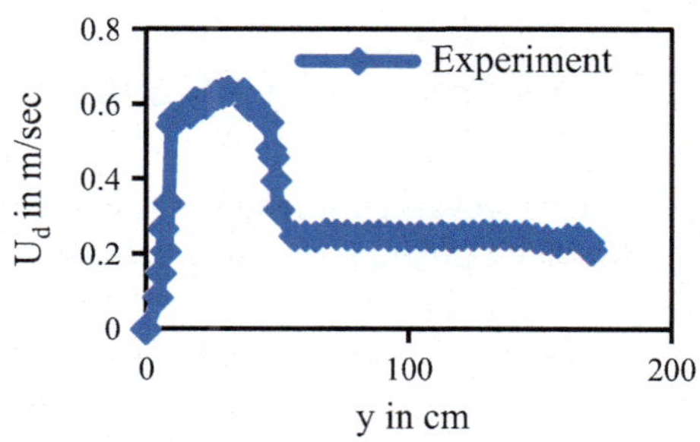

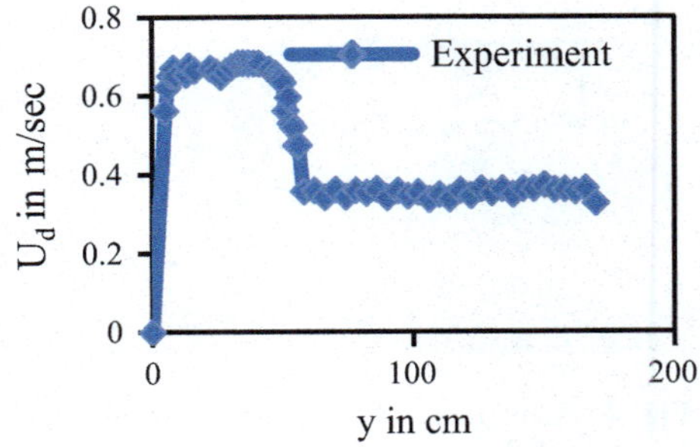

Fig. 6 For flow depth 14 cm

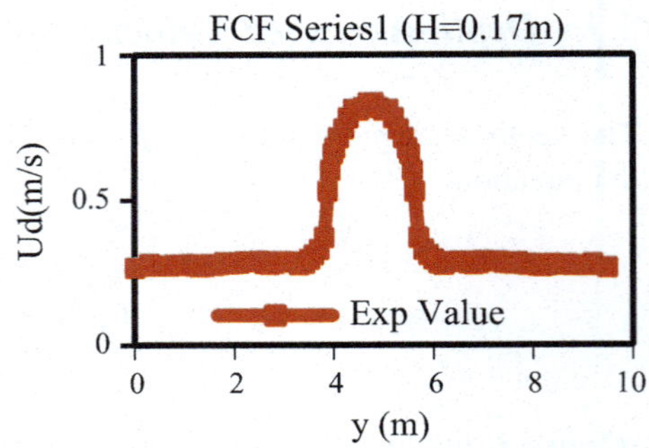

Fig. 7 For width ratio 12

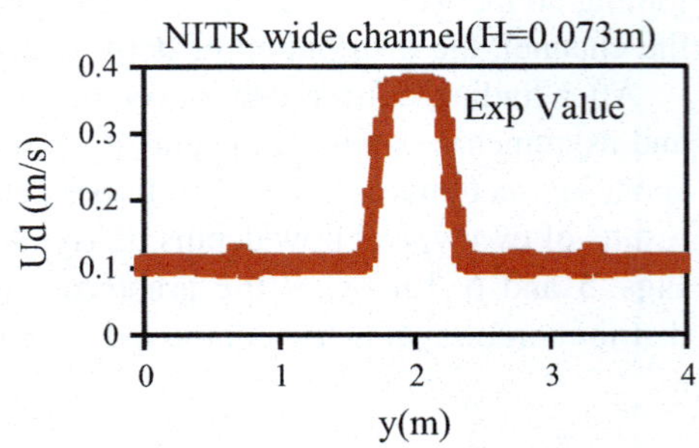

Fig. 8 For width ratio 6.67

3.2 Application of CES Software

1. The conveyance estimation system (CES) is developed by joint agency/DEFRA Research program on flood defence with contributions from the Scottish executive and the Northern Ireland Rivers Agency. HR Wallingford and the environmental agency, UK widely recommended it for the reliable prediction of conveyance. It is utilized to evaluate the river flood levels, flow capacities, depth-averaged velocities and boundary shear distributions of the channel.
2. This software consists of three parts: (1) The Roughness Advisor is assigned for roughness value, which is used for all roughness zones, (2) The Conveyance

Generator is used for determining the channel capacity based on both roughness and morphology, (3) The Uncertainty Estimator gives an indication of the uncertainty associated with CES outputs.

For steady uniform flows, the Reynolds-Average Navier–Stokes equation (RANS) is simplified as

$$\mathrm{v}\frac{\partial^2 \bar{u}}{\partial y^2} + \mathrm{v}\frac{\partial^2 \bar{u}}{\partial z^2} - \frac{\partial \overline{u'v'}}{\partial y} - \frac{\partial \overline{u'w'}}{\partial z} + g\left\{\frac{\partial h}{\partial x} - S_0\right\} = \frac{\partial \overline{u\mathrm{v}}}{\partial y} + \frac{\partial \bar{u}\bar{w}}{\partial z} \quad (2)$$

This generalized equation is applicable for obtaining the turbulent flow structure in different flow conditions. The RANS equation in X-direction (longitudinal flow direction) can be simplified as

$$\rho\left[\frac{\partial \overline{u\mathrm{v}}}{\partial y} + \frac{\partial \bar{u}\bar{w}}{\partial z}\right] = \rho g S_0 + \frac{\partial}{\partial y}\left(-\rho\overline{u'v'}\right) + \frac{\partial}{\partial z}\left(-\rho\overline{u'w'}\right) \quad (3)$$

where ρ the density of the water, S_0 the longitudinal bed slope, g the acceleration due to gravity, $\bar{u}, \bar{\mathrm{v}}$ and $\bar{w}$ are the component of the mean velocity, u', v' and w' are the fluctuations of the velocity components. Here the over bar represents a time-averaged parameters. The simplification of (3) has been done in SKM method. In Eq. (3), the first term is the secondary flow term consisting of lateral and vertical components of the velocity. The second term represents the weight component of water. The third and fourth terms account for the apparent shear or Reynolds shear stresses in vertical and horizontal planes, respectively.

Finally, Eq. (3) is simplified to

$$\rho\frac{\partial H(\overline{u\mathrm{v}})_d}{\partial y} = \rho \mathrm{Hg} S_0 + \frac{\partial}{\partial y}\left(\rho\lambda H^2\left(\frac{f}{8}\right)^{\frac{1}{2}} U\frac{\partial U}{\partial y}\right) - \frac{f}{8}\rho U^2\sqrt{1+\frac{1}{s^2}} \quad (4)$$

This is the simplified form of SKM. Here the first term of the left-hand side is due to secondary current (Γ). The first term of the right-hand side is gravitational term for a uniform flow, the second term is Reynold's shear stress and the third term is due to the bed shear. So we can say Eq. (4) is dependent upon three calibration coefficient f, λ, and Γ related to local bed friction, eddy viscosity and the secondary flow, respectively. The numerical results of depth-averaged velocity obtained from CES software for both symmetric and asymmetric compound channels are compared with the measured value found from experiments. The comparisons of the experimental results with CES have been demonstrated in Figs. 9, 10, 11, 12, 13 and 14 for various relative depths of asymmetric compound channel. Likewise, from four symmetric channel data sets, four typical cases (one from each channel) have been presented in Figs. 15, 16, 17 and 18.

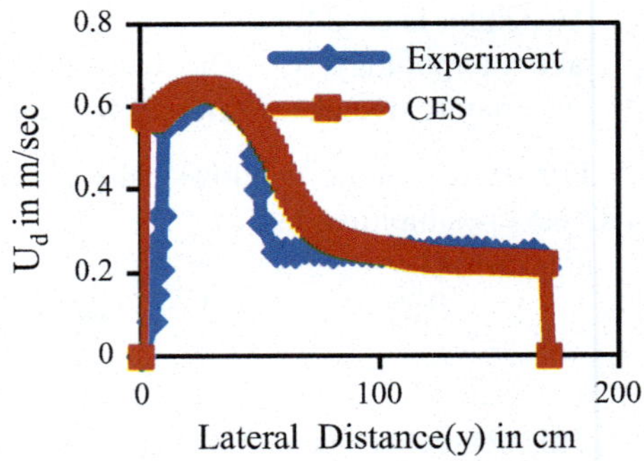

Fig. 9 For flow depth 12.5 cm

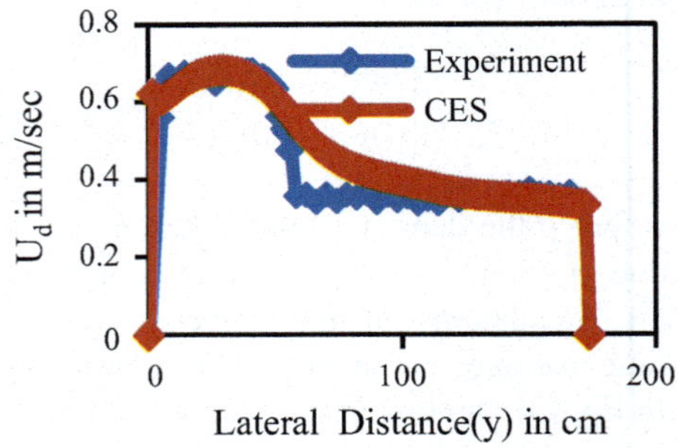

Fig. 10 For flow depth 14 cm

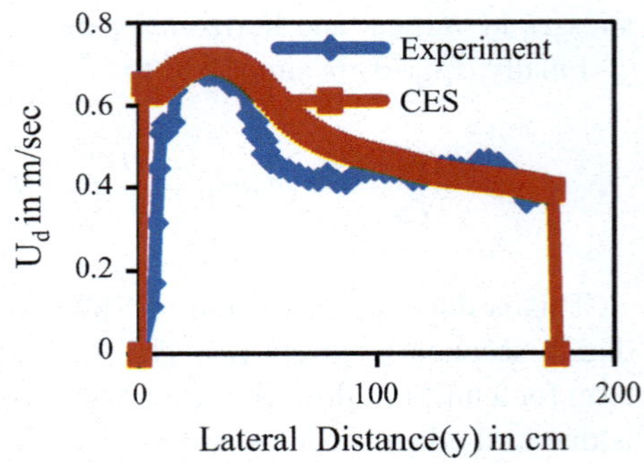

Fig. 11 For flow depth 15 cm

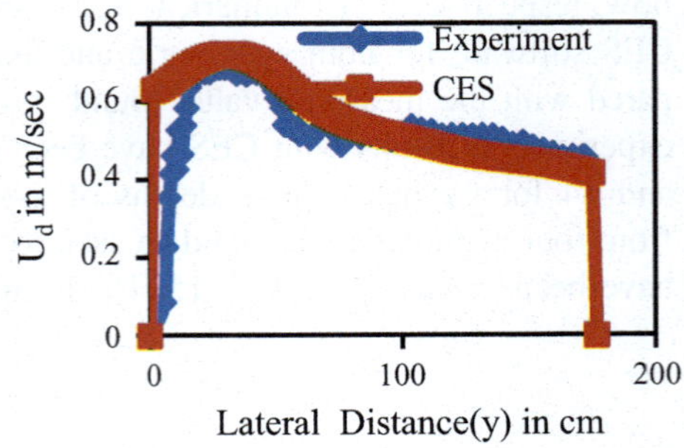

Fig. 12 For flow depth 15.5 cm

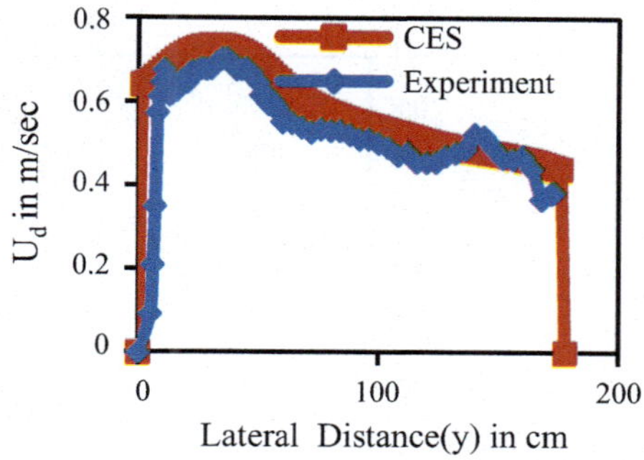

Fig. 13 For flow depth 16 cm

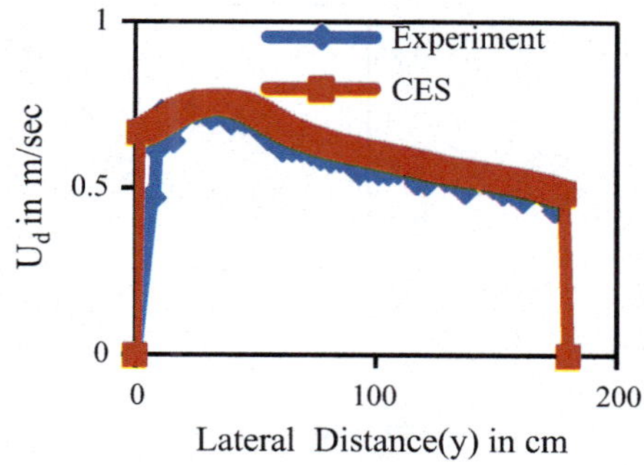

Fig. 14 For flow depth 17 cm

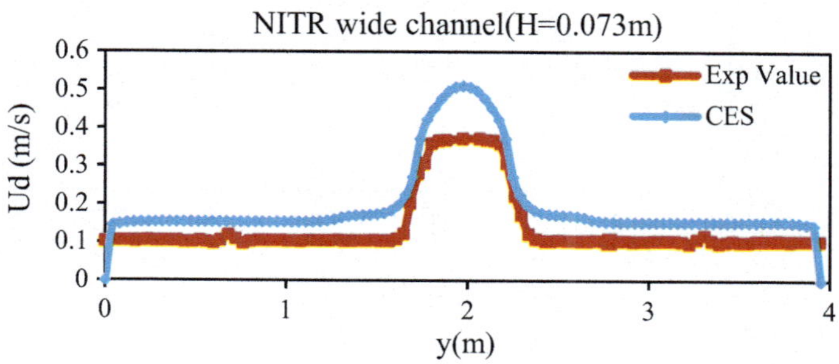

Fig. 15 For width ratio = 12

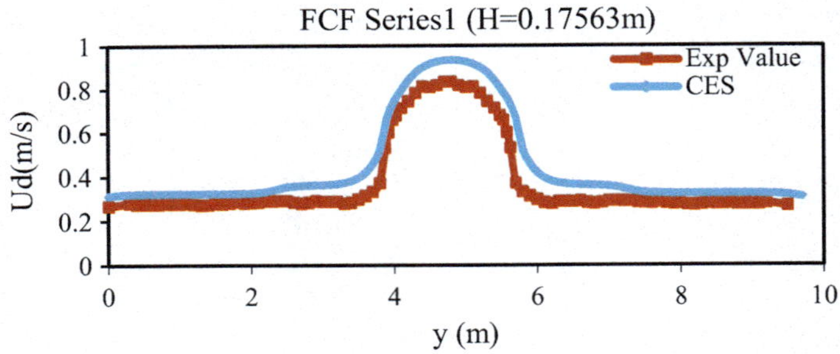

Fig. 16 For width ratio = 6.67

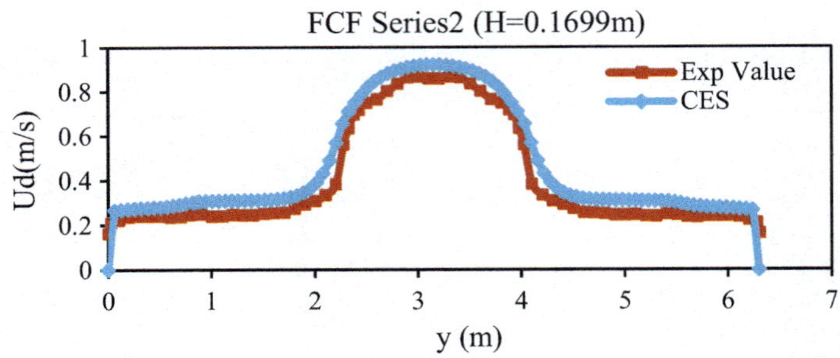

Fig. 17 For width ratio = 4.2

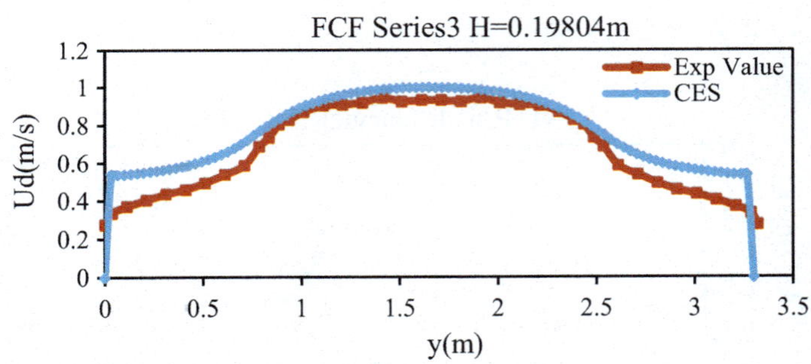

Fig. 18 For width ratio = 2.2

4 Results and Discussions

The comparisons of the experimental results with CES have been graphically presented for both symmetric and asymmetric compound channels (Figs. 9, 10, 11, 12, 13, 14, 15, 16, 17 and 18). From the results obtained by the most commonly used software CES, it has been noticed that due to the turbulence at interface, the

Table 1 Error analysis for asymmetric and symmetric compound channel

Channel type	Width ratio	Depth of flow(cm)	Percentage error (%)
Asymmetric channel (NITR)	5.1	12.5	11
		14	9.8
		15	8.45
		15.5	6
		16	5.15
		17	5
Symmetric channel (NITR)	12	7.3	15
Symmetric channel (FCF 1)	6.67	17.5	8
Symmetric channel (FCF 2)	4.2	17	5.79
Symmetric channel (FCF 3)	2.2	20	4.98

depth-averaged velocities have been over estimated at this interface for each flow depths. It can be clearly seen that especially for asymmetric compound channel the overestimation of depth-averaged velocity at junction is significant. For lower relative flow depths, the CES is providing erroneous values for entire junction (Figs. 9 and 10). For the higher-overbank cases, the less error has been observed in both symmetric and asymmetric compound channels.

This is due to higher values of momentum transfer at shallow depth as compared with higher-flow depths. At lower-flow depths, the effect of friction and interaction between the main channel and floodplain is more than that for higher-overbank flow depths. The results from CES have also been compared to their corresponding experimental values using error analysis. From error analysis, it is observed that in symmetric compound channels, 15% error has been found for high-width ratio channels and when width ratio is less, the error also decreased to 4%. In case of asymmetric compound channels, the error is more for lower-flow depth and subsequently, the minimum error has been obtained for higher-flow depths. The results of percentage error analysis are tabulated in Table 1. From the present analysis, it is recommended that the calibrating coefficients should be modified especially near the shear layer plateau.

5 Conclusions

Experimental data sets of the depth-averaged velocity in symmetric and asymmetric compound channels have been taken to study.

In this study, a series of laboratory experiments of different width ratio was considered for analysis. Flow characteristics in each model were observed, and corresponding error analysis is done for a wide range of depth-averaged velocities. From the error analysis of the results, it was observed that CES is providing good prediction for lower-width ratio channels and should be used to estimate discharges.

One should keep in mind that in interface region, this software is not predicting well due to the calibration of eddy viscosity coefficients, friction factor and secondary currents and it should be preferred carefully for depth-averaged velocity at transition region. From the present analysis, it is recommended that the calibrating coefficients used in this CES software should be modified especially near the shear layer plateau.

References

1. I.A. Al-Khatib, A.A. Dweik, M. Gogus, Evaluation of separate channel methods for discharge computation in asymmetric compound channels. Flow Meas. Instrum. **24**, 19–25 (2012)
2. I.A. Al-Khatib, H.A. Hassan, K.A. Abaza, Application and validation of regression analysis in the prediction of discharge in asymmetric compound channels. J. Irrig. Drainage Eng. **139** (7), 542–550 (2013)
3. D. Bousmar, N. Rivière, S. Proust, A. Paquier, R. Morel, Y. Zech, Upstream discharge distribution in compound-channel flumes. J. Hydraul. Eng. **131**(5), 408–412 (2005)
4. K.K. Khatua, Interaction of flow and estimation of discharge in two stage meandering compound channels. (Doctoral dissertation), 2007
5. K.K. Khatua, K.C. Patra, P.K. Mohanty, Stage-discharge prediction for straight and smooth compound channels with wide floodplains. J. Hydraul. Eng. **138**(1), 93–99 (2011)
6. D.W. Knight, M.E. Hamed, Boundary shear in symmetrical compound channels. J. Hydraul. Eng. **110**(10), 1412–1430 (1984)
7. P.K. Mohanty, Flow analysis of compound channels with wide flood plains. (Doctoral dissertation) 2013
8. B.C. Van Prooijen, J.A. Battjes, W.S. Uijttewaal, Momentum exchange in straight uniform compound channel flow. J. of Hydraul. Eng. **131**(3), 175–183 (2005)
9. K. Shiono, D.W. Knight, Mathematical models of flow in two or multi stage straight channels. *In Proc. Int. Conf. on River Flood Hydraulics* (Wiley, New York NY, 1990), pp. 229–238
10. H. Wang, K.J. Yang, S.Y. Cao, X.N. Liu, Computation of momentum transfer coefficient and conveyance capacity in compound channels. J. Hydrodyn. Ser. B **19**(2), 225–229 (2007)
11. P.R. Wormleaton, P. Hadjipanos, Flow distribution in compound channels. J. Hydraul. Eng. **111**(2), 357–361 (1985)
12. P.R. Wormleaton, J. Allen, P. Hadjipanos, Discharge assessment in compound channel flow. J. Hydraulics Div. **108**(9), 975–994 (1982)

Application of Lateral Distribution Method and Modified Lateral Distribution Method to the Compound Channel Having Converging Floodplains

Bhabani Shankar Das, Kishanjit K. Khatua and Kamalini Devi

Abstract This paper examines the use of lateral distribution method (LDM) and modified LDM in the computation of depth-averaged velocity distributions and boundary shear stress distributions of compound channel having converging floodplain. In two-stage channel flow, the main channel is influenced by the adjoining floodplains and the conveyance capacity is normally decreased. The many-sided quality of the issue rises progressively when dealing with a compound channel with non-prismatic floodplains. Due to change in floodplain geometry, water streaming on the floodplain now traverses in the main channel, resulting in increased interaction and momentum exchange. This additional exchange in momentum should also be considered in the flow modelling. In this research work, the modified LDM equation considers friction slope and LDM equation is discretized by finite difference scheme, and for solving those equations, MATLAB tool is used. Depth-averaged velocity distributions and boundary shear stress distributions obtained from LDM and MLDM are compared with the experimental data sets.

Keywords Depth-averaged velocity · Boundary shear stress · LDM
MLDM · MATLAB

B. S. Das (✉) · K. K. Khatua · K. Devi
Department of Civil Engineering, N.I.T, Rourkela, India
e-mail: bsdas7190@gmail.com

K. K. Khatua
e-mail: kkkhatua@yahoo.com

K. Devi
e-mail: kamalinidevi1@gmail.com

V. Nath (ed.), *Proceedings of the International Conference on Microelectronics, Computing & Communication Systems*, Lecture Notes in Electrical Engineering 453,
https://doi.org/10.1007/978-981-10-5565-2_26

1 Introduction

Flood in rivers often cause a danger to the population living side by it. Modelling of such type of flow is very critical due to complex exchange of momentum at the junction of river and its adjacent floodplain. So it causes difficulties in handling flood risk problems like flood routing and flood mitigation. During peak discharge time, river generally flows over their banks to expand their conveyance and gain extra storage capacity. This is a risk to the populace living on the surge fields. The water resources engineers are required to gauge surge hazard and to create relief plans. One-dimensional numerical models are normally utilized as a part of such studies, taking into account subsection division strategies. These models for the most part do not yet incorporate compound channel stream components, for example the momentum exchange because of velocity contrast between the main channel and the floodplains. The two-dimensional (2D) models could thus be preferred when the velocity and bed shear stress distribution across the section have to be estimated accurately, for example in the sediment transport studies. Such models are most costly in terms of data survey and numerical resources. Several authors have therefore suggested simplified methods aimed solely at the determination of the longitudinal velocity transverse distribution. All the versions of so-called lateral distribution method (LDM) are obtained by depth averaging the Navier–Stokes equation under a hypothesis of a steady uniform flow and reducing to a single ordinary differential equation. LDM is revisited by Wormleaton [1] and Knight et al. [2]. This is based on two-dimensional approach and developed for proper prediction of flow variables involving flood channel facility data. The basic LDM accounts for the bed friction and the lateral turbulent shear stress. This LDM incorporates three calibrating parameters, i.e. eddy viscosity, secondary flow and friction factor. The roughness is evaluated by Darcy–Weisbach friction factor [2–6].

Shiono and Knight [7] suggested that the secondary current effect can be modelled by constant parameters, whereas the Ervine et al. [8] stated that the secondary current term is proportional to the square of longitudinal velocity. Knight et al. [2] assumed the eddy viscosity as constant, and it is proportional to shear velocity. SKM (Shiono-Knight method) is an analytical method for the solution of lateral distribution methods which give good results for prismatic compound channel which is used in commercially available CES software to model the flow in compound channel with prismatic floodplains. This research examines whether the lateral distribution method is capable of predicting accurate depth-averaged velocity, boundary shear distribution, and stage discharge relationship for an overbank flow having converging floodplain. Figure 1 demonstrates the experimental compound channels having converging floodplain with different angles of convergence for the current research work.

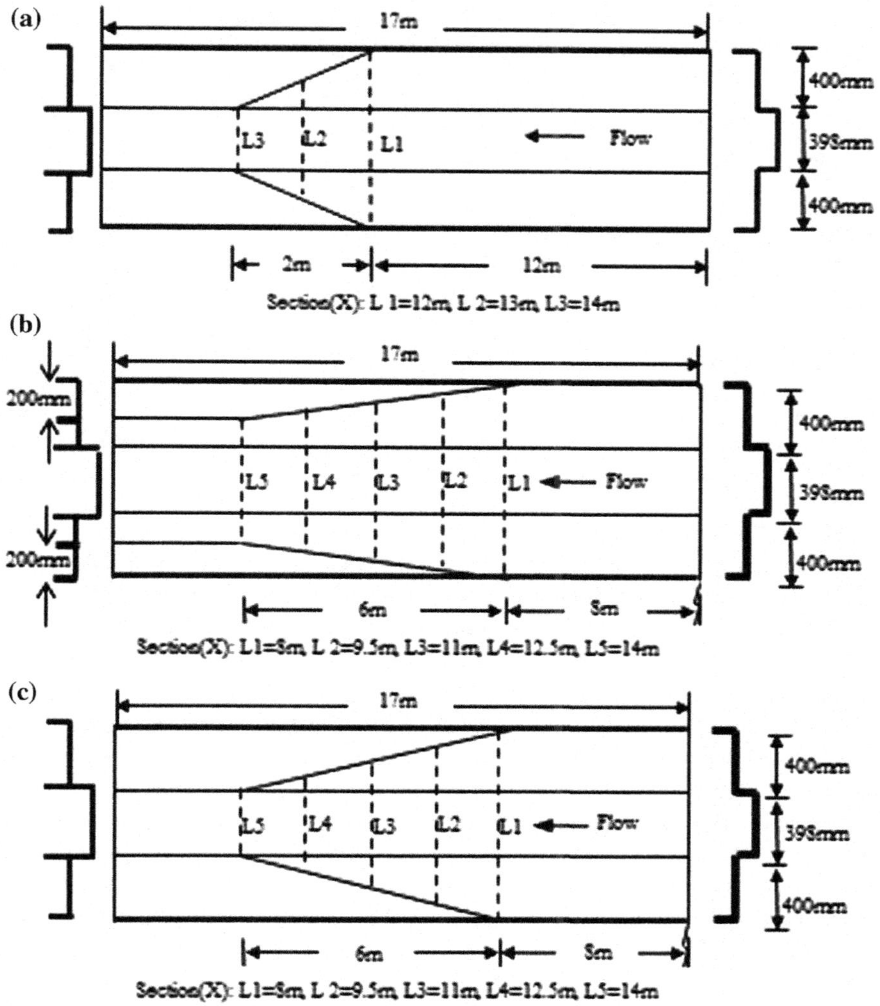

Fig. 1 Schematic view of compound channels with non-prismatic floodplains, converging from **a** 400 to 0 mm along a 2 m length (ONPC2-0), **b** 400 to 200 mm along a 6 m length (ONPC6-200), **c** 400 to 0 mm along a 6 m length (ONPC6-0)

2 Data Collection

For this research work, we consider the data of non-prismatic converging compound from Université Catholique de Louvain (UCL) flume [9]. The dimension of the experimental flume is shown in Fig. 1. There are three converging compound channels: ONPC2-0, floodplain width converges from 400 to 0 mm along a 2 m length (θ = 11.31°), and ONPC6-0, floodplain width converges from 400 to 0 mm along a 6 m length (θ = 3.81°), and ONPC6-200, floodplain width converges from 400 to 200 mm along 6 m length (θ = 1.91°)

3 LDM for Compound Channel

LDM equation Previous Work:

The lateral distribution method (LDM) is derived from a depth averaging of the Navier–Stokes momentum conservation equation in the stream-wise direction:

$$\rho\left(\frac{\partial\bar{u}}{\partial t}+\frac{\bar{u}(\partial\bar{u})}{\partial x}+\frac{\bar{v}(\partial\bar{u})}{\partial y}+\frac{\bar{w}(\partial\bar{u})}{\partial z}\right)=\rho F_x-\frac{\partial\bar{p}}{\partial x}+\mu\Delta\bar{u} -\rho\left(\frac{\partial\overline{u'u'}}{\partial x}+\frac{\partial\overline{u'v'}}{\partial y}+\frac{\partial\overline{u'w'}}{\partial z}\right) \quad (1)$$

where $(\bar{u},\bar{v},\bar{w})$ are Reynolds averaged local velocity components, respectively, in the x-(stream-wise), y-(lateral) and z-(normal to bed) directions; ρ is the density of water; F_x is the x-wise component of gravitational forces which equals the longitudinal bed slope S_0 time the gravity constant g; p is the pressure; μ is the molecular viscosity; and $\left(\overline{u'u'},\overline{u'v'},\overline{u'w'}\right)$ are the Reynolds turbulent shear stresses. From this momentum equation, the LDM equation is derived by taking various assumptions. Various assumptions made for the derivation of LDM for compound channel are as follows: (1) a permanent $(\partial/\partial t=0)$ and uniform $(\partial/\partial x=0)$ flow. (2) Viscous friction in regard to the Reynolds stresses is neglected. (3) There is no secondary current in the simple channel. (4) Water level is assumed to be horizontal $(\partial z_{\mathrm{w}}/\partial y=0)$ in the transverse direction as a consequence of a one-dimensional flow hypothesis. The RANS equation in x-direction (i.e. longitudinal flow direction) considering all the assumption for steady, uniform, incompressible fluid flow can be simplified as

$$\rho\left[\frac{\partial\bar{u}\bar{v}}{\partial y}+\frac{\partial\bar{u}\bar{w}}{\partial z}\right]=\rho g S_0+\frac{\partial}{\partial y}\left(-\rho\overline{u'v'}\right)+\frac{\partial}{\partial z}\left(-\rho\overline{u'w'}\right) \quad (2)$$

u', v' and w' are the fluctuation of the velocity components. Here, the over bar represents a time-averaged parameters. The simplification of (2) is done in LDM method. In Eq. (2), the first term is the secondary flow term consisting of lateral and vertical components of the velocity. The second term represents the weight component of water. The third and fourth terms account for the apparent shear or Reynolds shear stresses in vertical and horizontal planes, respectively. Considering the mean velocity component in z-direction is very negligible, then $\bar{w}$ is equal to zero and $\tau_{yx}=-\rho\overline{u'v'}$, $\tau_{zx}=-\rho\overline{u'w'}$, and integrating (2) in the normal direction over the total flow depth H, we have.

$$\rho g H S_0 + \frac{\partial}{\partial y} H T_{xy} - \tau_b \sqrt{1 + S_{0y}^2} = \frac{\partial}{\partial y} \int_0^H \rho \bar{u} \bar{v} \mathrm{d}z \tag{3}$$

where $\tau_b = \rho U_*^2$ is bed shear stress, U_* is the shear velocity, S_{0y} = lateral slope or transverse bed slope along the channel. The depth-averaged turbulent shear stress $T_{xy} = \rho \vartheta_t \frac{\partial U}{\partial y}$ is classically modelled using the Boussinesq eddy viscosity assumption, where U in Eq. (3) is depth-averaged longitudinal velocity and ϑ_t is the eddy viscosity. The right-hand term is known as secondary term. Knight et al. [1] used a Boussinesq eddy viscosity model for T_{xy} and assumed that (1) the eddy viscosity $\vartheta_t = \lambda H U^*$ is proportional to the water depth H and to the shear velocity U^*, where λ is the dimensionless eddy viscosity; (2) the bed shear stress τ_b can be evaluated using Darcy–Weisbach friction factor; and (3) the secondary current term is negligible. So, $\frac{\partial}{\partial y} \int_0^H \rho \bar{u} \bar{v} \mathrm{d}z = 0$, Now, Eq. (3) becomes

$$\rho g H S_{0x} + \frac{\partial}{\partial y} \left(\rho \lambda H^2 \sqrt{\frac{f}{8}} U \frac{\partial U}{\partial y} \right) - \rho \frac{f}{8} U^2 \sqrt{1 + S_{0y}^2} = 0 \tag{4}$$

Equation (4) can be solved by both numerically and an analytically. Using dimensionless eddy viscosity range $\lambda = 0.2–0.3$ for a natural river test case incorporates good estimates of longitudinal velocity distribution and total discharge.

A similar equation was developed by Wark et al. [10]. For the longitudinal unit flow $q = UH$, using the Manning's roughness n, good velocity distribution estimates were obtained for the large-scale flood channel facility (FCF) and for a natural river geometry, using realistic value of friction coefficient and dimensionless eddy viscosity in the range $\lambda = 0–0.24$ with distinct value in the main channel and floodplain. Lambert and Sellin also developed an alternative version of the LDM, using Prandtl mixing–length model to estimate the eddy viscosity. With appropriate value of friction factor, good estimates of the velocity distribution were obtained for the FCF experiments [11].

3.1 Modified Lateral Distribution Method

In prismatic channel cases, the flow generally remains uniform. But in non-prismatic channel, the flow is non-uniform; i.e., the depth of water varies along the length of the channel. Due to this, variation of depth in non-prismatic channel, the role friction slope or energy slope (S_e) comes into account in place of bed slope in Eq. (4). So the modified version of the LDM equation becomes

$$\rho g H S_{\mathrm{e}} + \frac{\partial}{\partial y}\left(\rho \lambda H^2 \sqrt{\frac{f}{8}} U \frac{\partial U}{\partial y}\right) - \rho \frac{f}{8} U^2 \sqrt{1 + S_{0y}^2} = \Gamma \tag{5}$$

It is known as modified SKM [12]. But here in modified LDM, we consider Manning's n value in place of Darcy's friction factor f. Thus, the Eq. (5) can be presented as

$$\rho g H S_{\mathrm{e}} + \frac{\partial}{\partial y}\left(\rho \lambda H^2 \sqrt{\frac{g n^2}{H^{1/3}}} U \frac{\partial U}{\partial y}\right) - \frac{\rho g n^2}{H^{1/3}} U^2 \sqrt{1 + S_{0y}^2} = \Gamma \tag{6}$$

3.2 Energy Slope Calculation

For a compound channel with non-prismatic floodplains, the depth is effectively non-uniform and the energy line slope can be evaluated by $S_{\mathrm{e}} = S_{0x} - \frac{\partial H}{\partial x} - g\frac{\partial U}{\partial x0}$ less than S_{ox} = 0.002003. A comparison between bed slope S_{e} and energy slope S_0 for non-prismatic compound channel at selected section is shown in Fig. 2.

Figure 2 indicates the discrepancies between S_{e} and S_0 at all sections of converging compound channel and also a negligible variation for skewed compound channel for relative depth (Dr = 0.5). Using the Manning's roughness coefficient n in modified LDM, the energy slope can be estimated. The energy slope S_e and the average flow velocity U are unknown. Thus, to calculate those variables, an iteration method is used as below.

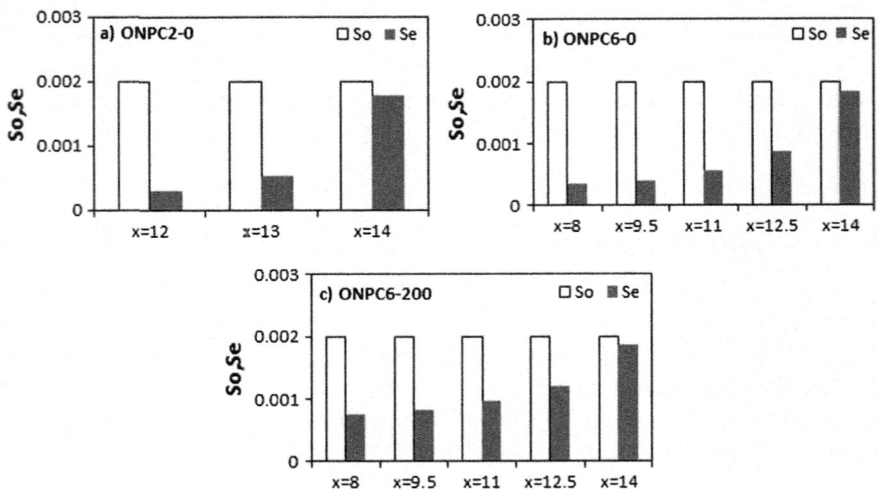

Fig. 2 Relationships between the bed slope S_o and energy slope S_e for **a** 2 m converging at 3 sections (11.3°) **b** 6 m converging at 5 sections (3.81°) and **c** 6 m converging at 5 sections (1.91°)

$$S_e \approx \left(\frac{n^2 U^2}{R^{4/3}}\right) \tag{10}$$

where Q = total discharge, A = cross-sectional area and R = hydraulic radius = (Area/wetted perimeter).

3.3 Solution for the Model

In the present research work, the LDM numerical solution is obtained by writing the differential equation in a discrete form, using the finite difference method (Fig. 2). The finite difference method is carried out by central difference method. This reduces the LDM equation to a set of quadratic algebraic equations linking together velocities U_i at each node of the mesh. At the boundaries, a no-slip condition is used by setting the velocity equal to zero along the walls. The so-defined set of equation is solved either by a Newton method, or by any other appropriate method.

$$A\frac{\partial^2(U^2)}{\partial y^2} + B(U^2) + C = \Gamma \tag{11}$$

where

$$A = \frac{1}{2}\rho\lambda\sqrt{\frac{gn^2}{H^{1/3}}}H^2,\quad B = \frac{\rho g n^2}{H^{1/3}}\sqrt{1+\frac{1}{s_{0y}^2}},\quad C = \rho g H S_{0x}$$

By substituting $U^2 = t$,
We have

$$A\frac{\partial^2 t}{\partial y^2} + Bt + C = \Gamma \tag{12}$$

The matrix corresponding to this set of equation is tridiagonal. Crout factorization [13] technique is used for solving the matrix, and the code is written in MATLAB tool. After solving the matrix, it gives $U_d = \sqrt{t}$; from the value of depth-averaged velocity (U_d), the boundary shear stress (τ_b) is found out by the relation (Fig. 3)

$$\tau_b = U_d^2 \frac{gn^2}{H^{1/3}} \tag{13}$$

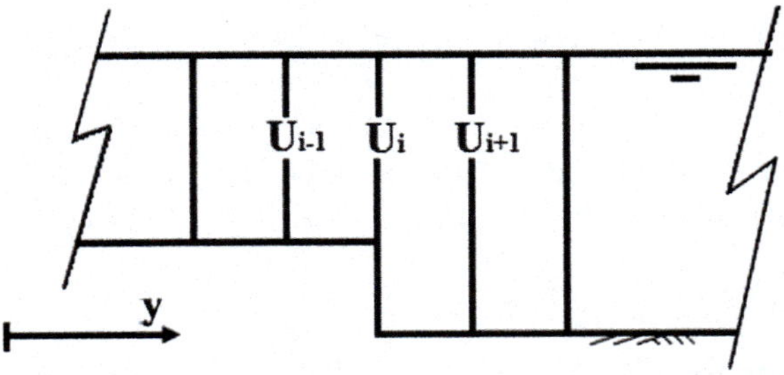

Fig. 3 Discrete meshes for LDM and MLDM numerical solving

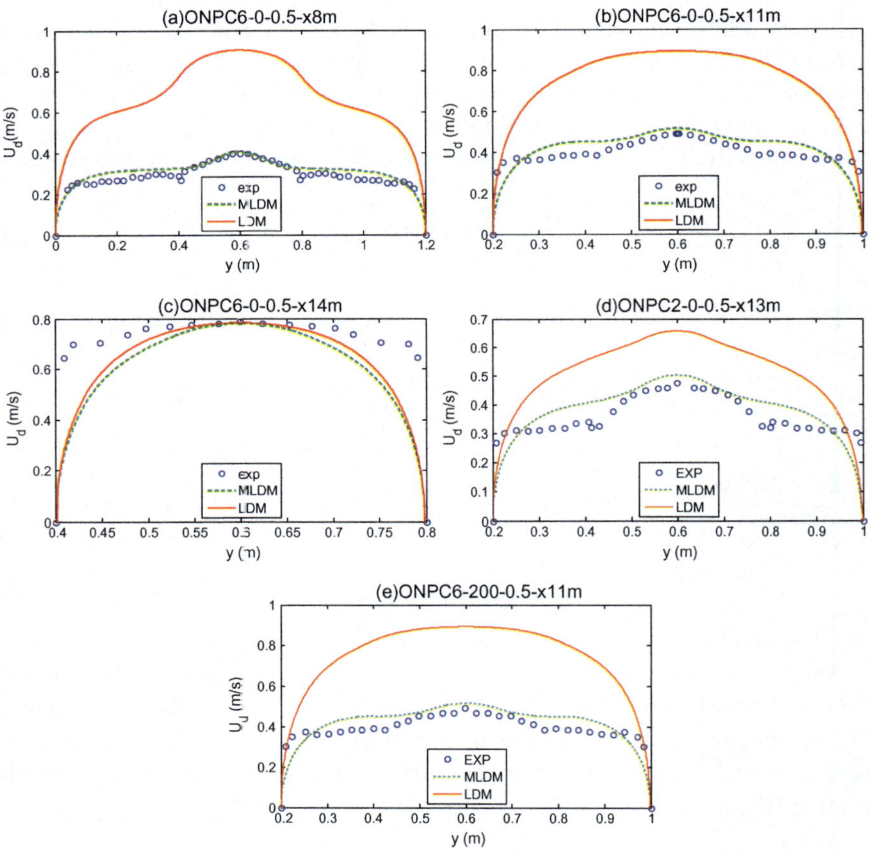

Fig. 4 Comparison of depth-averaged velocity (U_{d}) between experimental data, LDM and modified LDM for relative depth 0.5 with different channel geometries and sections **a** ONPC6-0-section 8 m, **b** ONPC6-0-section 11 m, **c** ONPC6-0-section 14 m, **d** ONPC2-0-section 13 m, **e** ONPC6-200-section 11 m

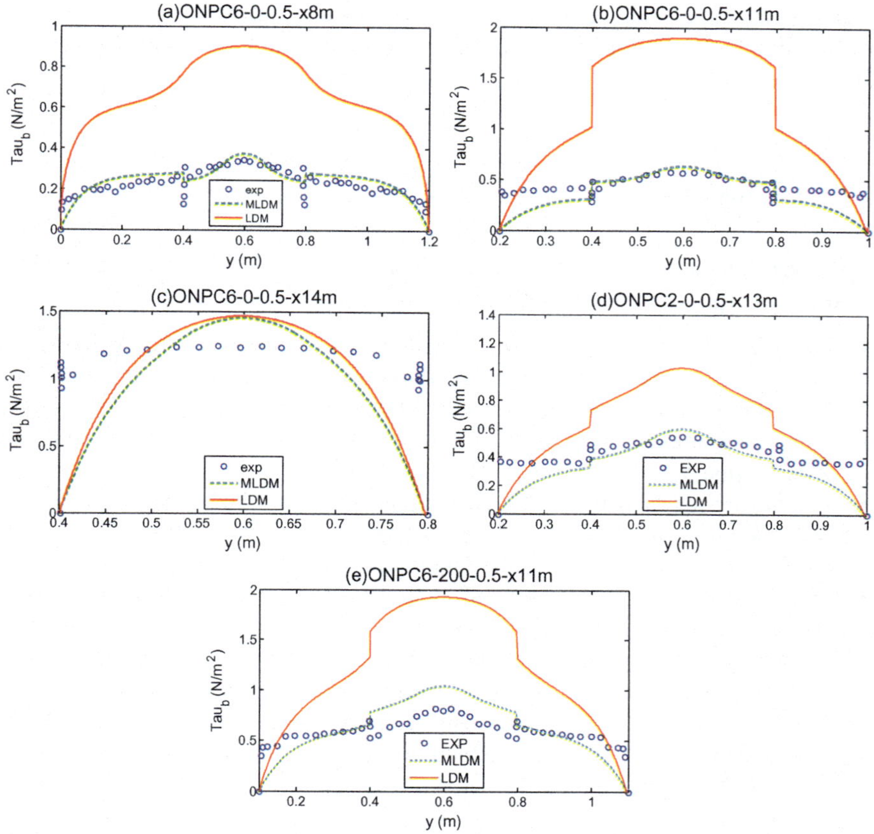

Fig. 5 Comparison of boundary shear stress (τ_b) between experimental data, LDM and Modified LDM for relative depth 0.5 with different channel geometries and sections **a** ONPC6-0-section 8 m, **b** ONPC6-0-section 11 m, **c** ONPC6-0-section 14 m, **d** ONPC2-0-section 13 m, **e** ONPC6-200-section 11 m

4 Results and Discussion

Figures 4 and 5 indicate, in general, for ONPC2-0, and ONPC6-0, ONPC6-200 that the usual LDM overestimates the both depth-averaged velocities and boundary shear stress distributions. However, there is good agreement with modified LDM and the experimental data sets. From Fig. 5b, d, e, it has been observed that for the midsection of the channel the numerical method MLDM does not model the boundary shear stress and velocity distributions at floodplain as accurately as main channel. For ONPC6-0, the developed numerical model has good agreement with the test results for section x = 8 m, for both depth-averaged velocity and boundary shear stress distribution. But at midsection of convergence and after that, i.e. at last section x = 14 m, the model is underestimated at floodplain and overestimated in main channel for both velocity and boundary shear stress distribution. This is mainly due to poor modelling of zonal friction factor and secondary current cells [12].

5 Conclusions

A numerical method based on lateral distribution method has been demonstrated to predict the depth-averaged velocity and boundary shear stress distributions.

- To model depth-averaged velocity and boundary shear stress distribution in compound channel with narrowing floodplain with angles 1.91°, 3.81° and 11.3°, the lateral distribution method (LDM) is modified.
- Converging impacts were dealt with by substituting the energy slope S_e in place of bed slope S_0.
- The inconsistency between the energy line slope and bed slope is remarkable and increases with the angle of convergence.
- Energy slope increases along the longitudinal direction of flow.
- The depth-averaged velocity and boundary shear stress distributions were modelled using LDM and modified LDM in compound channel with non-prismatic floodplains. The numerical solution of MLDM for depth-averaged velocity and boundary shear stress distributions for all sections of converging compound channel concurs well with test results when compared with LDM.
- Modified lateral distribution method results get improved with decrease in convergence angle from 11.3° to 1.91°.

References

1. P.R. Wormleaton, Determination of discharge in compound channels using the dynamic equation for lateral velocity distribution. In: Proceedings of the International Conference on Fluvial Hydraulics, pp. 98–103 Belgrade, Hungary (1988)
2. D.W. Knight, K. Shiono, J. Pitt, Prediction of depth mean velocity and discharge in natural rivers with the overbank flow. In: Proceedings of the International Conference on Hydraulic and Environmental Modelling of Coastal, Estuarine and River Waters, pp. 419–428. Gower Publishing (1989)
3. K.K. Khatua, K.C. Patra, P.K. Mohanty, Stage-discharge prediction for straight and smooth compound channels with wide floodplains. J. Hydraul. Eng. 138(1), 93–99 (2011) B. Rezaei, D.W. Knight, Application of the Shiono and Knight Method in compound channel with non-prismatic floodplains. J. Hydraul. Res. **47**(6), 716–726 (2009)
4. K.K. Khatua, Interaction of flow and estimation of discharge in two-stage meandering compound channels. Doctoral dissertation, (2007)
5. D.W. Knight, K. Shiono, Turbulence measurements in a shear layer region of a compound channel. J. Hydraul. Res. IAHR 28(2), 141–156 (1990)

6 K. Devi, K.K. Khatua, B.S. Das, A numerical solution for depth-averaged velocity distribution in an open channel flow. ISH J. Hydraul. Eng. 1–10 (2016)

7 K. Shiono, D.W. Knight, Turbulent open-channel flows with variable depth across the channel. J. Fluid Mech. 222, 617–646 (1991)

8 D.A. Ervine, K. Babaeyan-Koopaei, R.H.J. Sellin, Two-dimensional solution for straight and meandering overbank flows. J. Hydraul. Eng. 126(9), 653–669 (2000)

9 B. Rezaei, Overbank flow in compound channels with prismatic and non-prismatic floodplains. PhD Thesis, University of Birmingham, U.K (2006)

10 J.B. Wark, P.G. Samuels, D.A. Ervine, A practical method of estimating velocity and discharge in compound channels. In: Proceedings of an International Conference on River FloodHydraulics, Wiley, Chichester, pp. 163–172 (1990)

11 D. Bousmar, Y. Zech, Velocity distribution in non-prismatic compound channels. Proceedings of the Institution of Civil Engineers-Water Management 157 (2):99–108 (2004)

12 B. Rezaei, D.W. Knight, Application of the Shiono and Knight Method in compound channel with non-prismatic floodplains. J. Hydraul. Res. 47(6), 716–726 (2009)

13 B. S. Das, K. K. Khatua, K. Devi, Numerical solution of depth-averaged velocity and boundary shear stress distribution in converging compound channels. Arabian Journal for Science and Engineering, 42(3), 1305–1319 (2017)

A 0.533 dB Noise Figure and 7 mW Narrowband Low Noise Amplifier for GPS Application

Namrata Yadav, Mohd. Javed Khan, Jyoti Singh, Abhishek Pandey, Manish Kumar, Vijay Nath and L. K. Singh

Abstract In this research article, low noise amplifier (LNA) circuit is proposed. This circuit is most important block of receiver system. In wireless communication system, LNA is used in receiver front-end circuitry. It should be necessarily having high gain and minimum noise figure for optimum performance. This work is an attempt to develop the same without disturbing stability and linearity in the circuit. The proposed low noise figure LNA contains single-ended cascode topology including the input matching network and output matching network at input and output sides, respectively, so that minimum components are required when the circuit follows for LNA IC fabrication. The CMOS low noise amplifier is designed through Cadence spectre RF simulation in standard UMC 90 nm CMOS process. It is designed for 1.575 GHz frequency which seeks its application in GPS receiver. The parameters like gain, input matching, output matching, reverse isolation and stability are examined by S-parameters. The noise figure, 1-dB compression point IIP3 and power consumption are also examined for 1.5 V input LNA. The proposed LNA is compared with existing LNA for performance analysis using the above parameters.

N. Yadav · Mohd. Javed Khan · A. Pandey · V. Nath (✉)
VLSI Design Group, Department of ECE, Birla Institute of Technology Mesra, Ranchi 835215, Jharkhand, India
e-mail: vijaynath@bitmesra.ac.in

N. Yadav
e-mail: namratanushashi@gmail.com

J. Singh
Department of ECE, PES Institute of Technology, Bangalore, Karnataka, India

M. Kumar
Department of Electronics & Communication Engineering, MMMUT Gorakhpur, Uttar Pradesh, India

L. K. Singh
Department of Physics & Electronics, Dr. RML Avadah University Faizabad, Uttar Pradesh, India

V. Nath (ed.), *Proceedings of the International Conference on Microelectronics, Computing & Communication Systems*, Lecture Notes in Electrical Engineering 453,
https://doi.org/10.1007/978-981-10-5565-2_27

Keywords Cascode transistor · Gain · Impedance matching
Source degenerated LNA · Global positioning system (GPS)

1 Introduction

As there has come boom in market of wireless communication, the designing of optimum circuitry has become major focus. This has lead researchers to design good performing circuits. Ostensibly, one of the key components of any receiver circuitry is the low noise amplifier after antenna [1]. The signal received by antenna is very weak ~1μVp-p there is a necessity to amplify the signal which gives significance to LNA. Low noise is also one of the aspects which adds to this block. As the result, we get a significant signal without noise dominating it [2].

LNA is not a complex designed circuit. In fact, it is very simple in its topology. Transistors used are NMOS for having good gain results. The proposed circuit is simulated to have LNA on single chip. Making the block in single chip has largely helped in reducing the cost as well as increasing reliability. Because of scaling in VLSI process, the channel length of MOS has fallen to nanometres, rising the transit frequency to several gigahertz and power consumption to mill watts [3]. The challenges are continuous and imply motivation in exploration of RF architectures.

In recent years, a lot of effort has been put into the design of LNAs to explore the lowest possible noise figure while trying to have linearity and good gain [4]. The LNA designed has transistor working in sub-threshold region that is weak inversion. For linearity, we should have LNA in strong inversion. There are some design trade-offs that are found while simulation of the proposed circuit. It does not usually favour all performance criteria's. Thus, compromise might be required and performance criteria depend on application requirement. A classical narrowband LNA should have high voltage gain, low NF, good linearity, unconditional stability, low power consumption [5].

The paper is discussed in following way. Sect. 2 has description of cascade LNA along with input and output matching networks used in the circuit. Sect. 3 has simulation and verification. The results and discussion of parameters is summarized in Sects. 4 and 5 is conclusion.

2 Cascode LNA Circuit

2.1 Circuit and Component Description

The circuit is given the dc supply voltage of 1.5 V. There are two ports port 1 and port 2 as input and output port, respectively. Port 1 is fed by sinusoidal signal of 1.575 GHz, and port 2 is the output port which is connected to next block that is

Mixer. At input of port 1 capacitance (C1), gate inductor (Lg) and source degeneration inductance (Ls) are used for input matching of the LNA. Similarly, at output port 2 capacitance (Cg), inductor (Lg) is used as tuning circuit for output matching along with resistance (rd).

M0 is the common source transistor having high aspect ratio (W/L $\sim$ 1000) and large gm. M0 and M2 form current mirror, and it provides biasing to CS amplifier M0. The aspect ratio of transistor M2 is not as high as of transistor M0. The biasing is done so as to operate the transistor in sub-threshold region. M1 is a cascode transistor which isolates the input from output (Miller capacitance) providing stability and reducing harmonics. At microwave frequencies, the parasitic capacitance of transistor becomes significant. Thus, Cgd will be Miller capacitance at input and output terminal. This capacitance will create nonlinearity in the circuit. So, to minimize its effect, cascade M1 is implanted in circuit. Biasing of M1 is done by voltage divider biasing.

2.2 Matching Network

To have maximum power transfer in the circuit topology, input and output matching of LNA is the primary requirement. We need 50 Ω matching at input and output. Input impedance of LNA is given by Cgs, and for this input capacitive CMOS, we use inductors (gate inductor and source degenerated inductor) which create equivalent resistance with zero reactance. Matching network behaves like series resonance circuit for the input signal. This helps to have good gain (voltage gain) and also suppress noise by circuit [6].

Here in matching network if resistor would have been used in place of Ls, then in simulation noise figure of LNA has increased considerably, as resistors give thermal noise. Also if we use capacitance in place of Ls, then resultant Z_{in} derived below in small signal analysis would be negative real value. It means effect of oscillations is taking place, and it results in nonlinearity in the circuit. Therefore, the LNA is to be matched with 50 Ω is obtained using source degeneration inductor Ls and gate inductor Lg only, which acts as series resonance circuit (current amplification) for required resonant frequency fc = 1.575 GHz i.e. the passive LC matching network will provide gain. This relates to the fact that the LNA circuit proposed will be acting as a gm stage in the receiver circuitry.

2.3 Small Signal Model

The small signal model of proposed circuit is shown above in Fig. 2. Few components are being ignored to derive approximate mathematical values in the circuit.

After applying Kirchhoff's voltage law (KVL) at the input, it is found.

$$V_i = I_i j\omega L_g + \frac{I_i}{j\omega C_{gs}} + (I_i + g_m V_{gs}) j\omega L_s \tag{1}$$

$$V_{gs} = I_i \frac{1}{j\omega C_{gs}} \quad C = C_{gs} \tag{2}$$

So, the input impedance is given as

$$Z_{in} = \frac{V_{in}}{I_{in}} = j\omega(L_g + L_S) + \frac{1}{j\omega C} + \frac{g_m L_S}{C} \tag{3}$$

For impedance matching following condition should be met

$$\text{Re}[Z_{in}] = \frac{g_m L_S}{C} = Rs \tag{4}$$

$$\text{Im}[Z_{in}] = j\omega(L_g + L_S) + \frac{1}{j\omega C} = 0 \tag{5}$$

Real term need to be 50 Ω giving real part equal to 50.

If input matching is performed at centre frequency, imaginary part needs to be cancelled.

$$\omega_c = \frac{1}{\sqrt{C(L_g + L_S)}} \tag{6}$$

For lower values $\omega < \omega_c$, it becomes capacitive in nature and inductive if $\omega > \omega_c$.

2.4 Output Matching

At output, we have another resonant circuit. If we add a resistance at output, it would add noise to circuit. Therefore, a LC tank circuit is used which give pure resistance value at resonance for will output matching.

The small signal for cutput match is below

where

- C_{db} Drain to bulk capacitance,
- C_L load capacitance,
- gm effective transconductance, and
- r_d Matching resistance.

For this network,

$$f_c = \frac{1}{2\pi\sqrt{(c_d + c_L + c_{db})L_d}} \tag{7}$$

For good operation, we need to make sure $C_d \gg C_{db}$.

Since, we don't want transistor to vary due to parasitic capacitance.

Also rd resistance is used in output matching. Its value is produced after optimizing the S21 (gain) as follows.

$$r_d = 2\pi\sqrt{L_D/C_d} \tag{8}$$

2.5 *Stability Analysis*

The electronic circuits that are designed should meet stability for different source and load impedances. The electronic circuits that are designed and simulated need also be stable for becoming real-life circuits. This means that the source and load impedance of any value should not affect the output of the circuit. Stability is seen through S-parameters and alternate stability through K_f and $B1_f$.

$$K_f = \frac{1 + \Delta^2 - S_{22}^2 - S_{11}^2}{2S_{12}S_{21}} \tag{9}$$

For $K_f > 1$ and $B1f(\Delta) < 1$, the circuit is unconditionally stable. This means the circle does not produce oscillation for any source or load impedance. The alternate stability parameters for proposed circuit are shown in Figs. 1 and 2.

3 Simulation Result and Verification

The LNA is simulated and verified with [4]. The input and output matching for the LNA circuit is done differently in the paper [4]. There is a separate pi input network before the CS amplifier. Similarly, there also lies the output pi network after the LNA. It can be the limitation as it makes the circuit bulkier while fabricating the LNA since to many inductors would take more area on single chip. Moreover, the proposed paper with only three inductors could design the LNA having all parameters close to [4]. The parameter that makes this work worth is the noise figure (0.533 dB) (Figs. 4 and 5).

The linearity of the proposed LNA is derived through parameters like 1-db compression point and input intercept point IIP3. It is being tried to have positive value so that we can get linear and stable LNA.

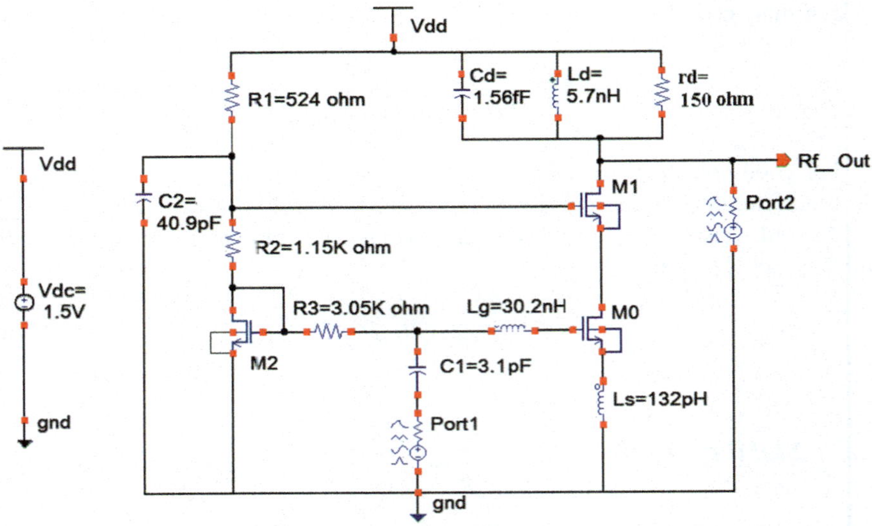

Fig. 1 Proposed low noise amplifier

Fig. 2 Small signal model of LNA

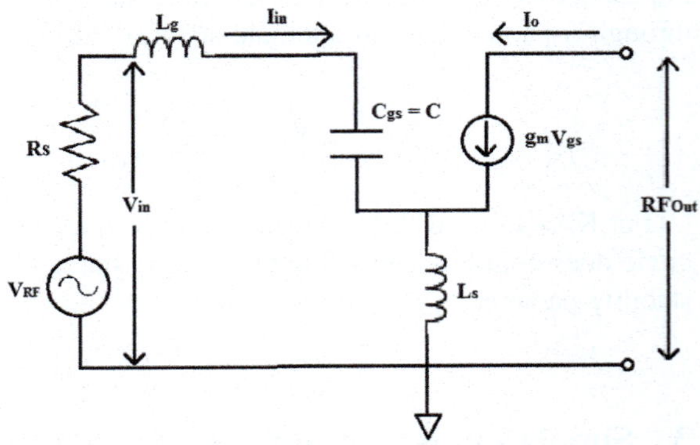

The comparison in results is done in the Table 1:

The LNA circuit is designed and simulated through cadence tool with help of UMC90 nm library. The circuit is operated at 1.5 V power supply. The proposed LNA is shown in Fig. 1. The small signal analysis of proposed circuit is shown Figs. 2 and 3.

As demonstrated in Figs. 6 and 7, the proposed LNA shows power gain by S21 (29.259 dB) and noise figure (0.533 dB). The S11 and S22 of the circuit are −9.9 and −11.86 dB, which are shown in Figs. 8 and 9, respectively. S12 (35.68 dB) is reverse isolation shown in Fig. 10.

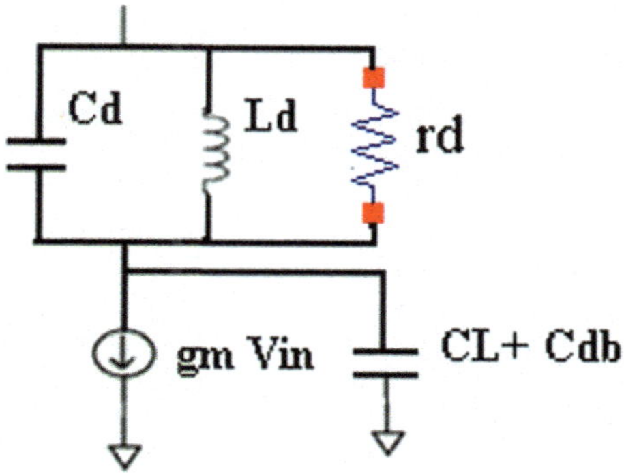

Fig. 3 Small signal model of LNA for output match

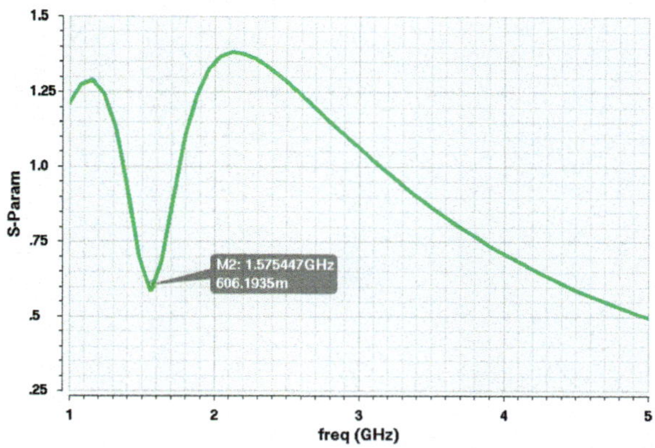

Fig. 4 Plot between B1f versus frequency

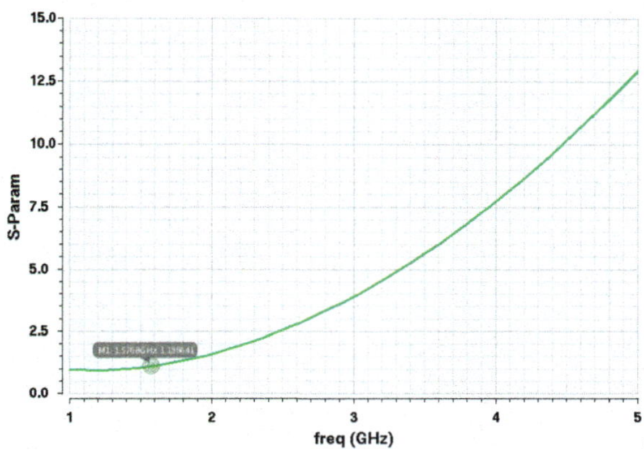

Fig. 5 Plot between Kf versus frequency

The 1-dB compression point of circuit is −16.9541 dBm as shown in Fig. 11. The performance summary of the proposed LNA was compared to other LNAs, and it is summarized in Table 2.

Table 1 Comparative Results

Specification	[4] 2014	This work
Technology	90 nm UMC	90 nm UMC
Operating frequency GHz	2.45	1.575
S11 (dB)	−9.19	−16.21
S12 (dB)	−38.03	−38.10
IIP3 (dB)	−5.70	2.9140
Noise figure	2.34	0.533
Power supply	1.2	1.5
Gain	31.53	29.25

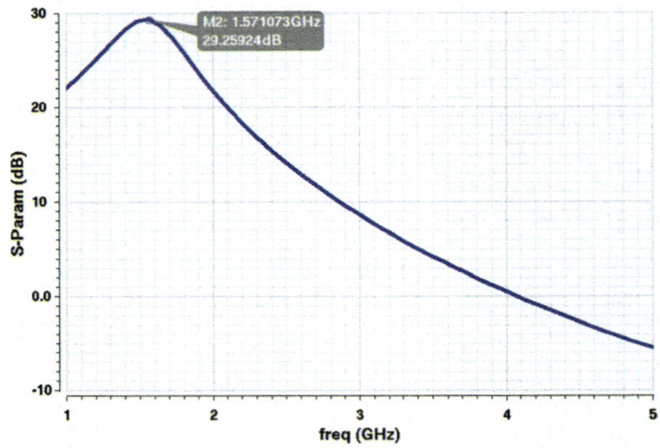

Fig. 6 Plot between Gain versus frequency (S21)

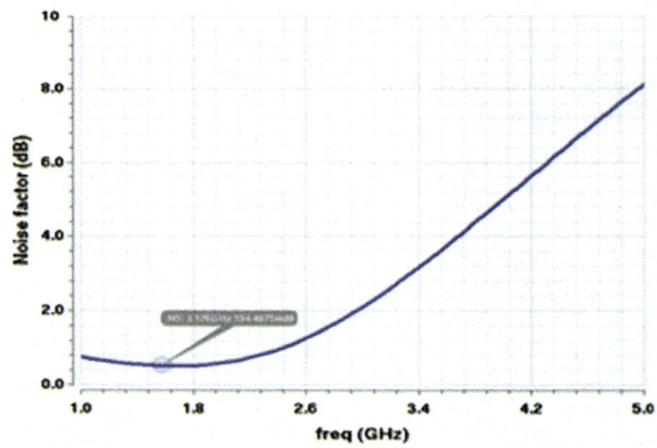

Fig. 7 Noise figure of circuit with of 1.5 V in voltage supply

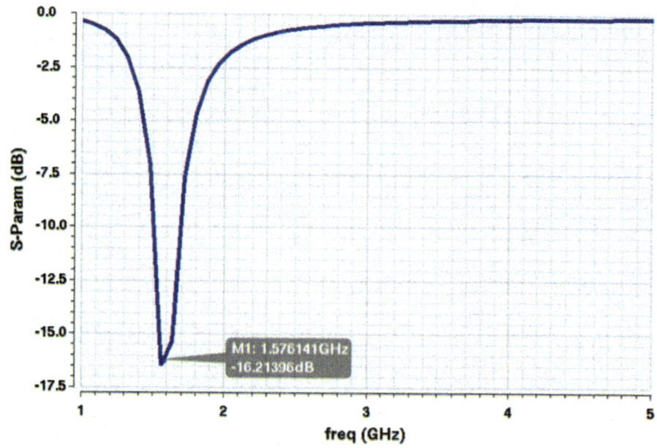

Fig. 8 Input return loss S11 V/s freq

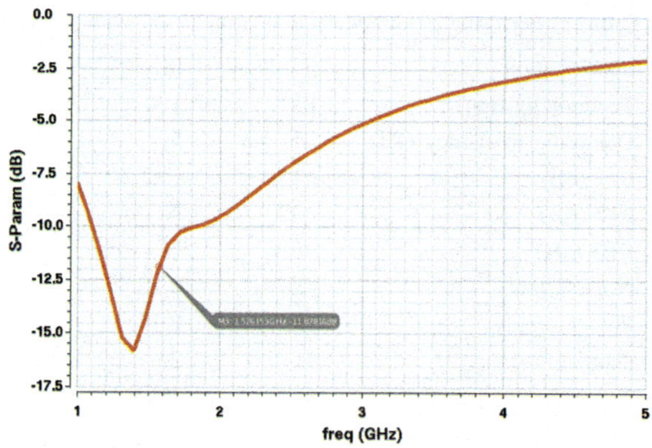

Fig. 9 Input return loss S22 V/s freq

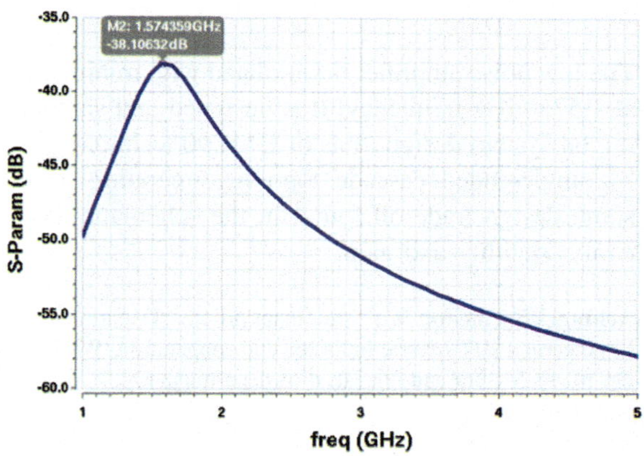

Fig. 10 Reverse isolation S12 versus frequency

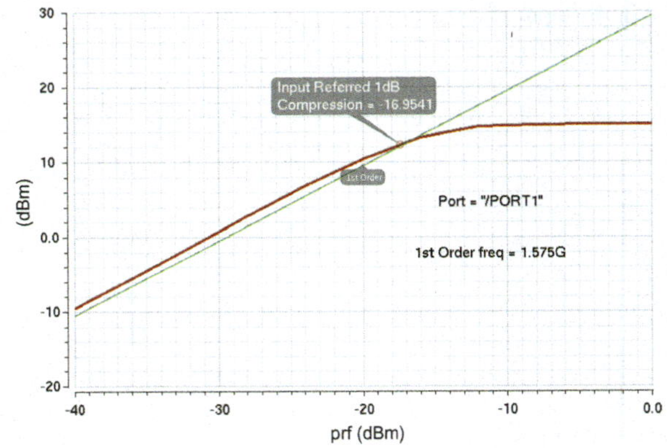

Fig. 11 1-dB compression point

Table 2 Performance summary and comparison with recent papers

Specification	[7] 2014	[8] 2012	[9] 2012	[4] 2014	This work
Technology (nm)	130	180	90	90	90
Operating frequency GHz	3.5, 5.8	5	1.575	2.45	1.575
S11 (dB)	−13.1	−12.2	−16.34	−9.19	−16.21
S12 (dB)	−32	N. A	−20	−38.03	−38.10
IIP3 (dB)	−5.4	N. A	1.655	−5.70	2.9140
Noise figure	4.2	2.22	0.933	2.34	0.533
Power supply	1.2	2.14	0.7	1.2	1.5
Gain	11	25.3	14.64	31.53	29.25
Power dissipation	9.36	N. A	35	8.5	8.7

4 Conclusion

The low noise amplifier is simulated with minimum noise figure and maximum gain at 1.575 GHz frequency. It is designed and simulated through cadence analog and digital system design tools of UMC90 technology. The increase in gain is achieved by single-ended cascade topology operating in sub-threshold region. There are some design trade-off found in the schematic designs like between gain and linearity, stability and gain.

Acknowledgements We are thankful to Department of Electronics & Communication Engineering, BIT Mesra, India for this cooperation. We are also thankful to our Vice-Chancellor, Dr. M. K. Mishra and our Head of department Dr. V. R. Gupta for their constant inspiration and encouragement.

References

1. B. Leung, VLSI for Wireless Communications, Prentice Hall Electronics and VLSI series Charles G.Sodini
2. T.H. Lee, *The Design of CMOS Radio Frequency Integrated Circuits, Cambridge* (Cambridge Univ. Press, U.K., 1998)
3. S. Ziabakhsh, H. Alavi-Rad, M.C.E. Yagoub, A high-gain low-power 2–14 GHz ultra-wide band CMOS LNA for wireless receivers. Int. J. Electron. Commun. (AEU) **66**, 727–731 (2012). doi:10.1016/j.aeue.2012.01.002
4. R. Kundu, A. Pandey, V. Nath, A CMOS low noise amplifier based on common source technique for ISM band application. Microsyst Technol. DOI 10.1007/s00542-015-2550-Accepted 16 Apr 2015 © (Springer-Verlag, Berlin Heidelberg, 2015)
5. M.T. Hsu, Y.C. Chang, Y.Z. Huang, Design of low power UWB LNA based on common source topology with current-reused technique. Microelectron. J. **44**, 1223–1230 (2013). doi:10.1016/j.mejo.2013.08.008
6. B. Razavi, CMOS technology characterization for analog and RF design. IEEE J. Solid-State Circuits **34**, 268–276 (1999)
7. A. Zokaei, A. Amirabadi, A 0.13 μm dual-band common-gate LNA using active post distortion for mobile WiMAX. Microelectron. J. **45**, 921–929 (2014). doi:10.1016/j.mejo.2014.03.012
8. R. Kumar, Design and noise optimization of rf low noise amplifier for ieee standard 802.11a WLAN. Int. J. VLSI Design Commun. Syst. (VLSICS) **3**(2) Apr 2012
9. B. Najeemulla, D.S. Chandu, B. Satish, Design and analysis of a CMOS 0.7 V low noise amplifier for GPS L1 band. Int. J. Eng. Innovative Technol. (IJEIT) **2**(5) Nov 2012

Design of Ultra-Low-Power CMOS Class E Power Amplifier

Jyoti Singh, Megha Agarwal, Vinita Mardi, Madhu Ray, Deepak Prasad, Vijay Nath and Manish Mishra

Abstract This paper is proposed to design an ultra-low-power CMOS class E power amplifier circuit to analyze its power gain and output power in periodic steady state (PSS) response. A technique is presented to facilitate the control power of the class E power amplifier (radio frequency power amplifier). The basic circuit of RF PA is designed which has different switching actions as different values of capacitors are taken into account. A driver F stage has been added in the basic circuit which increases the switching action considerably. When the voltage becomes high, the current decreases and when the voltage decreases, the current becomes high. Therefore, by using above technique, the power consumption is minimized upto great extend. The efficiency of power amplifier obtained is 78% and power gain is 60 dB.

Keywords Complementary metal oxide semiconductor field effect transistor (CMOS) power amplifier · Power amplifier · Figure of merit · Power gain

1 Introduction

An increase in the popularity of wireless communication has resulted in an increasing demand for compact, low power portable transceivers. The manufacturing cost was high in earlier days, but now the cost of fabrication and formation is much low due to mass production. The solution proposed was use of single-chip

J. Singh (✉)
Department of ECE, PES Institute of Technology, Bangalore, Karnataka, India
e-mail: jyoti6242@gmail.com

M. Agarwal · V. Mardi · M. Ray · D. Prasad · V. Nath
VLSI Design Group, Department of ECE, Birla Institute of Technology Mesra, Ranchi, Jharkhand, India
e-mail: prasaddeepak007@gmail.com

M. Mishra
Department of Electronics, DDU University, Gorakhpur, UP, India

V. Nath (ed.), *Proceedings of the International Conference on Microelectronics, Computing & Communication Systems*, Lecture Notes in Electrical Engineering 453,
https://doi.org/10.1007/978-981-10-5565-2_28

radio transceivers [1]. The maximum power consumption is within the transmitter; if this drawback is fixed, we can lower the power consumption rate and increase its efficiency [2]. The power amplifiers are the large signal amplifiers which supply greater ac power to load because it internally converts the dc power obtained from biasing supply into ac power [3]. Its output current and voltage vary by large amount which makes possible to obtain greater ac output current and voltage. Here, output current lies in ampere while voltage lies in 10's of volt. The performance of power amplifier is measured in terms of efficiency, figure of merit, and amount of harmonic distortion. Conversion efficiency is a measure of ability of power amplifier to convert dc power into ac power [4–6].

Mathematically expressed as,

$$\%\,\text{efficiency} = \frac{\text{ac power supplied to load}}{\text{dc power obtaied from bias supply}} \times 100 \tag{1}$$

Figure of merit (FOM) is the ratio of maximum power dissipated in the transistor to the maximum ac power which can be supplied to load. Ideally figure of merit (FOM) should be zero.

$$\text{FOM} = \frac{P_{D,\max}}{P_{\text{ac},\max}} \tag{2}$$

where

$P_{D,\max}$ is the maximum power dissipated and $P_{\text{ac},\max}$ is the maximum ac power which can be supplied to load.

In a large signal amplifier, transistor behaves as nonlinear element due to which harmonic or nonlinear distortion will appear in output. The total ac power supplied to load is

$$P_{\text{out}} = P_1 + P_2 + P_3 + \cdots \tag{3}$$

where

P_1,P_2, P_3 are the power supplied at fundamental 1st, 2nd, and 3rd harmonic frequency

$$\begin{aligned} P_{\text{out}} &= \left[\frac{B_1}{\sqrt{2}}\right]^2 R_L + \left[\frac{B_2}{\sqrt{2}}\right]^2 R_L + \left[\frac{B_3}{\sqrt{2}}\right]^2 R_L + \cdots \\ P_{\text{out}} &= \left[\frac{B_1}{\sqrt{2}}\right]^2 R_L \left\{1 + \left[\frac{B_2}{B_1}\right]^2 R_L + \left[\frac{B_3}{B_1}\right]^2 R_L + \cdots\right\} \\ P_{\text{out}} &= P_1\{1 + D_{\text{H2}}^2 + D_{\text{H3}}^2 + \cdots\} \end{aligned} \tag{4}$$

where

$$D_{Hn} = \frac{B_n}{B_1}$$

$$P_{out} = P_1\left\{1 + D_H^2\right\}$$

where

$$D_H = \sqrt{D_{H2} + D_{H3}} + \cdots$$

D_H is called amount of harmonic distortion.

Class A power amplifier has operating point approximately at the center of the load line. It is mostly used as small signal amplifier. When ac input is applied both output current and voltage can vary symmetrically w.r.t dc value. Hence, output current and voltage waveform will be undistorted. Its figure of merit (FOM) is 4. The advantages of class A power amplifier is to get distortion less output while large power dissipation is its biggest drawback.

In class B power amplifier, operating point is located in edge of cutoff or saturation region. If operating point lies in cutoff region, output current varies during positive cycle of input but remains zero during negative cycle, and if operating point lies in saturation region, then output current varies during negative cycle of input but remain constant during positive cycle of input. Its power dissipation is negligible but its output waveform is half sinusoidal which is considered as distorted waveform. It is used as untuned (register works as load) power amplifier [7, 8]. Class C power amplifier is operated such that output current varies for less than one half cycle of the input signal. Practically, its efficiency is not more than 90% and figure of merit (FOM) is equal or less than 0.25.

In case of class D, E, and S, the transistors act as switch and not a current source. Since an ideal switch has zero voltage across it or zero current through it, the switches dissipate no power. Class E power amplifiers have a single switching transistor connected to passive load network. Because class E uses capacitance shunting across the switch to shape the voltage and current waveforms, it avoids the power loss due to charging and discharging the capacitance, thus achieving a better efficiency than other power amplifiers [2].

2 Methodology

The basic circuit of Class E power amplifier composes of a single-pole switch, the driver stage, load network, and the load shown in Fig. 3. The load consists of a series LC resonant circuit, DC drain power supply, and a DC shunt capacitor. The network preceding the load network is the switch. From Fig. 1a and b, it is clear

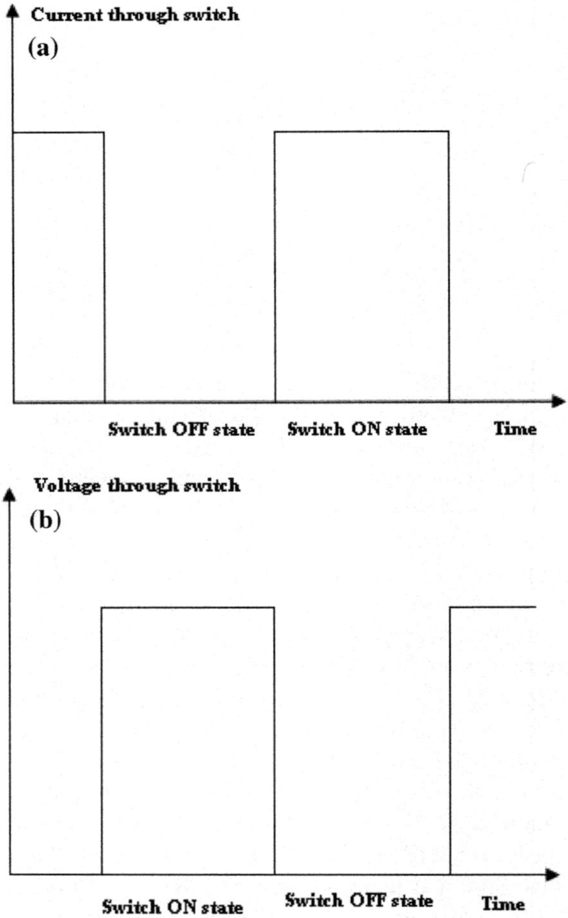

Fig. 1 **a** Ideal switching action of amplifier, **b** ideal switching action of amplifier

that when the switch is in ON stage, the current flows with no voltage drop across the transistor whereas when the switch is OFF, there is no current flow but an induced voltage. Waveforms describing the relationship between voltage and switch and current and switch suggest that at the rise of one, the other should fall.

The supply voltage of this amplifier is given by the following equation.

$$V_{CC} = \frac{\mathrm{B.V_{CEV}}}{3.56}.\mathrm{SF} \tag{5}$$

where SF is the safety factor and BV_{CEV} is the breakdown voltage of the MOSFET device [4].

The load resistance can be calculated as follows:

$$R_L = \left\{\frac{V_{CC}}{P_{out}}\right\}^2 0.576801\left(1.001245 - \frac{0.451759}{Q_L} - \frac{0.402444}{Q_L^2}\right) \quad (6)$$

The shunt capacitance can be given by the following formula:

$$C_1 = \frac{1}{2\pi fR\{\frac{\pi^2}{4}+1\}\frac{\pi}{2}}\left(0.99866 + \frac{0.91424}{Q_L} - \frac{1.03175}{Q_L}\right) + \frac{0.6}{(2\pi f)^2 L_1} \quad (7)$$

Or,

$$C_1 = \frac{1}{34.2219fR}\left(0.99866 + \frac{0.91424}{Q_L} - \frac{1.03175}{Q_L}\right) + \frac{0.6}{(2\pi f)^2 L_1}$$

L1 can be calculated from the equations that follow:

$$X_{L1} > 30 * X_{C1}$$

Or,

$$\omega L_1 > \frac{30}{\omega . C_1}$$

Or,

$$\omega L_1 > 30.26845$$

Or,

$$2\pi f_0 L_1 > 805.353$$

Or,

$$L_1 > 128.18n\text{H}$$

The equation for calculating C_2 is given as follows:

$$C_2 = \frac{1}{2\pi fR}\left(\frac{1}{Q_L - 0.104823}\right)\left(1.00121 + \frac{1.01468}{Q_L - 1.7879}\right) - \frac{0.2}{(2\pi f)^2 L_1} \quad (8)$$

The driver F stage is added to provide suitable switching action to the basic design. The circuit consists of two LC circuits, two bypass capacitors at the input and output, a depletion mode MOSFET. The inductor is used to change the dc value of the class F output to suit the value of the CMOS.

3 Power Control Techniques

Power control technique is one of the most important techniques used to improve the efficiency of the power amplifier which is shown in Fig. 2. Supply voltage power control technique (SVPCT) is conventionally used in case of switch mode power amplifier.

A 24–34 dB power control range is needed by the GSM900 standard (GSMK modulation) for a mobile station specified by European Telecommunications Standard Institute (ETSI) while in case of DCS1800 and PCS1900 frequency bands; standard requires 24–36 dB power control range (PCR) [5].

$$\mathrm{PCR}[\mathrm{dB}] = P_{\mathrm{out,max}} - P_{\mathrm{out,min}} = 20\log_{10}\frac{V_{\mathrm{dd,max}}}{V_{\mathrm{dd,min}}} \tag{9}$$

where

$$P_{\mathrm{out,max}} = \text{maximum average output power in dBm}$$

$$P_{\mathrm{out,min}} = \text{minimum average output power in dBm}$$

There are few drawbacks like limited output power control range, high sensitivity to load variations which create negative aspect.

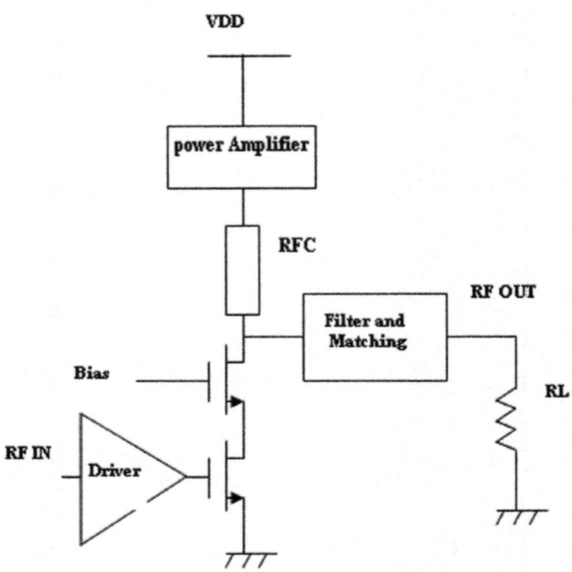

Fig. 2 Power control technique

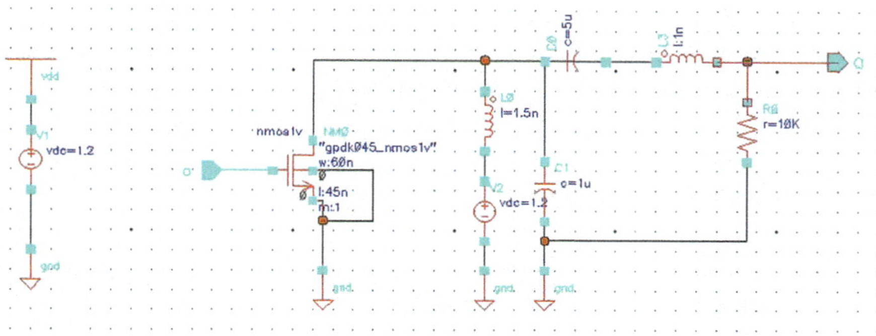

Fig. 3 Basic class E amplifier circuit

4 Proposed CMOS Class E Amp Design

The circuit shown in Fig. 4 shows the switching action of the amplifier with time. The vertical arrow indicates the time of transistor turn on.

The time response depends on the Q on the network. The class E power amplifier is a nonlinear device and observes a trade-off between efficiency and output power. In standard cases, the Q of the output network must be sufficiently small.

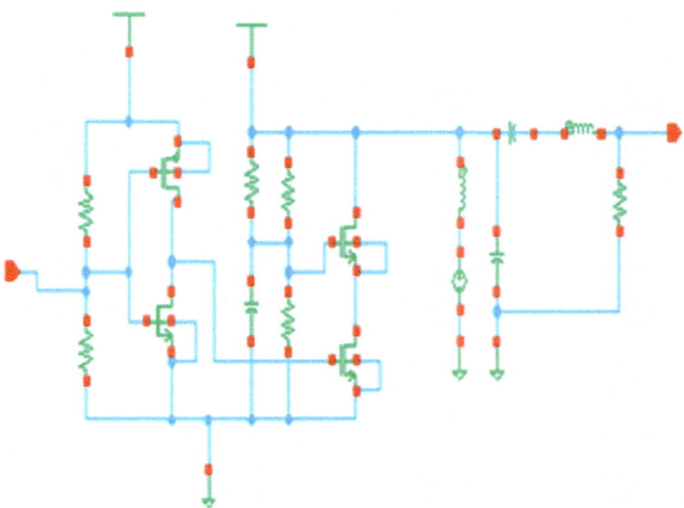

Fig. 4 Class E power amplifier with driver stage

The two LC networks resonate at third and fifth harmonics of the input frequency. The inductance of the final inductor is made to resonate with input capacitance so there is no drop in the signal.

5 Result and Discussion

The operation of the power amplifier in terms of power amplifier efficiency and output power is shown in Figs. 5, 6, and 7.

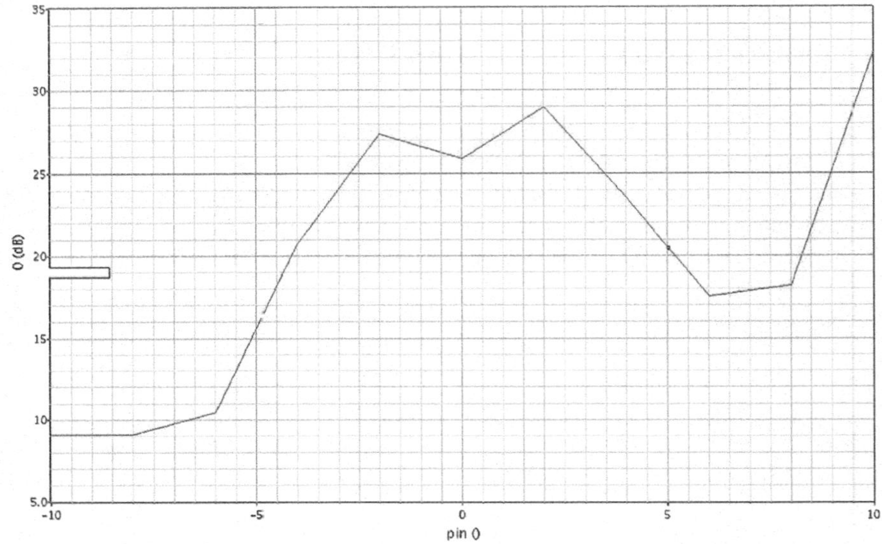

Fig. 5 Power added efficiency of proposed class E power amplifier

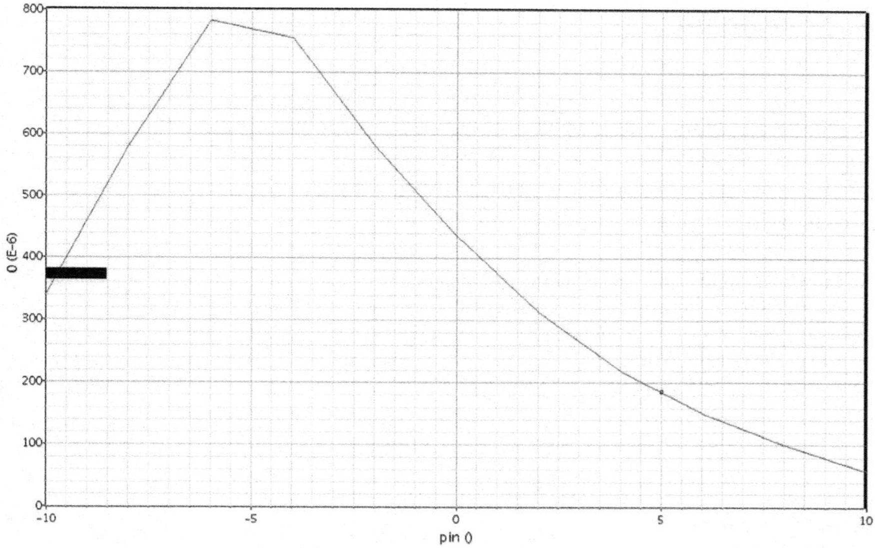

Fig. 6 Periodic steady state response of the amplifier

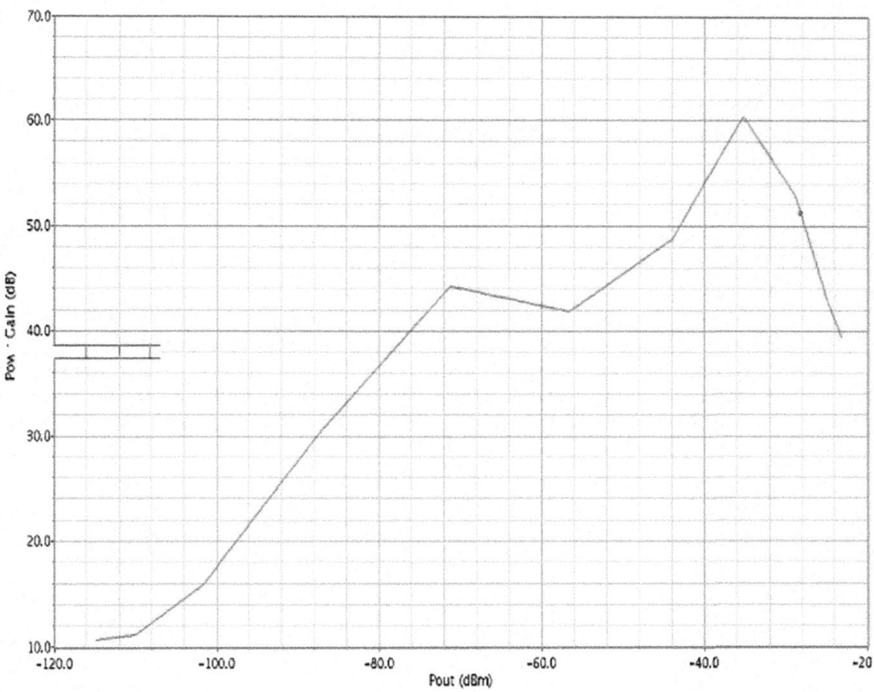

Fig. 7 Power gain versus Pout of the amplifier

6 Conclusion

The design proposed here has better output power gain and efficiency as compared to the works earlier. The proposed design has low power consumption and hence is applicable in radio and wireless communications, biomedical applications. The efficiency of power amplifier obtained is 78% which is shown in Fig. 5 and power gain is 60 dB which is shown in Fig. 6.

Acknowledgements We are thankful to Department of Electronics and Communication, BIT, Mesra, India for kind cooperation. We are also thankful to Vice Chancellor Dr. M. K. Mishra and head of department Dr. V Gupta for encouragement and support. We are also thankful to Principal, PES Institute of Technology Dr. K. S. Sridhar and Head of Department, Electronics and communication, PES Institute of Technology Dr. T. S Chander for providing us with suitable resources.

References

1. S.C. Cripps, *RF Power Amplifier For Wireless Applications*, 2nd edn. (Artech House, Boston, MA, 2006)
2. B. Razavi, CMOS Technology characterization for analog and RF design. IEEE J. Solid State Circuits **34**, 268–276 (1999)
3. R. Kumar, V. Nath, *Design of Low Power CMOS Amplifier*, National Conference on Frontiers Electronics, Communication & Instrumentation Technology (FECIT-2011) organized by Dept. of ECE, ISM Dhanbad, India from 3–4th Nov 2011
4. J. Tan et al., Design of efficient class E power amplifiers for short distance communications. IEEE Tran. Circuits Syst.-1, Regular Papers **59**(10) Oct 2012
5. S. Sivakumar, A. Eroglu, Analysis of class-E based RF power amplifier using harmonic modelling. IEEE Trans. Circuits Syst. I, Regular Papers **57**(1), 299–311 (2010)
6. R. Brama, L. Larchaer, A. Mazzanti, F. Svelto, A 30.5 dBm 48% PAE CMOS class E power amplifier with integrated balun for RF Sapplications. IEEE J. Solid State Circuits **43**(8), 1755–1762 (2008)
7. Y. Song, S. Lee, J. Lee, S. Nam, *A 29 dBm, CMOS Class E Power Amplifier with 63% PAE Using Negative capacitance.* IEEE Custom Integrated Circuits Conference, INMC School of Electrical Engineering and Computer Science, Seoul National University, Seoul Korea 2009
8. P. Maniknandan, R. Mathew, *Design of Class E Power Amplifiers for WLAN and Bluetooth applications.* IEEE International Conference Devices Circuits and Systems, Coimbatore, Mar 2012

Effect of Temperature on Dark Current in QWIP for Unmanned Aerial Vehicles

Vishal Kumar and R. K. Lal

Abstract This paper deals with results of optimizing the structure and temperature effects leading to dark or noise current mitigation in quantum well IR photodetector (QWIP) using mathematical modeling. The quantum wells are formed by heteroepitaxial process where a narrow E_{GAP} material between wide bandgap materials. Results show that the fine tuning of aluminum (Al) mole fraction and well-width helps in achieving high responsivity for the both near and far IR wavelength. Low noise operation, as well as comparative study, is done between the experimental and theoretical value for temperature analysis. The modeled QWIP detector consists of GaAs quantum wells and $Al_xGa_{(1-x)}As$ barriers. The temperature-based operation and analyses show the cause of band splitting, and reduction of noise is observed in MQW IR sensor structure. This type of QW finds application in broadband sensors used in unmanned aerial vehicles (UAV).

Keywords Quantum well infrared photodetector QWIP · Dark current
Temperature · Optical efficiency · Quantum confinement · Minibands

1 Introduction

In the last 10–20 years, there have been many developments in research and funding activities due to advancement in growth and fabrication of quantum well-based infrared sensors [1–4]. The IR detectors have large commercial and defense applications including application in the field of astronomy. The working of quantum well infrared photodetectors (QWIPs) is based on the excitation of carriers mainly electrons in a quantum well by near to far IR signals. A repetition of well-barrier gives array or superlattice sensors. For sensitive QW detectors, doping lattice-matched quantum well is used create carriers in the dormant state using

V. Kumar · R. K. Lal (✉)
Department of E.C.E, B.I.T, Mesra, Ranchi, Jharkhand, India
e-mail: rklal@bitmesra.ac.in

V. Nath (ed.), *Proceedings of the International Conference on Microelectronics, Computing & Communication Systems*, Lecture Notes in Electrical Engineering 453,
https://doi.org/10.1007/978-981-10-5565-2_29

either n-type (for the conduction band well) or if p-type sensor (for the valence band well) is used.

The importance of QWIPs in a wireless network sensor system can be understood by considering a system which requires a highly sensitive receiver device. In military UAV, wireless network is made up of several components which include transmitter, antennas, and receiver. For high sensitivity and selectivity, a QWIP lens is used in the receiver system as a peripheral device for the detection of a signal as it can detect a multitude of wavelengths with accuracy and lesser noise as compared to conventional detectors like MCT. QWIPs have proved to be a quintessential element of target tracking systems of aerodynamics and surveillance systems for military applications.

The dark current is an un-desired parameter in photodetectors, and the consequence of minimization of dark current showed the enhanced detectivity over the entire operating range.

For temperatures, under study, above 45 K, the dark or noise current of the QWIP is almost genuinely dominated by classical or bulk effects such as thermionic or thermodynamic emission of ground state electrons from well into the energy continuum (Fig. 1).

Among several theoretical techniques studies for of minimizing its effects [3], one important method is the band gap engineering of the top or bottom (for CB or VB) excited state to align with the kink in the band gap or Eg. The work is based on the temperature modeling of the quantum well devices in which we will analyze the system under consideration for a varied range of temperature. The reported work utilized mathematical tools MATLAB and Atlas TCAD.

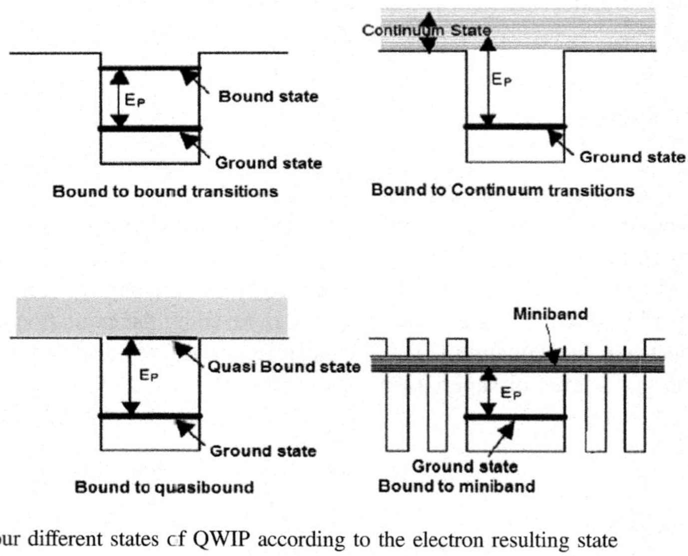

Fig. 1 Four different states of QWIP according to the electron resulting state

2 Device Structure

The classical quantum well structure forms a crystalline wurtzite lattice, which is naturally occurring, whereas under certain growth conditions can form a zinc-blend structure. In present work, for GaAs/AlGaAs, the following approach is applied.

i. An electron in a sensor QW is the particle-in-a-well example, and its solution is given by Schrodinger's equation. Such a square QW is desirable in the lattice-matched GaAs/AlGaAs material family by using a layer of GaAs between two layers of A1GaAs. The depth of the potential well (=the height of the potential barrier) is normally controlled by using the aluminum mole fraction (x) in the AlGaAs barrier/cap layers.
ii. The quantum well stack of Fig. 2 consists of GaAs quantum well layers and $Al_xGa_{(1-x)}As$ barrier layers. The combination of well and barrier layers is repeated for thirty times. The well-width is 59.4 Å, and barrier width is 300 Å with aluminum mole fraction x as 0.22. This choice is the result of reported papers and simulations done for this purpose. If layers numbers are increased, then the device footprint becomes larger and lower number results in lesser 2DEG and mobility.
iii. The design consists of deposition of repeated sequence of $Al_{0.22}Ga_{0.78}As$/ GaAs (59.4 Å)/$Al_{0.22}Ga_{0.78}As$. The silicon doping concentration in GaAs well is assumed to be 1.53×10^{18} cm^{-3}, which is derived from optimal value as higher values can lead to degenerate doping and carrier punch through.
iv. Table 1 shows the design of the structure, and Fig. 2 shows the multilayered format in which a calculated number of 2DEG cloud is formed by implanting the GaAs well with Si (an n-type dopant) during the MBE growth cycle. As layer by layer implant provides Gaussian distribution tapering off at deeper levels.

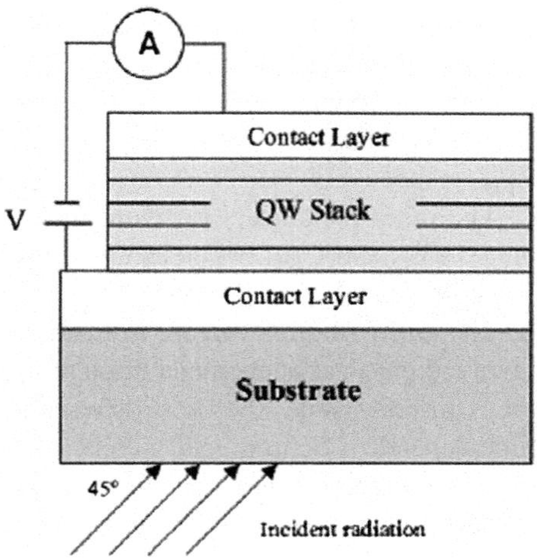

Fig. 2 Structure of proposed format in stacked form

Table 1 Typical QW parameters value for proposed structure

Typical QW	Parameter
Contact layer	$x = 0.22$, mole fraction
GaAs	300 Å
AlGaAs	59.4 Å
GaAs	300 Å
Contact layer	$n = 1.53 \times 10^{-18}\ \text{cm}^{-3}$, doping
Substrate	Sapphire

v. Several quantum wells (interspaced between barrier layers) are generally grown one on top of each other to increase sensitivity for photon collection. The higher range of this number for a commer detector structure in this material family is around 50–75, about the number of well layers that a photoelectron can pass through device internal electric field without being captured by another deep level well downstream from the well which the electron was emitted leading to reduction of quantum efficiency.

vi. The entire MQW is fabricated between heavily doped top (emitter) and bottom (called the collector) GaAs epilayers which provide electrical contacts to detector array.

vii. Photon absorption excites an electron from the ground state to the first excited state close well top, where it drifts out into the continuum (continuous energy levels in device) aided by an external voltage bias, leading to a sufficient photocurrent above dark/noise current level.

Normally, one nanoscale layer (well) is fabricated between two or more layers of wider E_{GAP} bandgap, and it gives rise to confinement of carriers. As mentioned in the previous section, the charge and carrier profile are defined by the band offsets/minibands and the formation of quantized energy levels, which leads to confined e^- or h^+ inside the wells, whereas continuum states are normally closer to contacts (outside well array).

QW structures and superlattices allow for the band gap modification of quantum confinement effects that are reported to be very useful in electronic devices (QWIP) and MODFET heterostructures. For an ideal schematic representation of the electrostatic potential, made up of multiple quantum wells (MQWs) is illustrated in Fig. 3. The barriers surrounding wells are highlighted with bandgaps, $\mathbf{E_{gB}}$, and $\mathbf{E_{gW}}$, respectively.

The other parameters in figure include valence and conduction band discontinuity, VB_O and CB_O, whereas the valence and conduction bands confined energy levels $\mathbf{E_{vi}}$ and $\mathbf{E_{ci}}$, respectively.

The utility of this structure from device point of view requires knowledge of the involved physical phenomena through mathematical models [5, 6]. The behavior of the carriers inside the heterostructure is calculated by modification of the time-dependent Schrodinger's equation along with the appropriate boundary conditions. Complementarily, density of states, subband carrier population (useful in sensors), charge distribution potential (for array type devices or superlattices),

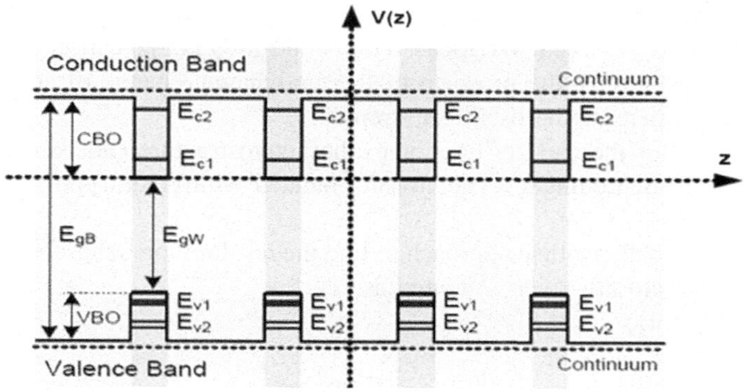

Fig. 3 Schematic diagram of a heterostructure potential profile V(z)

transition rates, and effects of external fields (cause significant in polarization between layers), among others, have to be worked out in order to allow for practical simulation, comprehension and required control in designing devices and planning fabrication procedures.

Although noise can be caused by interference of spectral patterns, under ideal MQW growth conditions with sidewall passivation, the structure itself causes reduction in it. The confinement due to separation by barrier layer between two groups of MQ wells of quantum well structure also reduces interference.

3 Temperature Simulation Methodology

The behavior of QW IR sensors under various conditions of operation has been modeled through simulations and for improvement of the performance of QW IR sensors, and formulated temperature-based performance was partially achieved.

The temperature analysis of the QWIP has shown the increase of the operating temperature from as low as 77 to 180 K without compromising the on device characteristics by implementing dynamic resistance.

This novel device shows that the alteration of the mole fraction along with the well-width, and barrier height can show higher sensitivity and selectivity of the device, which is of utmost importance in military applications where two color or multispectral resolution has been reported eight.

Temperature-dependent emissivity variations for reported QWIP translate to contrast effecting color bands for typical visible image, revealing objects at different temperatures and if there is relative motion, the IR contrast color translates to differential motion. Reducing the dark current is very important to the results of the IR sensor since it enables the highly sought after; increased temperature detector or sensor operation. Through our study of various material properties of MCT and GaAs–AlGaAs system [8], we found much scope for improvement in AlGaAs-based QW systems.

In our modeling, we used broadband multiquantum well structure and the photocurrent was observed. We focused on bound-to-Q bound carrier transmission, as it helps to minimize the prospect of thermodynamic change of state (to some extant), which increase the oscillator strength.

The activity of the carriers interior to the heterostructure can be described by modification of Schrodinger's equation formulated with grade-appropriate boundary conditions.

The general mathematical approach to find the result of the Schrodinger equation used with standard nth-layer is written as:

For even parity:

$$\mathrm{KL}/2 = [n\pi] + a \tan\left[\frac{m_\mathrm{w}\,\alpha}{m_\mathrm{b}\,k}\right] \qquad n = 0, 1, 2$$

For odd parity:

$$\mathrm{KL}/2 = \frac{\pi}{2} + [n\pi] + a \tan\left[\frac{m_\mathrm{w}\,\alpha}{m_\mathrm{b}\,k}\right] \qquad n = 0, 1, 2$$

where

$$\mathrm{K} = \sqrt{(2m_\mathrm{w}{*}\mathrm{E})}/\hbar \quad \text{and} \quad \alpha = \sqrt{(2m_\mathrm{b}{*}(\mathrm{V}-\mathrm{E}))}/\hbar$$

The effective mass for well region and barrier regions is taken as $m_\mathrm{w} = 0.067\ m_\mathrm{o}$, $m_\mathrm{b}^* = (0.0665 + 0.0835\times)\ \mathrm{m_o}$, where m_o = rest mass of electron.

While the higher-frequency region of the electromagnetic spectrum comprising of gamma rays, X rays, and ultraviolet rays (UV rays) on the far side the blue end of the visible range, and IR rays (bridging a wide wavelength region from 0.3 μm to 1 mm) and microwave signals after the red wavelength thus light sensing element operating in the 4–20 μm wavelength range are particularly of commercial significance. A working temperature of 105 K or well below it is not uncommon for externally cooled or heat dissipating industrial QWIPs [9]. The dark current is a function of device noise, ambient temperature, while internal temperature affects carriers so while the outcome remains in essence, unchanged but device characteristics need to be designed for better temperature management.

By using the broadband QWIPs, the dynamic resistance of the QW is used to find a point of intersection of graph. The graphs clearly depict that as the temperature is increasing the dark current which is increasing marginally slowly, and the $\mathrm{R_oA}$ factor is reducing which helps in minimizing the dark current.

It should be noted that interwell tunneling, which could be neglected due to the designed barriers, involves electrons scattering from the un-excited state in one quantum well (QW) and flowing to the next. Second, thermionic emission (TE) refers to thermal excitation of electrons from the upper part (with energy greater than the barrier height) of the ground state to the non-confined continuum on the top of the barrier.

The analysis and methodology validate the reduction of dark current up to half of its original value which has an effect in the detectivity of the Q well under discussion.

4 Sensor Current Analysis

In addition to the detectable optical current, all infrared region (near IR and far IR) detectors including reported detector array produce a un-desired current known as dark or noise current, which should be minimized so as to produce high-functional parameters. *The dark or noise current can be defined, current passing within the detector when no excitation is present (i.e., with no photons are incident on it) and is normally assumed to be zero* (Fig. 4).

Types of dark current:

a. thermionic dark current,
b. sequential dark current,
c. thermally assisted dark current.

Method of reduction of dark current:

i. A standard procedure to reduce the dark current due to thermal or temperature-dependent emission, for a range of ambient temperature, so as to optimize the performance of reported MQW array in reported detector is to utilize ***bound-to-quasi-bound intersubband absorption*** **(such as for the first excited state in QW to be in resonance with the top of the well)**.
ii. This mode of carrier generation is used to maximize the intersubband or internal QW absorption while maintaining higher-electron mobility for better 2DEG transport.
iii. The thickness and the dimensions of the QW IR array are designed in such a way that the QW detector contains only first bound level and the first excited level matches with the top of the barriers between the MQWs.

Fig. 4 Variation of RoA versus Temperature

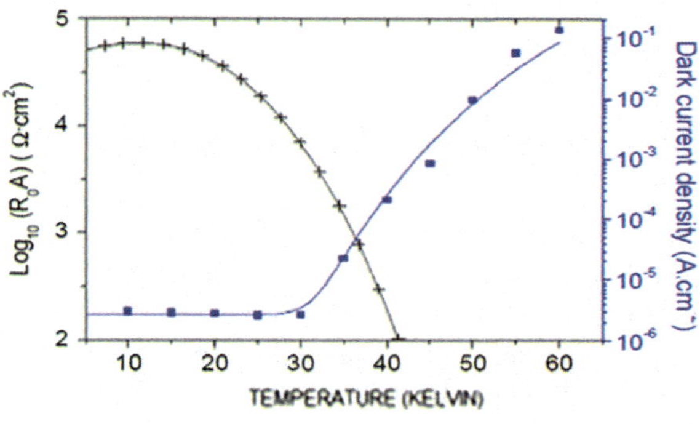

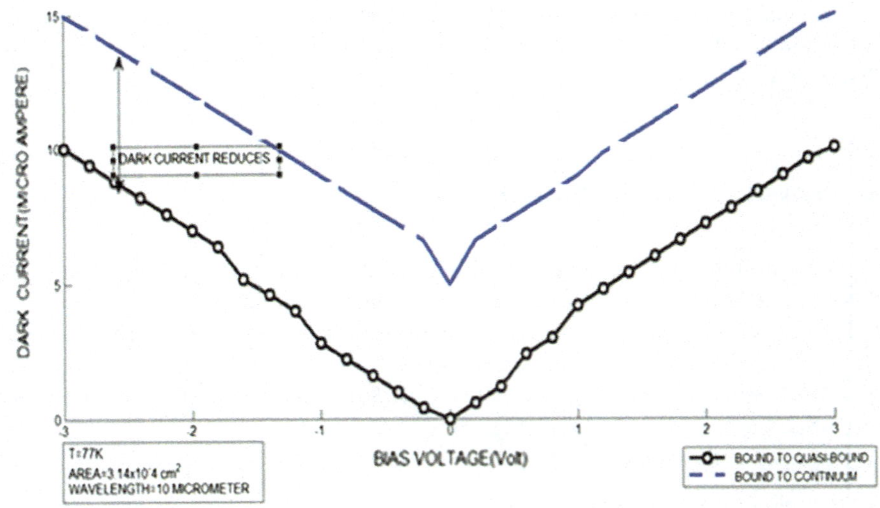

Fig. 5 Experimental comparison of dark current as a function of bias voltage in bound-to-continuum and bound-to-quasi-bound structure

These states, also known as continuum bound states or above barrier states, have properties of strong energetic and spatial localization similar to those of the discrete energy levels.

The graph of Fig. 5 clearly depicts that the dark current reduces substantially when the bound-to quasi-bound intersubband absorption is considered. The value of dark current decreases from 12 to 6 μA which shows the minimization of dark current to half the original value.

Figure 6 shows the effect of dark current reduction on detectivity of the device which shows considerable enhancement in the device performance.

a. The experimentation validates the reduction of dark current up to half its original value which has an effect in the detectivity of the sample.
b. The simulation of detectivity in Fig. 5 shows the result of optimization of dark current.
c. The experimental result shows the peak at 25 Jones instead of 15 Jones formulated before.

5 Temperature Characterization of QWIP

An operation temperature of 65 K or further cooled operation is not uncommon for QWIPs. The dark current is explicit function of the temperature while the output remains more or less unchanged [10].

So the dark current is minimized by simulation of dynamic resistance of BB-QWIP and validating the results. Here we note that density of states causes

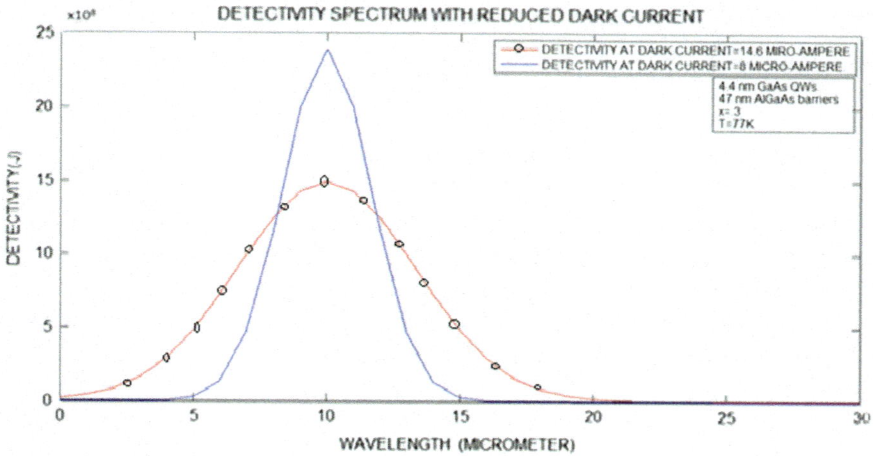

Fig. 6 Detectivity increases from 15 × 10^8 Jones to 25 × 10^8 Jones as a consequence of the minimization of dark current

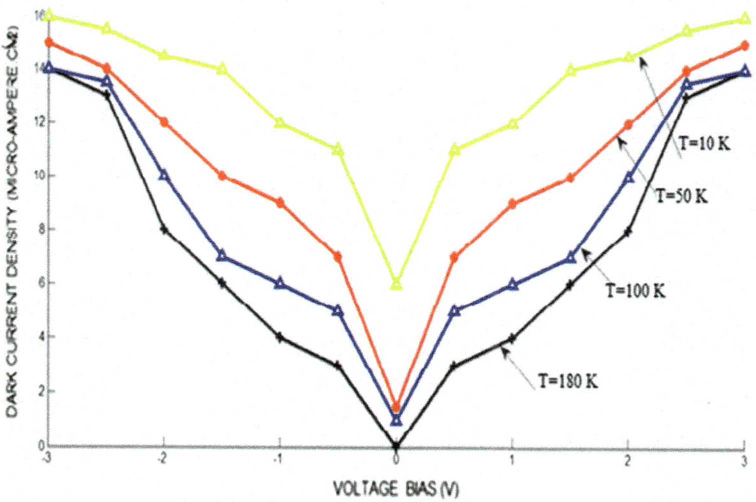

Fig. 7 Dark current as a function of the applied bias for different temperatures

splitting of bands into minibands as predicted by quantum mechanics, which requires multiple quantum well structure. The crystalline structure is assumed to be standard wurtzite of GaAs–AlGaAs system.

Figure 7 shows the simulated values of dark current at various temperatures using the effect of dynamic resistance. It is observed that at a voltage bias of 1 V, the minimum value of dark current is observed at 180 K.

Table 2 Dark current density at various temperatures

Temperature (K)	Dark current density (μA/Cm^2)
10	12.5
50	7.9
100	5.5
180	2.8

The experimentation of these temperature-dependent values shows the ability of tweaking the working temperature from 10 to 180 K successfully.

Simulation is carried out, and the results are tabulated in Table 2.

On using the BB-QWIP, the peak of the graph was observed within the range which shows enhancement in the device performance.

6 Conclusion

The region of QW IR sensors and device engineering is very large stretching from defense to space and astronomical applications. The advances in fabrication, material sciences, and semiconductor scaling technology through diverse and new fabrication methodologies have produces some novel devices for bandgap engineering to cover, all IR (near IR and far IR regions) with real-time tunability and suitable for mobile and stationary object segregation. Specifically, in the area of near room temperature operation or un-cooled operation using quantum well arrays, much research and analyses were done in the past 15–25 years to make these commercially possible and viable for mass production and several applications. The commercialization of technology is dependent on application, but cost plays major role. The quantized 2DEG from intersubband level is basically bound overall device mobility, and 2DEG continuum was altered to bound-to-quasi-bound which has given significant results which validated the minimization of dark current. The consequence of minimization of dark current showed the enhanced detectivity over the range of 10–15 μm. In temperature range, the limit of 180 K can be improved by suitable device cooling to reduce dark current in IR sensors used UAVs.

The temperature analysis of the QWIP has shown the increase of the operating temperature from as low as 77 to 180 K without compromising the on device characteristics by implementing dynamic resistance and broadband multiquantum well structure. This reduces additional cooling requirements in sensor applications such as used in unmanned aerial vehicles.

References

1. V. Ryzhiia, Characteristics of quantum well infrared photodetectors, vol. 62, ed. by H.C. Liu, F. Capasso, *Semiconductors and Semimetals, IEEE Xplore, 1997* (Academic Press, 2000), p. 197
2. A. Rogalski, J. Phillips, P. Bhattacharya, S.W. Kennerly, D.W. Beekman, M. Dutta, Infrared detectors: an overview. IEEE J. Quantum Electron **35**, 936–943 (2001)
3. D. Saeedkia, *Handbook of Terahertz Technology for Imaging, Sensing, and Communications*, Woodhead Publishing Series in Electronic and Optical Materials (Woodhead Publishing Limited, 2013)
4. A. Majumdar, E.L. Ginzton, K.-K. Choi, K.-M. Leung, T. Tamir, C. Monroy, Quantum well tera-hertz range infrared photo-detector in resonant cavity. IEEE J Quant. Electron. **40**(2), 130–142 (1999)
5. F.F. Sizov, V.V. Zabudsky, S.A. Dvoretsky, V.A. Petryiakov, A.G. Golenkov et al., MCT as sub-terahertz and infrared detector, in *Proceedings SPIE 9483, Terahertz Physics, Devices, and Systems IX: Advanced Applications in Industry and Defense, 94830V* (2015)
6. H. Schneider J. Fleissner, QWIP focal plane array (FPA) for the 8–12 and 3–5 μm regimes, NATO ASI Series Vol. E 270, *Quantum Well Intersubband Transition Physics and Devices*, ed by H.C. Liu, B.F. Levine, J.Y. Andersson (Kluwer Academic Publishers, Dordrecht, 1994), p. 55; 1999
7. J. Li, C.Z. Ning, S.D. Gunapala, S.V. Bandara, A. Singh, N.Q. Tran, J.D. Vincent, C.A. Shott, J. Long, P.D. LeVan, Microscopic theory and simulation of quantum-well intersubband absorption. Proc. SPIE **3698**, 687 (2001)
8. H.C. Liu, J.C. Cao, Terahertz quantum-cascade lasers based on a three-well active module. Appl. Phys. Lett. **90**(4), 041112 (2007)

Design of Circular Disc Monopole Antenna for UWB Application

Md Maqubool Hosain, Sumana Kumari and Anjini Kumar Tiwary

Abstract A new proposed circular disc monopole (CDM) antenna is designed for ultra-wideband (UWB) application. The dielectric substrate is used to print the antenna which is fed by 50 Ω CPW on the same substrate side. A stub is introduced across the feed line for improving the performances. In addition, the bandwidth (BW) is enhanced using modified ground plane. The simulated outcome shows that the antenna can give an impedance bandwidth of 2.58–12 GHz having reflection coefficient less than −10 dB. Also, the voltage standing wave ratio (VSWR) is less than 2 and the peak gain antenna is up to 4.5 dBi. A fine conformity is obtained between the simulation and the experiment.

Keywords Ultra-wideband (UWB) · Circular disc monopole (CDM) CPW-fed · Printed antennas · Stub · Bandwidth (BW) enhancement

1 Introduction

The revolutionary growth in the field of wireless communication has been experienced in last few years. This happened due to the invention of many wireless products and services such as wireless local area network (WLAN), Global Positioning System (GPS), mobile phone, Bluetooth. The lightweight, robustness and easy integration with planar circuits are the first requirements to lead the designers for communications systems. Also, in the last few years, the wireless communication system has increased significantly due to the development of UWB technology. Therefore, various modern wireless communication systems have

M. M. Hosain (✉) · S. Kumari · A. K. Tiwary
Department of ECE, Birla Institute of Technology, Ranchi, Jharkhand, India
e-mail: himaqubool@gmail.com

S. Kumari
e-mail: sumana_keya@rediffmail.com

A. K. Tiwary
e-mail: aktiwary@bitmesra.ac.in

V. Nath (ed.), *Proceedings of the International Conference on Microelectronics, Computing & Communication Systems*, Lecture Notes in Electrical Engineering 453,
https://doi.org/10.1007/978-981-10-5565-2_30

developed due to this technology. The broadband monopoles have certain advantages, such as wide bandwidths, simple structures, having satisfactory radiation properties and simple fabrication [1–3]. Since the radiators are perpendicular to its ground plane, their structures are not planar. Finally, their integration with the printed circuit board is not suitable. The realization of planar UWB monopoles is done by using either a microstrip line or CPW feeds [4–12]. A printed circular disc monopole [4, 5] has been proposed which is fed by microstrip line. The antenna operation affects the parameters in terms of characteristics for frequency domain. It has been examined both theoretically and experimentally to understand the antenna operation. It has been also explained that this type of antenna for optimum design can obtain a wide bandwidth for improved properties of radiation. In [6–12], a CPW-fed circular disc monopole antenna is presented which leads to complete the requirements of UWB characteristics. In this projected antenna, the effects of various parameters are studied. This antenna offers better flexibleness for circuit integration and has larger impedance bandwidth.

In this paper, the CPW-fed circular disc monopole is taken. The optimal operation of an antenna is analysed for design parameters. For this antenna, the performance and characteristics are also studied, both simulated and experimental. This type of antenna for optimal design shows satisfactory radiation patterns. This circular disc antenna offers not only the larger bandwidth of impedance but also a better tractability for integration of the circuit. In addition that, a stub is introduced across the feed line for matching the impedance and good performances. To extend the impedance bandwidth of an antenna, the ground plane (GP) is further modified.

2 Antenna Design and Performance

In Fig. 1, the structure parameters of CPW-fed CDM antenna are shown. The dielectric substrate FR4 is used for printing the antenna having a thickness (t) = 1.6 mm, loss tangent ($\tan\delta$) = 0.016 and relative permittivity (ε_r) = 4.4. The feed line width (W_f) = 3.05 mm which represents to a characteristic impedance of 50 Ω and gap (g) = 0.33 mm. Here, the feed gap between the GP and the disc is h. Also, L (10 mm) and W are the length and width of the GP, correspondingly. The Method of Moments based IE3D simulation software tool from Zealand, USA, is used for the simulation purpose.

Since, circular disc monopole (CDM) antenna structure has similar with planar circular antenna. The lower edge cut-off frequency is given by [13]

$$l = 0.24 \times \lambda \times F,$$

Fig. 1 Structure of CDM antenna with CPW-fed

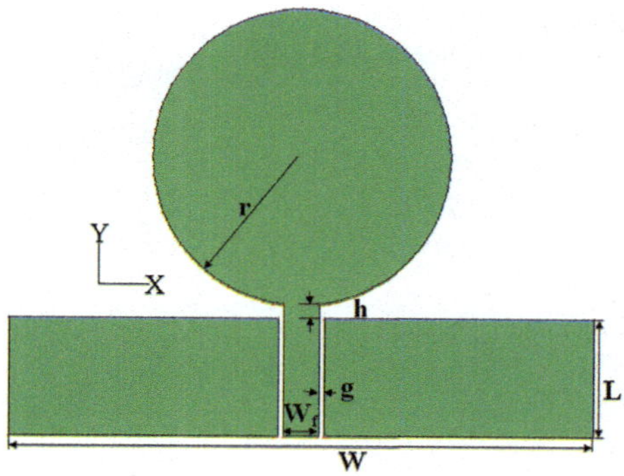

where

$$F = \frac{(l/r)}{(1 + l/r)}.$$

The resonant frequency from above equations is given by

$$f = \frac{c}{\lambda} = \frac{(30 \times 0.24)}{(l + r)} \text{GHz},$$

where l and r are in mm.

The parameters of circular disc monopole antenna are r = 12.5 mm, h = 1.1 mm, W = 47 mm and g = 0.33 mm as shown in Fig. 1, respectively.

Since, the circular disc monopole is exactly reckoned on the radius (r) of the disc, feed gap (h), the gap between ground plane and stripline (g) and also depending on the width of the ground plane (W). So, for getting maximum impedance bandwidth, these parameters are optimized.

3 Antenna Characteristics

3.1 The Variation of Feed Gap h

Figure 2 shows the replicated reflection coefficient graphs with different feed gaps (h = 1.0, 1.1, 1.2 and 1.3 mm) by keeping W, r and g as constant at 47, 12.5 and 0.33 mm, correspondingly.

Figure 2 plots the reflection coefficient curves which have almost identical shape for three distinct gaps, but with the variation of h, the antenna bandwidth changes drastically. It is also observed that the bandwidth increases with decreasing the h. The preeminent feed gap is found to be at h = 1.1 mm. In fact, the strengthening and weakening of capacitive coupling between the circular disc and GP depend upon the increase and decrease of h, respectively. This affects primarily at resonant frequencies of the disc.

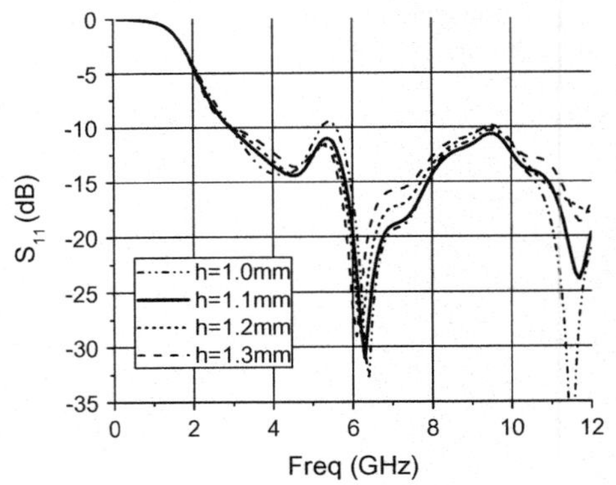

Fig. 2 Reflection coefficient graphs for different feed gaps

3.2 *The Variation of Width W of the Ground Plane*

Figure 3 plots the replicated reflection coefficient curves with the variation of width of coplanar ground plane (W = 45, 47, 49 and 51 mm) by keeping r = 12.5 mm, h = 1.1 mm and g = 0.33 mm as constant, respectively.

Figure 3 plots the reflection coefficient curves which vary to a great extent and no more have similar shapes. With the variation of W, the higher resonant frequency varies considerably, but the first resonant frequency almost constant. When W varies from 45 to 51 mm, it covers an ultra-wideband frequency for getting −10 dB bandwidth. From the simulation graph, the best result is found at W = 49 mm and its frequency ranges from 2.7 to 12 GHz. The impedance of GP is mainly determined by its width W. This is mainly due to the distribution of current along the y-direction because for a circuit, the ground plane (GP) serves as matching of impedance.

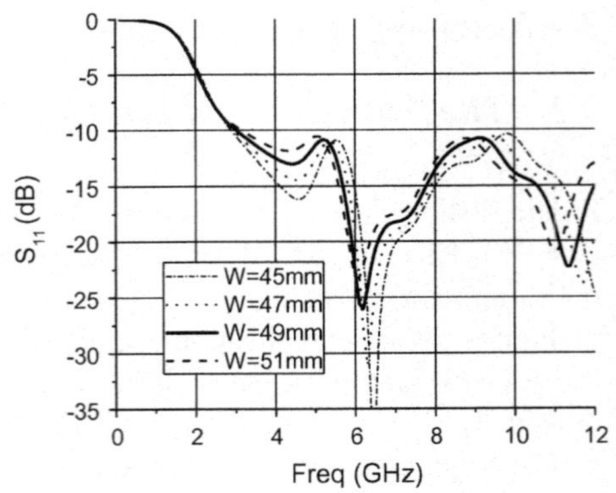

Fig. 3 Simulated reflection coefficient curves for variation of ground plane W keeping h = 1.1 mm constant and r = 12.5 mm

3.3 The Variation of Radius r of Circular Disc

The reflection coefficient graphs with the dissimilar radius of the disc ($r = 10.5$, 12.5, 16.5 and 21.5 mm) are shown in Fig. 4 by keeping $h = 1.1$ mm, $W = 49$ mm and $g = 0.33$ mm as constant, respectively.

From above curve, it is clear that the first resonant frequency decreases with increasing the radius r of the disc. It is also noticed from the simulation graph that at $r = 12.5$ mm, an ultra-wide frequency band covers only −10 dB bandwidth, i.e. from 2.7 GHz to more than 12 GHz.

Here, for the proposed antenna, the dimensions r and h are selected in such a manner so that the first resonant frequency is at about 3.5 GHz, as shown in Figs. 2 and 4, respectively.

3.4 The Variation of g

Figure 5 plots the simulated curves for the reflection coefficient with the variation of the gap between the coplanar GP and stripline and ($g = 0.33$, 0.35, 0.37, 0.39 and 0.41 mm) by keeping $h = 1.1$ mm, $r = 12.5$ mm and $W = 49$ mm as constant, respectively.

Figure 5 plots the simulated graph for the variation of gaps g which are in similar shape, but with increasing the gap, the first resonant frequency decreases and also noticed that all the graphs cover an ultra-wide frequency band for −10 dB bandwidth. The BW ranges from 2.7 GHz to more than 12 GHz at $g = 0.33$ and $g = 0.39$ mm, while at $g = 0.35$ and $g = 0.37$ mm, the BW ranges from 3.02 GHz to more than 12 GHz.

The CPW-fed line and the ground plane coupling effect may usually affect the impedance matching condition. It means the effect of the feed line gap 'g' provides impedance matching for the projected antenna. In addition, it is found that the overall bandwidth is increased if the feed line gap is increased.

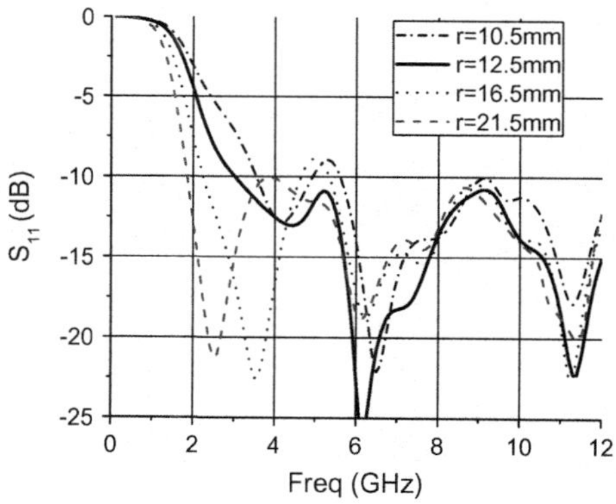

Fig. 4 Reflection coefficient curves for variation of r of the disc

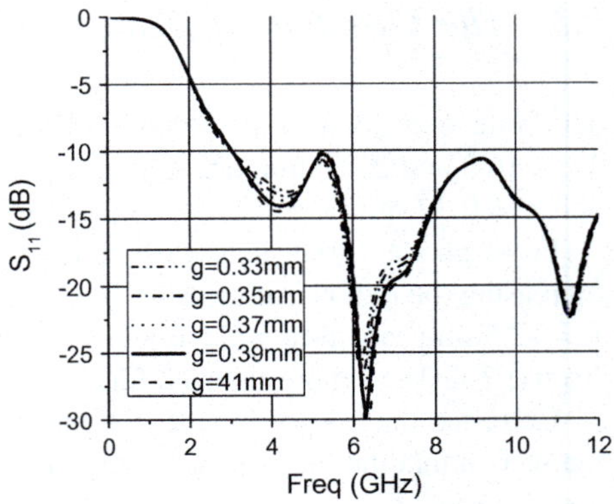

Fig. 5 Simulated reflection coefficient graphs for variation of gap *g* without stub

Thus, the optimized design parameters are as follows, i.e. $h = 1.1$ mm, $W = 49$ mm, $r = 12.5$ mm and $g = 0.39$. To improve the performances, a stub of length $Ls = 6$ mm and width $W_f = 3.06$ mm is introduced across the feed line of circular disc monopole antenna as shown in Fig. 6.

Different variations of gap g are observed and are plotted in Fig. 7 by keeping other parameters, i.e. *h*, *r*, *W* and *Ls*, constant. It is evident from Fig. 7 that at $g = 0.39$ mm, the bandwidth of an impedance varies from 2.8 GHz to more than 12 GHz. A comparative study is carried for the two structures shown in Figs. 1 and 6 and is tabulated in Table 1.

Table 1 shows that if the value of *g* increases in without a stub, the resonant frequency decreases, whereas the increase of *g* in the case of with stub, the resonant frequency increases. In addition to that, the higher the resonant frequency, the lower the 3 dB cut-off frequency. So, bandwidth increased.

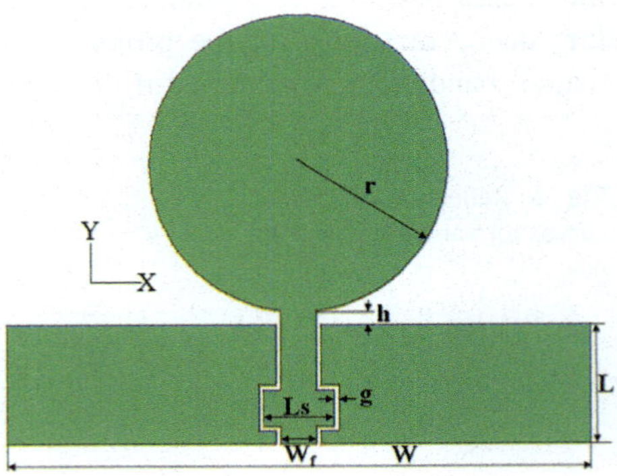

Fig. 6 Geometry of circular disc monopole antenna with stub

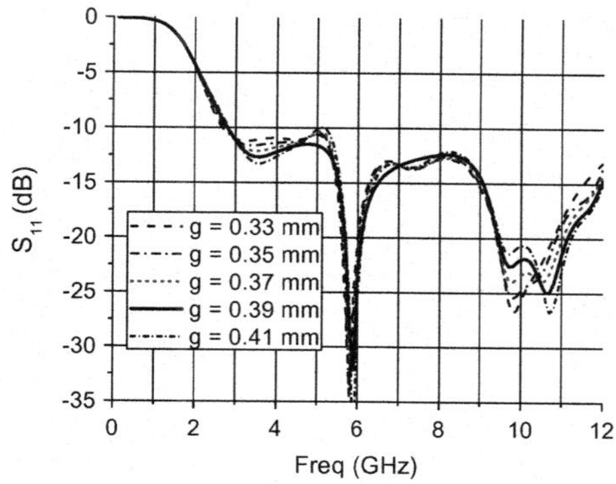

Fig. 7 Reflection coefficient graphs for variation of gap *g* with stub

Table 1 Comparison table for the structure without stub and with stub

Without stub				With stub		
Parameter variation (mm)	No. of resonant. frequencies (GHz)	Reflection coefficient (dB)	Bandwidth range (GHz)	No. of resonant frequencies (GHz)	Reflection coefficient (JB)	Bandwidth (GHz)
$g = 0.33$	(i) 4.4	−13.04	3.02 GHz to more than 12 GHz	(i)3.21	−11.21	2.68 GHz to more than 12 GHz
	(ii) 6.2	−26.00		(ii) 5.8	−34.27	
	(iii) 11.31	−22.26		(iii) 9.8	−26.85	
$g = 0.35$	(i) 4.39	−13.36	3.03 GHz to more than 12 GHz	(i)3.39	−11.61	2.73 GHz to more than 12 GHz
	(ii) 6.2	−27.09		(ii) 5.8	−34.45	
	(iii) 11.3	−22.44		(iii) 9.8	−25.43	
g = 0.37	(i) 4.3	−13.70	3.03 GHz to more than 12 GHz	(i) 3.5	−12.13	2.77 GHz to more than 12 GHz
	(ii) 6.29	−27.77		(ii) 5.8	−31.24	
	(iii) 11.29	−22.49		(iii) 9.8	−23.98	
				(iv) 10.5	−23.94	
$g = 0.39$	(i) 4 2	−14.08	3.04 GHz to more than 12 GHz	(i) 3.58	−12.67	2.8 GHz to more than 12 GHz
	(ii) 6.29	−29.80		(ii) 5.89	−31.98	
	(iii) 11.29	−22.39		(iii) 9.7	−22.56	
				(iv) 10.69	−24.94	

4 Modified Ground Plane Design (Proposed) and Performance

By using the above best possible design parameters, i.e. $r = 12.5$ mm, $h = 1.1$ mm, $W = 49$ mm, $g = 0.39$ mm and $Ls = 6$ mm, further an inverted triangular slot is cut from the left and right top ground plane to improve the bandwidth as shown in Fig. 8.

Figure 9 shows the simulated reflection coefficient for the variation of G (2, 3, 4, 6 and 8 mm). The first fundamental frequency decreases as the value of

Fig. 8 Proposed antenna

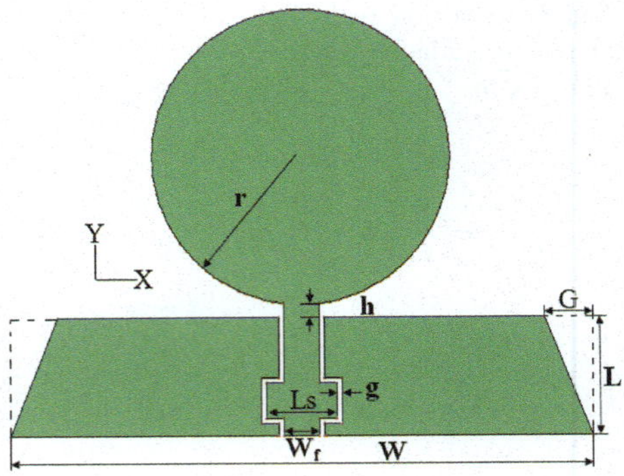

Fig. 9 Simulated reflection coefficient curves for variation of cutting ground plane *G*

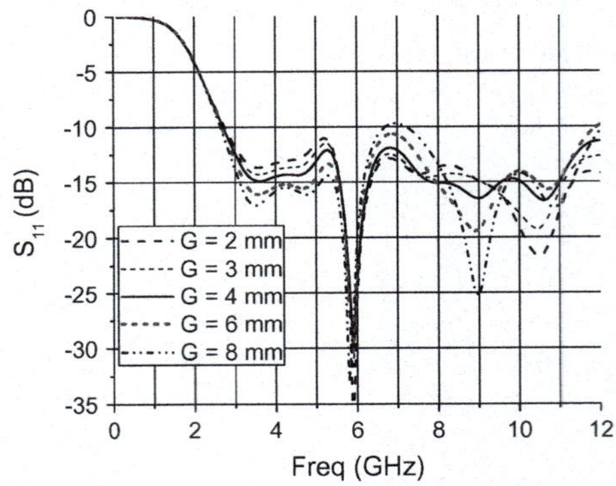

G increases, which covers an ultra-wide frequency band. The best result is found to be at $G = 4$ mm in which the −10 dB BW ranges from 2.58 to more than 12 GHz.

The simulated uniqueness of the three antennas as shown in Figs. 1, 6 and 8 is compared and is shown in Fig. 10 which shows that the projected antenna has an enhancement in the bandwidth with respect to other two configurations.

To know the occurrence behind this multi-resonance performance, the input impedance of the proposed antenna structures that were studied in Fig. 8 on a Smith chart is shown in Fig. 11. Introducing a stub in the ground plane and due to modification of ground plane, the upper frequency bandwidth is affected and shown in Fig. 11.

CDM antenna with CPW-fed and proposed antenna is compared and is shown in Table 2. In the proposed antenna, higher impedance bandwidth is achieved, i.e. 22% increment is there.

The proposed antenna with exact design was built and is shown in Fig. 12. The simulated and measured reflection coefficient for the proposed antenna is shown in Fig. 13. Figure 14 shows that the simulated and measured VSWR is below than 2 from 2.5 to 12 GHz.

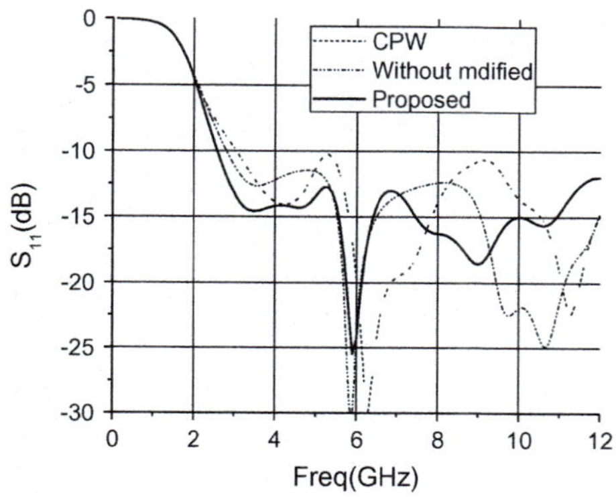

Fig. 10 Simulated characteristics of the three structures

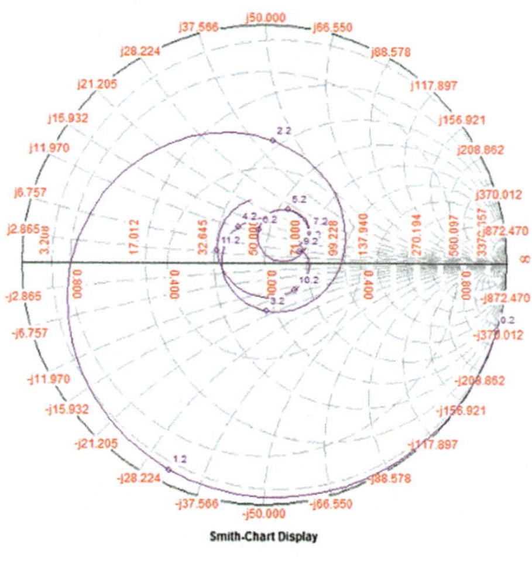

Fig. 11 Simulated Smith chart characteristics of proposed antenna

Table 2 Comparison table for the structure stub without modified and stub with modified ground plane

Stub without modified ground plane				Stub with modified ground plane		
Parameter variation (mm)	No. of resonant frequencies (GHz)	Reflection coefficient (dB)	Bandwidth range (GHz)	No. of resonant frequencies (GHz)	Reflection coefficient (dB)	Bandwidth (GHz)
$g = 0.39$	(i) 3.58	−12.67	2.8 GHz to more than 12 GHz	(i) 3.5	−14.60	2.58 GHz to more than 12 GHz
	(ii) 5.89	−31.98		(ii) 5.9	−25.36	
	(iii) 9.7	−22.56		(iii) 9	−18.60	
	(iv) 10.59	−24.94		(iv) 10.6	−15.64	

Fig. 12 Proposed fabricated structure

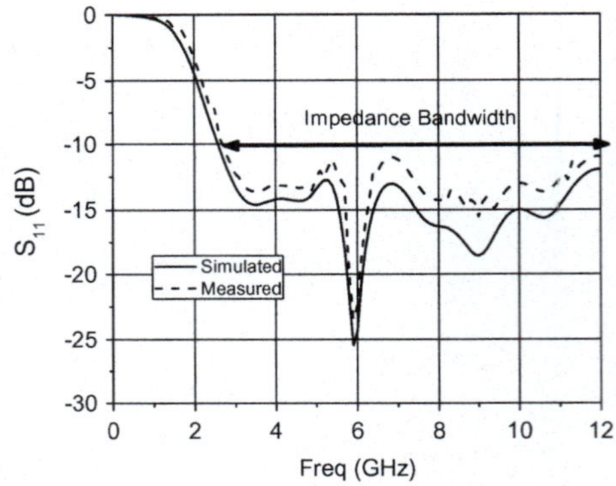

Fig. 13 Simulated and measured reflection coefficient curve for $g = 0.39$ mm, $h = 1.1$ mm, $W = 49$ mm, $r = 12.5$ mm, Ls = 6 mm and $G = 4$ mm

4.1 Current Distribution

Figure 15a–d shows the distribution of current at 3.5, 5.9, 9 and 10.6 GHz, respectively. It has also been clear that if the SMA port is detached from simulation, the simulated distributions of current at these two frequencies will not change. So, Fig. 15c, d explains two more complex patterns of current at resonant frequencies 9 and 10.6 GHz, consequent to the harmonics of third and fourth order, respectively. From Fig. 15, it is clear that along the edge of the disc, the current is frequently spread. It is also noticed that the current is distributed mainly towards the *x*-direction on the upper edge of the ground plane and it is maximum at the stub, which explicates the execution of an antenna is mostly reliant on the width *W* of the ground plane, the gap between the coplanar GP, the metal strip g and slotted ground plane *G*.

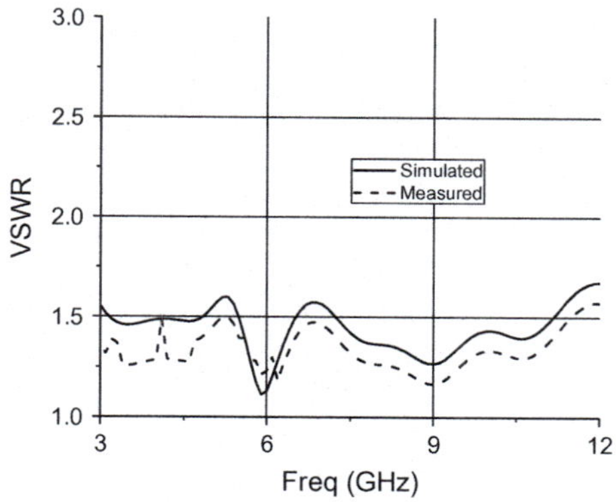

Fig. 14 Simulated and measured VSWR

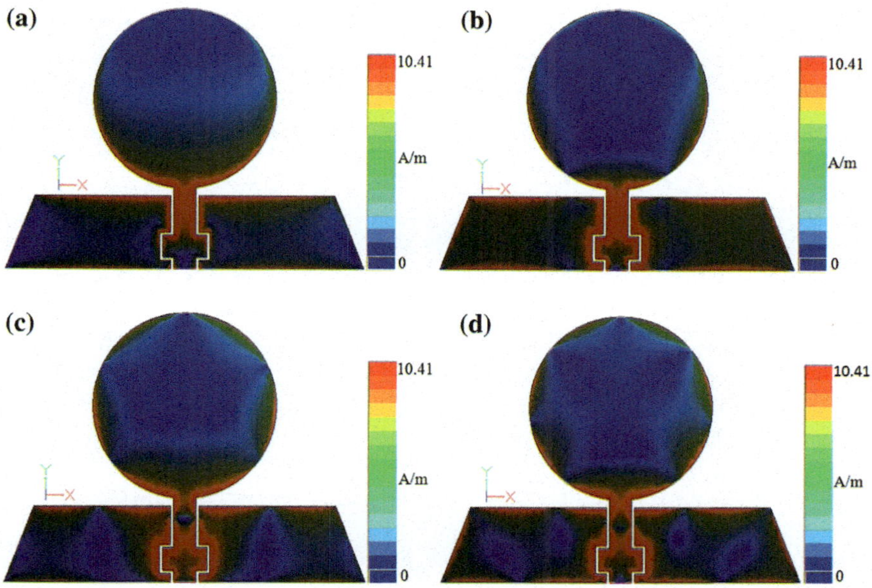

Fig. 15 Replicated current distributions of CDM with h = 1.1 mm, r = 12.5 mm, W = 49 and g = 0.39 mm at **a** 3.5 GHz **b** 5.9 GHz **c** 9 GHz **d** 10.6 GHz

4.2 Radiation Pattern and Gain

The radiation patterns and co-polarization pattern at 3.5, 5.9, 9 and 10.6 GHz are plotted in Fig. 16a, b severally. It is noticed that at all the resonant frequencies in

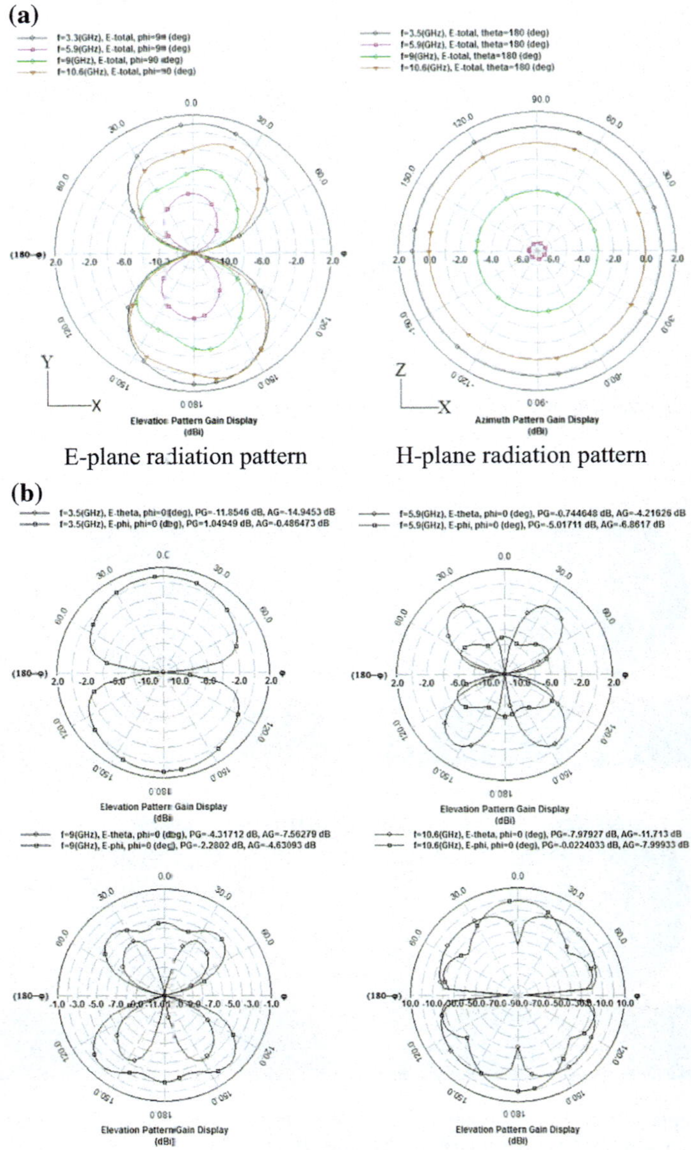

Fig. 16 **a** Simulated patterns with $h = 1.1$ mm, $r = 12.5$ mm, $W = 49$ mm, $g = 0.39$ mm, $Ls = 6$ mm and $G = 4$ mm at (i) 3.5 GHz (ii) 5.9 GHz (iii) 9 GHz (iv) 10.6 GHz, **b** co-polarization at different resonant frequency

the simulated E-plane (*x*-*y* plane) pattern are like a conventional monopole and the H-plane (*x*-*z* plane) pattern are near omnidirectional. Generally, the H-plane pattern shapes communicate fine to the patterns of current on the circular disc as shown in Fig. 16 at dissimilar resonant frequencies, correspondingly.

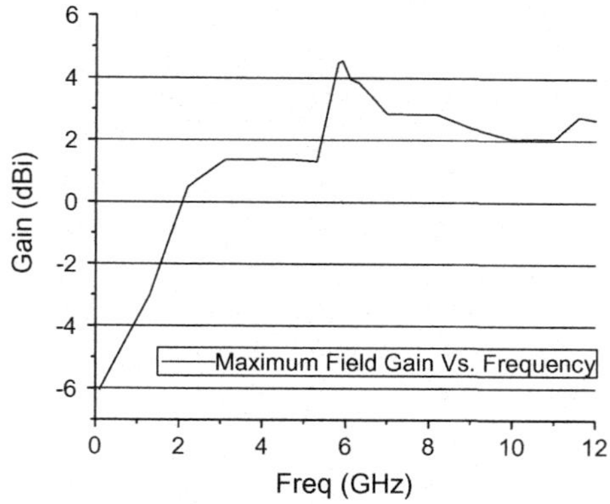

Fig. 17 Simulated peak gain of proposed antenna

Table 3 Comparison with other literatures

Antenna type	Gain	Size (mm^2)	Bandwidth
Printed circular disc monopole antenna [5]	6.7 dBi	42 × 50 = 2100	2.78–9.78 GHz
CPW-fed planar with frequency band notch function [6]	5 dBi	22 × 31 = 682	2.8–10.6 GHz Notch band 5.15–5.35 GHz
CPW-fed circular disc monopole antenna [12]	6 dBi	35 × 47 = 1645	2.73–12 GHz
Bandwidth increasing technique using modified ground plane [14]	–	30 × 50 = 1500	2.95–12.615 GHz
Proposed antenna	4.5 dBi	36.1 × 49 = 1768.9	2.58–12 GHz

At different frequencies, the replicated gain of the projected antenna is shown in Fig. 17. It is obvious that the utmost attainable peak gain is 4.5 dBi from 1.65 to 12 GHz. The poor gain comes from the pattern shape, poor match and internal losses.

A comparative study is done with the proposed work with the previously published work and is tabulated in Table 3. From the Table 3, proposed antenna shows wider bandwidth.

5 Conclusion

In this paper, the printed CDM antenna fed by CPW is examined. The performances of the antenna have been shown that they are frequently reliant on the width of the GP, feed gap, the gap between the strip and the coplanar GP and radius of the disc. The dimensions of the CDM are directly depending upon the first resonant frequency because the current is largely spread on the boundary of the disc. A stub is introduced across the feed line for improvement of matching and good performances. For tuning the input impedance, the GP works as impedance for matching circuit. Thus, the operating bandwidth can be improved by changing g, *h* and *W*. In addition to that, further the bandwidth is improved using slotted ground plane. It is also explained that the proposed printed CDM covers FCC-defined frequency band. It is also noticed that the antenna's emission pattern are almost omnidirectional over the entire bandwidth. Finally, the result shows that the proposed antenna is appropriate for future UWB applications.

References

1. M.J. Amman, Z.N. Chen, Wideband monopole antennas for multi-band wireless systems. IEEE Antennas Propag. Mag. **45**(2), 146–150 (2003)
2. N.P. Agrawall, G. Kumar, K.P. Ray, Wide-band planar monopole antennas, in *IEEE Transactions Antennas Propagations, February 1998*, vol. 46, pp. 294–295
3. J. Liang, C.C. Chiau, X. Chen, J. Yu, Study of a circular disc monopole antenna for ultra wideband applications, in *International Symposium Antennas Propagations, Sendai, Japan, August 2004*, pp. 17–21
4. J. Liang, C.C. Chiau, X. Chen, C.G. Parini, Printed circular disc monopole antenna for ultra wideband applications. Electron. Lett. **40**(20), 1246–1248 (2004)
5. J. X. Liang, C. C. Chian, X. D. Chen, and C. G. Parini, "Study of a printed circular disc monopole antenna for UWB systems," in *IEEE Transactions Antennas Propagation, November 2005*, vol. 53, pp. 3500–3504
6. Y. Kim, D.H. Kwon, CPW-fed planar ultra wideband antenna having a frequency band notch functions. Electron. Lett. **40**(7), 403–405 (2004)
7. W. Wang, S.S. Zhong, S.B. Chen, A novel wideband coplanar-fed monopole antenna. Microw. Opt. Technol. Lett. **43**(1), 50–52 (2004)
8. S.Y. Suh, W. Shutzman, W. Davis, A. Waltho, J. Schiffer, A novel CPW-fed disc antenna, in *IEEE Antennas Propagation Society Symposium, June 2004*, vol. 3, pp. 2919–2922
9. T. Yang and W. A. Davis, Planar half-disk antenna structures for UWB communications, in *IEEE Antennas Propagation Society Symposium, June 2004*, vol. 3, pp. 2508–2511
10. H. Yoon, H. Kim, K. Chang, Y. J. Yoon, and Y. H. Kim, A study on the UWB antenna with band-rejection characteristic, in *IEEE Antennas Propagation Society Symposium, June 2004*, vol. 2, pp. 1784–1787
11. K. Chung, T. Yun, J. Choi, Wideband CPW-fed monopole antenna with parasitic elements and slots. Electron. Lett. **40**(17), 1038–1040 (2004)
12. J. Liang, L. Guo, C.C. Chiau, X. Chen C.G. Parini, Study of CPW-fed circular disc monopole antenna for ultra wideband applications. IEEE Proceedings-Microw Antennas Propagation. **152**(6), 520–526 (2005)
13. C.A. Balanis, *Antenna Theory: Analysis and Design* (Harper and Row, New York, 1982)
14. N. Prombutr, P. Kirawanich, P. Akkaraekthalin, Bandwidth enhancement with diagonal edge. IETE J. Res. **55**, 196–200 (2009)

Cryptosystem for AVK-Based Symmetric Algorithms and Analysis Using Cryptic Pattern Mining

Shaligram Prajapat

Abstract This work introduces the basic concepts of Cryptic Mining, that is a specialized area of data mining discipline for cryptic text processing. This chapter explores the various tasks, models, and techniques that are used in Cryptic Mining in order to understand useful patterns and information from large and unorganized captured cipher logs.

Keywords Automatic variable key (AVK) · Cryptic Mining discipline
Cryptosystem · SGcrypter · Enciphering/deciphering time

1 Introduction

The importance of low power devices and device-to-device communication is the central demand in framework of Internet of Things (IOT). The security of these devices together with a balance with power efficiency will decide the long-term sustainability of the system. Automatic variable key is also gaining pace and finding its applicability in low power devices [1]. The feature of short life period of key and small size makes it a promising candidate for energy-efficient secure communication. Parameterized key-based cryptosystem will add one more level of security in the design of efficient cryptosystem. This article presents the concept of automatic variable key and its enhanced framework of key exchange by parameters only along with its importance and benefits. We have also presented the investigation of AVK-based framework from hacker's/cryptanalyst's perspective and evolution of Cryptic Mining discipline. This specialized study will be useful for auditing AVK-based cryptosystem [2, 3].

S. Prajapat (✉)
International Institute of Professional Studies, Devi Ahilya University, Indore, India
e-mail: shaligram.prajapat@acm.org; shaligram.prajapat.in@ieee.org
shaligram.prajapat@iips.edu.in

V. Nath (ed.), *Proceedings of the International Conference on Microelectronics, Computing & Communication Systems*, Lecture Notes in Electrical Engineering 453,
https://doi.org/10.1007/978-981-10-5565-2_31

2 Effect and Consequences of Increasing Key Length

There are many symmetric key algorithm for performing cipher generation and processing. Hence, question comes into our mind that which one is best and up to what extent? Which one is most secure and energy efficient? The state-of-the-art approach to increase the security is by increasing the key size from 32 to 48 bit or to 56, 64,128, or 256 bit or beyond, but it compromises the computation time and storage cost. In order to choose efficient or optimum symmetric key algorithm from a list of symmetric key algorithms {DES, 3DES, Blowfish, Twofish, …}, comparative analysis of these algorithms may help one to get the best out of them. One may use online tool to analyse or compare them for decision making with parameters such as encryption/decryption time, key-size, data size, data-type, power consumption etc. [4, 5].

2.1 Web-Based Cryptic Algorithm Analysis Tool SGcrypter

SGcrypter is customized Web tool (Shaligram and Gaurav crypter tool) [2] created for analysis of cyptic algorithms (symmetric key based). To analyze SGcrypter generates statistics of encryption or decryption algorithms under different cases. SGcrypter supports the user to choose a cryptic algorithm from a list of conventional algorithms and to analyze or compare these algorithms to choose best parameters such as encryption/decryption time, key size, data size, power consumption [2–5]. **Step-1**: On clicking the button for any category, the user is redirected to the concerning page where one may asked to provide data files, key, key files, etc. Depending on the type of category, the user has selected. **Step-2**: Feed all the information and get the analysis report [2].

Analysis result is in the form of tables and graphs. With various combinations of input of different sizes and different key lengths, we present one example. For illustration purpose, we take random set of input files and key files with different sizes for elucidation. The four cases/categories of graphs supported by SGcrypter are Category I: Study of *input Data size with execution time*, Category II: Study of *input Data size on execution time*, Category III: Study of *variation in Key size on execution time*, and Category IV: Study *of variation of Key size versus execution time* [6–8].

Performance metrics of SGcrypter: *Key length, data size, key size* are the prime parameters considered for efficiency comparison using SGcrypter, where *Key Length*—key size is directly proportional to security level. It directly influences encryption/decryption time. *Data size*—It is the size of input information file. Larger data size may result in increased encryption/decryption time. *Enciphering/Deciphering time*—It is the conversion time that depends on the complexity of algorithm, processor speed, RAM, and other factors of the operating environment. Algorithm with lesser encryption/decryption time is considered to be better. *Throughput*—It is the ratio of total plaintext in MB (to be encrypted/decrypted) and total encryption/decryption time for each algorithm. The greater the throughput, the lesser the power consumption. The technique with maximum throughput will be most efficient [9].

2.2 *Limitation and Scope for SGcrypter*

The results shown in Table 1 are suggestive and random. Since the operations are in some milliseconds, so each time the result may slightly vary. Since different devices have different loads at different time that can affect the speed of encryption/decryption, so the results given here are not reference but provide hints about behavior. It can be considered as an average and random case, and in this way, the analysis seems to be legit. Moreover, SGcrypter can only analyze DES, 3DES, Blowfish, and Twofish algorithms on the already discussed parameters. Also, SGcrypter cannot operate upon input data with size above 2 MB for now. The input data files provided to SGcrypter should not exceed 2 MB in size. Also, the maximum key size supported is 56 bytes [2].

3 Summary of Analysis of Cryptic Algorithms Using SGCrypter

The analysis of variation of file size and key length on performance of symmetric key cryptosystem has been demonstrated here. Outcome generated through SGcripter tools [2, 3] indicate that if we choose the speed as a deciding factor of efficiency, conventional techniques to encrypt/decrypt data has Encryption time, Decryption time, Throughput and memory utilization as prime parameters that must be taken under consideration. These parameters will determine the future of symmetric key-based cryptosystem [10–12].

1. Time taken by TWOFISH is minimum (for both encryption and decryption) for all size combinations of data and key.
2. Efficiency of TWOFISH and BLOWFISH has a very close competition.
3. Time taken by 3DES is maximum among all cases and so it is least efficient on encryption/decryption time parameter among these algorithms.
4. Whether we are encrypting or decrypting input file of variable size with the same key (fixed key size) or in the case when we encrypting or decrypting input file of fixed size with variable length keys or keys with different sizes. The order of encryption/decryption time taken by these techniques in ascending order is TWOFISH < BLOWFISH < DES < 3DES.
5. The throughput for TWOFISH is maximum, and hence, TWOFISH will consume least processing power and recommended for low power devices and of course for IOT. On the other hand, 3DES with maximum processing power with minimum throughput has to think for alternative solutions.

Table 1 Response of SGCrypter for encryption and decryption time input file [2, 3]

Comparative analysis response	Key size and corresponding execution time
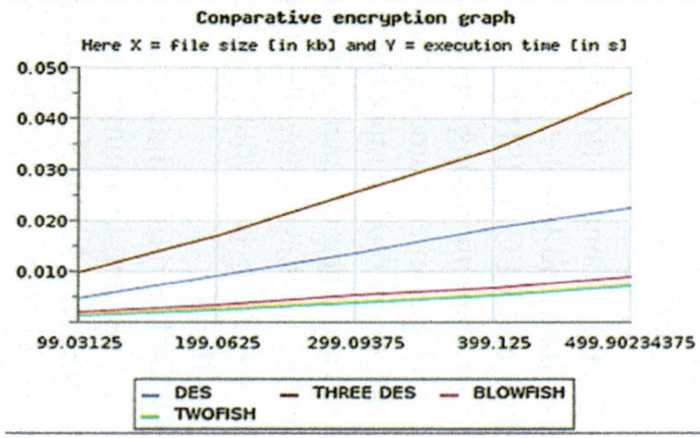	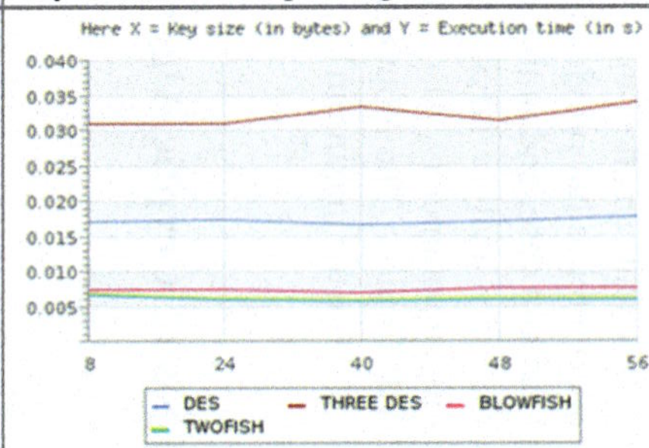

3.1 The Automatic Variable Key Approach (AVK-Based Cryptosystem)

Automatic variable key (AVK) coined by C.T. Bhunia presented with XOR operation opens a new hope in this direction. It says that for secure information exchange using symmetric key cryptography is to reduce the lifetime of key and restrict it to one session only. It must be noted that in the design of efficient cryptosystem the biggest challenge of the designer is to make key unbreakable, whereas the challenger threats to break the key. Vernum also proposed that it would be impossible to break the key if the key is made time-variant or dynamic. The AVK can be implemented by changing key from session to session (a time-variant key). The AVK schema is illustrated in the table below that focuses on two sessions between Alice (sender) and Bob (receiver), whereby they, respectively, exchange data between 34 and 78. In original AVK schema, the key is made variable with data. That is, if K_0 = initial secret data, then next key will be computed from following recurrence relation $K_i = K_{i-1}$ XOR D_i for all $i > 0$ where D_i = data in ith session [13, 14].

Figure 1 indicates two keys K_{i} and $K_{\mathrm{j},}$ respectively, of session i and j. Message of session I is encrypted/decrypted by key $K_{i,}$ and similarly, K_j key will be used for messages M_j. In this way, keys of session i will not be useful for session j, and thus, the security of information system will not be compromised. Further, the key length will not be needed to extend further. Once the key length is fixed (sufficient key length—6 characters), it would be changed from session to session [3]. In this model, (shown below) Node 1 and Node 2 (can be extended to node-n) are communicating with each other by sharing parameters instead of key exchange. The model also demonstrates that for same parameters different approaches may generate same key. Thus, additional level of security may be achieved by parameterized model. The two approaches for computation of key from parameters have been demonstrated by approach-1 and approach-2 [15].

3.2 Elucidation of Working of AVK-Based Cryptosystem

Figure 2 demonstrates that both Alice and Bob use different keys for different sessions and both are able to communicate securely [1, 3, 23, 24].

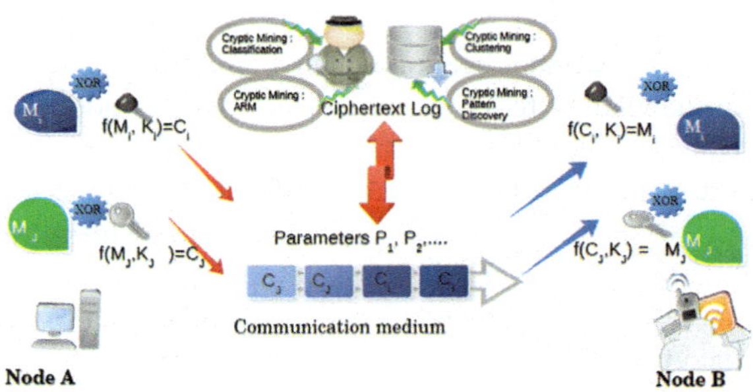

Fig. 1 Framework of AVK-based cryptosystem

AVK in Symmetric Key Cryptography

Session ID	Alice sends	Bob receives	Bob sends	Alice receives	Remarks
1	Secret key (say 2)	2	Secret key (say 6)	6	For next slot, Alice will use 6 as key and Bob 2 as key for transmitting data.
2	Alice sends Bob first data as: $3 \oplus 6$	Bob gets back original data as: $(3 \oplus 6) \oplus 6 = 3$	Bob sends first data as: $7 \oplus 2$	Alice gets back original data as: $(7 \oplus 2) \oplus 2 = 7$	Alice will create new key $6 \oplus 7$ for next slot. Bob will create new key $(2 \oplus 3)$.
3	Alice sends next data as: $4 \oplus (6 \oplus 7)$	Bob gets back original data as: $((4 \oplus (6 \oplus 7)) \oplus (6 \oplus 7)) = 4$	Bob sends next data as: $8 \oplus (2 \oplus 3)$	Alice recovers data as: $((8 \oplus (2 \oplus 3)) \oplus (2 \oplus 3)) = 8$	Thus Alice and Bob respectively exchange data 34 and 78.

Fig. 2 Working of AVK-based symmetric cryptosystem

4 Summary of AVK Model of Symmetric Cryptosystem

Communication among two nodes works well in the model shown in Figs. 1 and 2. Node A (Alice) has message M_i in session i, and enciphered with the key of that session, say $K_{i,}$ and transmits it as ciphertext $C_{i.}$ The ciphertext travels over communication medium and received by recipient Node B (Bob) where the plaintext information would be recovered as the reverse process done at node A. Now, for the session j, A (Alice) has new message M_j in session j, and enciphered with the key of that session, say $K_{j,}$ and transmits it as ciphertext $C_{j.}$ The ciphertext travels over communication medium and is received by recipient Node B (Bob) where the plaintext information would be recovered as the reverse process done at Node A and successfully recovers the transmitted plaintext information [16].

4.1 Parameters Only Scheme for AVK Approach

Over the communication channel, entire key may not be exchanged; instead, it is exchanged in terms of parameters only to add additional key security also. The ciphertext log may be captured by hacker or cryptanalyst that can be mined to extract patterns from the logs, and variety of tools and methods can be applied on it to exploit the weakness or harvest the secure information. The framework also highlights the mechanism of key computation using parameters only. Both Alice and Bob will compute keys using parameters exchanging over communication medium. This can illustrate using following algorithms. Following algorithms of

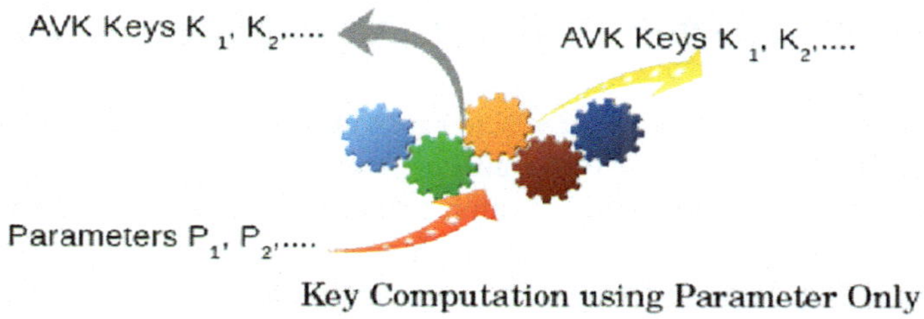

Fig. 3 Key computation using parameters only

Alice and Bob will demonstrate the procedure of key computation using parameters only (Fig. 3) [10, 19, 20].

Without exchanging entire key, Node 1 and Node 2 will securely communicate with each other. Both the nodes are computing the same key using different functions, which in turn enhance the level of security.

Hence, even if parameters or method at any one node for key computation is known, it will not work for next node or parameter set [21–23].

Algorithm-1 **parameters4Key-Alice** (*parameters* p_1,p_2)

{

1. Sense parameters p_1, p_2;
2. Compute the key for information exchange by: $key_i = (p_1 * p_2)^{1/2}$;
3. Sense the information to exchange= D_i ;
4. If (mode==transmit)

   Generate Cipher text C_i =f_Encrypt(D_i, key_i);
   Transmit Ci;
5. else

   Receive Plain text P_i = f_Decrypt (D_i, key_i);
   Use P_i;

}

Algorithm-2 **Parameters4Key-Bob** (*parameters* p_1, p_2)

{

1. Sense parameters p_1,p_2;
2. Compute the Arithmetic Mean A.M.= ($p_1 + p_2$)/2;
3. Compute the Harmonic Mean H.M.-2* $p_1 * p_2$ /($p_1 + p_2$)
4. Compute the $Key_i = (A.M. * H.M)^{1/2}$
5. If (mode==transmit)

   Generate Cipher text C_i =f_Encrypt (D_i, key_i);
   Transmit C_i;
6. else

   Receive Plain text P_i = f_Decrypt (D_i, key_i);
   Use P_i;

}

4.2 Summary of Parameterized AVK Model

Parameterized only key exchange schema provides additional security with keys generated and exchanged using of parameters only. That is, there is no need to share entire key over the public communication medium. Instead of it, only parameters will be shared that can be utilized by both sender and receiver of Node A and Node B for the generation of key. The parameterized model will be discussed in next unit. This has been implemented very recently and still under development phase for AVK approach [1, 15–17].

1. AVK scheme can be extended further to explore alternative approach. Two methods have been discussed to demonstrate how AVK-based cryptosystem can be developed. Both methods use some parameters to construct key. Fibonacci method (for a particular session, with given n and p values computations can be done for F_{n-1}, F_n and F_{n+1}).
2. Sparse matrix (location coordinate (i, j) will act as parameter for encryption/decryption)-based approach can be modeled for automatic variability of key for secure information exchange.

For these AVK-based cryptosystems, parameters (n, p) or location (i, j) can vary from session to session. So, even if the intruder gets unwanted access to the key of session i, it would not be valid for original message extraction in session $i + 1$ onward. In this model, key is not transmitted in the data transfer. So, it becomes highly difficult to interpolate any information regarding plaintext or key.

4.3 Analysis of AVK-Based Cryptosystem from Hacker's Perspective

Investigation of application of techniques of data mining in cryptography domain forms the basis of Cryptic Mining. Cryptic Mining tools are the set of techniques for following activities. Originally, data mining methods are concerned with information extraction at application level or for business and commercial need of individual or organization. The term "Cryptic-Mining" is used for low-level information domain. This knowledge area increases the security level of information and power of cryptic algorithms by helping cryptanalyst. In order to strengthen the cryptosystem, automated tools can be developed that intelligently exploit patterns among ciphertext, plaintext, key size, key life time, and log of partially recovered plaintext–ciphertext-derived knowledge. Cryptic Mining domain assumes that ciphertexts present in the network or stored encrypted files/logs are not 100% random and exhibit some patterns. These patterns may be useful to exploit weakness using mining algorithms [17].

Cryptic pattern discovery, cryptic classification, cryptic ARM, cryptic clustering, cryptic forecasting together form Cryptic Mining system. This mining is a set of techniques to provide/present probable relationships in plaintext, ciphertext, or both (not based on cause–effect relationships). Moreover, Cryptic Mining methods explore significant relationships, not just between key and key length, but also among number of parameters used for key computation, key strength, ciphertext–plaintext relationship, and running encryption/decryption key cryptanalyst have to find out correlations and determine what is significant. Some set of techniques are available in the literature in initial phase, such as cryptic classification and inference using perception, inference of parameter or key information using Bayesian belief networks, cryptic association rule mining, cryptic clustering [17–20]. The work of Claudia presents one typical usage of extracting key size information from clustering of ciphertext which is given in Algorithm 4.1 to Algorithm 4.4. It assumes that ciphertext is a regular document written in an unknown language. Now perform 3-Step procedure as follows:

Step-1: Create group from collection of ciphers (based on similarity)
Step-2: Compute some kind of index from largest most frequent prime factor from the distance between repeated n-grams. The most frequent index will act as label.
Step-3: Measure the distance between labels and known key length of groups previously classified.

```
Algorithm4.1 CipherClustering4Kasiski (input cipher text)
{
//Premise: Pattern in plaintext will manifest in the cipher text.
1) Identify repeated patterns of trigrams or longer n-grams.
2) For each pattern, write down the all possible instances of the pattern.
3) Compute the differences between starting position of successive instances.
 4) Determine all the factors of these differences.
5) Key Length factor appearing more often.
}
```

```
Algorithm-4.2 CipherClustering4Friedman (input cipher text)
{
        1.      // compute Index of coincidence(IC), for this Let C1 and C2 are two random ciphers with probability P1
        and P2 and if P1=P2 then, C1 is coincident with C2 and P1=P2=P =IC.
        2.      // The range of IC lies in interval [0.038, 0.066 ] ,For English Text with mono alphabetic ciphers,
        IC(Plaintext) = IC(Cipher)
        3.      //Drawback: Precision decreases with key increase in key length.
}
```

```
Algorithm-4.3CipherDocument Clustering (input cipher text)
{
//Ciphers/cryptogram is a vector of words delimitation (space) if it is preserved otherwise standard word length.
// create clusters.
```

```
}
Algorithm-4.4 CipherDocument categorization (input cipher text)
{
//Ciphers/cryptogram is a vector of words delimitation (space) if it is preserved otherwise standard word length.
// it attempts to assign cipher text into 2 or more predefined categories. Training of algorithm will be done to produce knowledge for categorization algorithm.
//Group documents/ciphered together based on similarity
}

Algorithm-4.3 Vector space model (input cipher text)
{

//Input: A collection of ciphers and a Ciphers/cryptogram is a vector of words delimitation (space) if it is preserved otherwise standard word length.

//it attempts to assign cipher text into 2 or more predefined categories. Training of algorithm will be done to produce knowledge for categorization algorithm.
Group documents/ciphered together based on similarity
}
```

5 Future Work and Extension

In future extension work, we may introduce the concept of cipher generation based on fuzzy logic. Cipher generation can be possible when we divide the original message in frames and assign a fuzzy value corresponding to each packet [21–23].

6 Conclusion

Cryptic Mining gains base from data mining. It tries to apply and use analysis of data from session logs and identifies patterns related to attacks. Finding the preintimation of an attack can help to develop good prevention tool and techniques, and seeing the action associated with an attack can help to locate vulnerabilities to control possible damages. This chapter presents novel schemes of secure information exchange over the network that may be useful for wired and wireless systems. AVK approach is claimed to be secured, but parameter-based communication would add extra security feature in the system. Association rule for predicting probable parameters from parameter space using association rule may provide hints for future parameters to predict key. But since both number of parameters and key of session are variable and changing from session to session, so the security of the system would not be compromised with the automatic variable scheme. For improvement and learning, Cryptic Mining gains training database for mining from various sources. It tries to apply and use analysis of data from session logs and identifies patterns related to attacks. Finding the preintimation of an attack can help to develop good prevention tool and techniques, and seeing the action associated with an attack can help locate vulnerabilities to control and possible damages.

References

1. S. Prajapat, D. Rajput, R.S. Thakur, Time variant approach towards symmetric key, In Proceedings of IEEE Science and Information Conference (SAI), 2013, p. 398–405
2. S. Prajapat, G. Parmar, R.S. Thakur, Towards investigation of efficient cryptosystem using SGcrypter. Special Issue Int. J. Appl. Eng. Res. (IJAER) **10**(79), 853–858 (2015)
3. S. Prajapat, R.S. Thakur, Cryptic mining for automatic variable key based cryptosystem. Elsevier Proc. Comput. Sci. **78**, 199–209 (2016)
4. S. Prajapat, R.S. Thakur, Realization of information exchange with Fibo-Q based symmetric cryptosystem. Int. J. Comput. Sci. Inf. Secur. (IJCSIS) **14**(2), 216–223 (2016)
5. C. Oliveira, A. José, A.C. Carlos, Clustering and categorization applied to cryptanalysis. Cryptologia **30**(3), 266–280 (2006)
6. S. Prajapat, R.S. Thakur, Cryptic mining: apriori analysis of parameterized automatic variable key based symmetric cryptosystem. Int. J. Comput. Sci. Inf. Secur. (IJCSIS) **14**(2), 233–246 (2016)
7. D.S. Elminaam, H.M. Abdual, M.M. Hadhoud, Evaluating the performance of symmetric encryption algorithms. Int. J. Netw. Secur. **10**(3), 216–22 (2010)
8. P. Chakrabarti, B. Bhuyan, A. Chowdhuri, C. Bhunia, A novel approach towards realizing optimum data transfer and automatic variable key (AVK) in cryptography. Int. J. Comput. Sci. Netw. Secur. **8**(5), 241 (2008)
9. P. Chakrabarti et al., Application of automatic variable key (AVK) in RSA. Int. J. HIT Trans. ECCN **2**(5), 304–311 (2007)
10. P. Chakrabarti et al., Various new and modified approaches for selective encryption (DES, RSA and AES) with AVK and their comparative study. Int. J. Trans. ECCN, **1**(4), 236–244
11. C.T. Bhunia, Application of AVK and selective encryption in improving performance of quantum cryptography and networks. United Nations Educ. Sci. Cult. Organ. Int. Atomic Energy Agency **10**(12), 200–210 (2006)
12. C.T. Bhunia, New approaches for selective AES towards tracking error propagation effect of AES. Asian J. Inf. Technol. Pak **5**(9), 1017–1022 (2006)
13. Prajapat, Shaligram, Ramjeevan Singh Thakur "Cryptic-Mining: Association Rules Extractions Using Session Log". In proceedings of Computational Science and Its Applications–ICCSA 2015. Springer International Publishing. pp. 699–711., 2015
14. S. Prajapat, A. Thakur, K. Maheshwari, R.S. Thakur, Cryptic mining in light of artificial intelligence, in *Proceedings of Second International Conference on Advances in Computing, Control And Networking* (ACCN 2015), 2015, p. 131–135
15. S. Gaurav, H. Karnik, A. Manindra, in Classification of ciphers using machine learning, Thesis IIT-K, 2008. http://www.security.iitk.ac.in/contents/publications/more/ciphers_machine_learning.pdf
16. P. Maheshwari, in The classification of ciphers, Thesis IIT-K, 2001. http://citeseerx.ist.psu.edu/viewdoc/download?doi=10.1.1.4.9410&rep=rep1&type=pdf
17. C. Girish, in *Classification of modern ciphers*, Thesis IIK-2002. http://www.security.iitk.ac.in/contents/projects/cryptanalysis/repository/girish.pdf
18. M.B. Rao, in Classification of RSA and IDEA Ciphers, Thesis IITK-2003. http://www.security.iitk.ac.in/contents/projects/cryptanalysis/repository/anoopjain.pdf
19. S. Prajapat, A. Thakur, K. Maheshwari, R.S. Thakur, Cryptic mining in light of artificial intelligence. IJACSA **6**(8) (2015)
20. S. Prajapat, R. Chaudhary, A. Khan, S. Raikwar, R.S. Thakur, Various ANN approaches towards crypt analysis, in *Proceedings of the International Conference on Innovations in Computer Science and Information Technology (ICICSIT-2015)*. ISBN 978-93-85100-08-6
21. S. Prajapat, R.S. Thakur, Various approaches towards crypt analysis. Int. J. Comput. Appl. **127**(14) (2015). doi:10.5120/ijca2015906518

22. S. Prajapat, Parameterised key diffusion approach of AVK based cryptosystem. Covenant J. Inf. Commun. Technol. **5**(1) (2017)
23. S. Prajapat, A. Sharma, R.S. Thakur, AVK based cryptosystem and recent directions towards cryptanalysis. J. Inter. Comput. Serv. (JICS) **5**, 97–110 (2016)
24. S. Prajapat, R.S. Thakur, Towards parameterized shared key for AVK approach. Patt. Data Anal. Healthcare Settings, p. 61 (2016)

Silicon–Germanium Channel Heterostructure p-MOSFETs

Tara Prasanna Dash, Sanghamitra Das and Rajib K. Nanda

Abstract In this paper, we study the heterostructure p-MOSFETs with Silicon–Germanium channel. This chapter deals with the hole confinement in the SiGe well and the design trade-off for $Si_{1-x}Ge_x$ p-channel MOSFET devices. Also the selection of gate electrode, optimization of SiGe channel width and profile, Si cap and gate oxide thicknesses and the method of threshold voltage adjustment have been addressed.

Keywords Heterostructure · p-MOSFET · Hole confinement · Design trade-off

1 Introduction

Advances in integrated circuit speed depend upon the continued technological evolution and miniaturization of transistors and the interconnect networks between them. One device central to modern IC design is the Metal-Oxide-Semiconductor-Field-Effect Transistor (MOSFET). This device was first proposed by Lilienfeld in 1930, but it was not until 1960 that an operational MOSFET was demonstrated by Kahng and Atalla [1]. The technological and economic forces driving the semiconductor industry are compelling the designers to increase both the number and complexity of the components in integrated circuits leading to a decrease in device size.

A major thrust in the IC community is to achieve symmetrical electrical operation from equivalently sized n- and p-MOSFETs for increased packing densities in Complementary MOS (CMOS) circuits. The channel mobility (100–150 cm^2/V-s) of PMOS device is lower than that of the NMOS device (300–450 cm^2/V-s) in a CMOS technology. Hence to maintain constant drive current the size of PMOS has

T. P. Dash (✉) · S. Das · R. K. Nanda
Department of Electronics and Communication Engineering,
Siksha 'O' Anusandhan University, Bhubaneswar 751030, Odisha, India
e-mail: taradash@soauniversity.ac.in

V. Nath (ed.), *Proceedings of the International Conference on Microelectronics, Computing & Communication Systems*, Lecture Notes in Electrical Engineering 453,
https://doi.org/10.1007/978-981-10-5565-2_32

to be 2–3 times larger in comparison to NMOS. This limits the integration level, and the circuit speed degrades due to larger parasitic.

The ITRS 2013 edition has reported that to achieve sufficient current drive, various major technological innovations like introduction of new materials and device structures are necessary [2]. Efforts have been made to enhance the speed and drive current of FETs through device scaling and the use of materials possessing high carrier mobilities (e.g. GaAs). Recently many reports indicate that specific materials like Germanium [3] or III–V semiconductors [4] can be chosen for high-mobility channel region. But it puts severe challenges to the fabrication process and yields serious issues like gate leakage, tunnelling, etc.

Substrate induced strain (Biaxial) or process-induced strain (uniaxial) are the popularly used technologies which can serve as mobility boosters in long channel devices. The ongoing development in the deposition of heteroepitaxial silicon–germanium ($Si_{1-x}Ge_x$) alloys on Si has made it possible to apply the band gap engineering techniques to semiconductor devices based on silicon. Very exciting and promising results from recent development in SiGe materials and novel high performance devices [5] have opened up a new aspect in the area of VLSI which had previously been solely based on Si homojunction devices. Most of the SiGe work done to date has focused mostly on HBTs. In contrast, progress in SiGe-based field-effect devices has been limited.

The CMOS field-effect transistor is the workhorse of modern VLSI industry. In heterojunction field-effect transistor (Hetero-FET) area, SiGe channel p-MOSFETs have drawn a lot of attention due to its possible applications in VLSI. There has been considerable interest in the SiGe PMOS device, since perhaps this technology might be incorporated into CMOS design, where packing density and circuit speed are limited by the intrinsically poorer PMOS device. Advantages of band gap engineered devices are as follows:

- Increased transistor speed/transistor,
- Reduced power consumption/transistor,
- Better analog circuit functionality,
- Lower mask costs,
- Less expensive process,
- Strained SiGe saves because it extends the utilization of existing fabrication facilities (wafer and transistor).

In this review paper, we shall present the current status of strained SiGe heterostructure FETs. A brief introduction of the importance of heterostructure MOSFETs, the band offsets at the SiGe/Si heterostructure and band alignment in the strained SiGe material is presented in Sect. 2. In Sect. 3, the design trade-off of a SiGe quantum-well p-MOSFET obtained using a 1-D Poisson solver is demonstrated. Sect. 4 outlines the conclusion and recommendations for future investigations.

2 Heterostructure MOSFETs

Strained SiGe alloys exhibit smaller fundamental band gaps as compared to Si due to the larger lattice constant (see Fig. 1). In the presence of compressive strain, the band gap is further reduced. For epitaxial $Si_{1-x}Ge_x$ films grown on Si, the band gap difference resides mostly in the valence band. As a result, this heterostructure combination is well suited for the confinement of holes. For a sufficient degree of hole carrier confinement (~200 meV), Ge mole fractions in excess of 30% are required. There is a very small conduction band discontinuity (~20 meV) between strained SiGe and cubic Si.

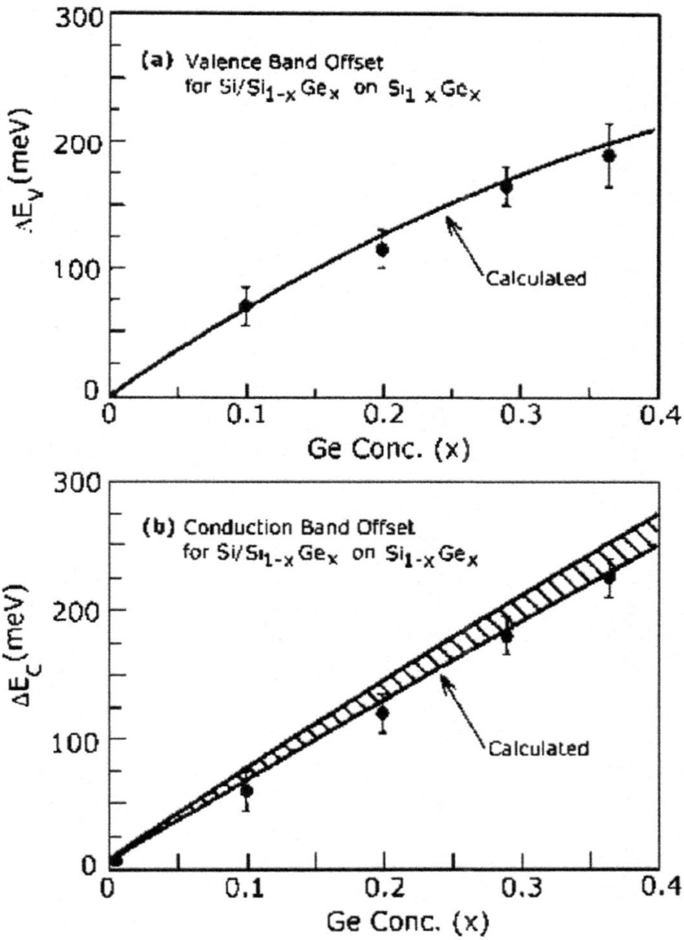

Fig. 1 Valence and conduction offsets in strained SiGe films [6]

In surface channel Si p-MOSFET, the channel mobility is limited to 0.25–0.5 of the bulk value because of scattering at the oxide/substrate interface. The surface scattering on the carriers is reduced appreciably in case of a buried channel SiGe p-MOSFET. The mobility of holes in the strained SiGe alloy is affected by the presence of strain. The compressive strain in these films breaks the heavy hole (HH) and light hole (LH) valence band degeneracy and leads to a reduction of the hole effective mass—the primary sources for the predicted hole transport improvement in these films. The in-plane velocity of holes in a strained SiGe alloy is higher than that of bulk Si and increases with Ge concentration at low and high fields due to decrease in transverse hole effective mass [7]. A smaller mass leads to a reduction in the density of states and a decrease in the scattering rate. The high-field hole mobility of a strained $Si_{0.6}Ge_{0.4}$ layer was found to be the same as that of bulk Ge. Since carriers in ultra small electronic devices attain such velocities with small biases, it is essential to understand high electric field carrier transport. Reduction of Coulombic scattering of the ionized impurities also arises due to the undoped nature of the SiGe alloy. This will enhance the channel mobility at low temperature.

Hole-phonon scattering in strained SiGe quantum wells was characterized using standard deformation potential theory within the framework of the momentum conservation approximation. It was found that confinement of the carriers does not significantly reduce the hole-phonon scattering rate since the two-dimensional density of states (DOS) is similar to the three-dimensional DOS value. Furthermore, holes confined within a SiGe well can scatter into a large number of sub-bands, which contributes to the overall scattering rate. However, an increase in hole saturated drift velocity is expected in strained $Si_{1-x}Ge_x$ quantum wells with increasing Ge mole fraction since the optical phonon spectrum retains a high energy character while the carrier effective mass decreases. This aspect of hole transport may lead to significant improvements in the high-frequency performance and current drive capability of short-channel p-MOSFETs.

Nayak et al. [8] first fabricated quantum-well SiGe p-MOSFET with higher channel mobility and saturation current. Carrier confinement in the quantum well was observed at room temperature and at low temperature. Other workers also confirmed the advantages of a SiGe p-MOS transistor over that of a Si p-MOS transistor. p-MOSFET fabrication on SiGe can directly improve the packaging density and speed in VLSI applications. The channel hole mobility is two times higher in the dual channel structure (SiO_2/Si/SiGe/Si) than that in submicron Si p-MOSFETs. Nayak et al. [9] fabricated a high-mobility quantum-well SiGe p-MOSFET on a SIMOX substrate with higher mobility and improved carrier confinement. p-MOSFET fabrication directly on strained SiGe/Si was first demonstrated in 1994 which has useful applications in high speed electronics and optoelectronics. Few reports are also available for n-channel SiGe MOSFET [5].

It has been difficult to realize SiGe channel MOSFETs due to the formation of either a pure SiO_2 or GeO_x mixed with SiO_2 layer at the oxide interface. A thin Si cap layer with SiGe channel overcomes this problem. Using this idea, the first fabrication of a quantum-well SiGe p-MOSFET was demonstrated by Nayak et al. [8]. It was found that SiGe p-MOSFET devices exhibit higher channel mobility and saturation current drive compared to an identically processed bulk Si p-MOSFET.

Design considerations for pseudomorphic Si/SiGe/Si heterojunction p-MOSFETs on Si substrates have been addressed by several authors [10, 11]. These studies have been based on the use of the 1-D Poisson simulator or self-consistent Schrodinger-Poisson solver or analytical techniques. "Schrodinger-Poisson solver" can be used to optimize the p-MOSFET design parameters, but it is a long duration process, because iteration is used to find the solution. To simplify device studies, a charge control semi-analytical model has also been reported in p-MOSFETs [12] that compute the charge distribution within the device for any arbitrary gate voltage without iterative loops. The analytical description as well as the quantum mechanical descriptions for hole charges in the SiGe well and Si cap, respectively, have been taken care in this model. In the following, the pseudomorphic silicon–germanium SiGe p-channel quantum-well MOSFET (p-MOSFET) design trade-off is presented.

3 Strained SiGe Channel p-MOSFET Simulation

The SiGe MOSFET design objective is to obtain maximum device transconductance and hence mobility. For this, the density of holes at the Si/SiO_2 interface has to be minimized to get maximum hole density in SiGe channel. The important $Si_{1-x}Ge_x$ channel MOSFET design factors are the selection of material used for gate, Ge concentration, $Si_{1-x}Ge_x$ well thickness, thickness of silicon cap and gate oxide layers. To obtain the hole densities in the SiGe and the parasitic Si channels, simulations have been performed. The SiGe channels exceeding 50 Å are considered for simulation, and hence, the quantum effects are neglected. A 1-D MOSCAP simulator (modified to include SiGe material parameters) was used in the present study to evaluate the effects of variation of different parameters on the device performance. Basically theoretical or experimental data are considered for strained SiGe material else a linear interpolation method can also be adopted. Figure 2 illustrates the SiGe MOS structure under investigation having the cited sequence: (100) oriented n-type Si substrate, Si buffer layer, SiGe well, Si cap layer, oxide and gate. It is assumed that the semiconductor region is uniformly doped throughout the layer.

3.1 Gate Material Selection

The material used in gate of SiGe p-channel MOS devices is predominantly influenced by the hole confinement amount. For n^+-poly gate SiGe p-channel MOSFET, Fig. 3 illustrates the distribution hole in SiGe and Si cap layer. At a situation where gate voltage almost close to threshold, the quantum well generally

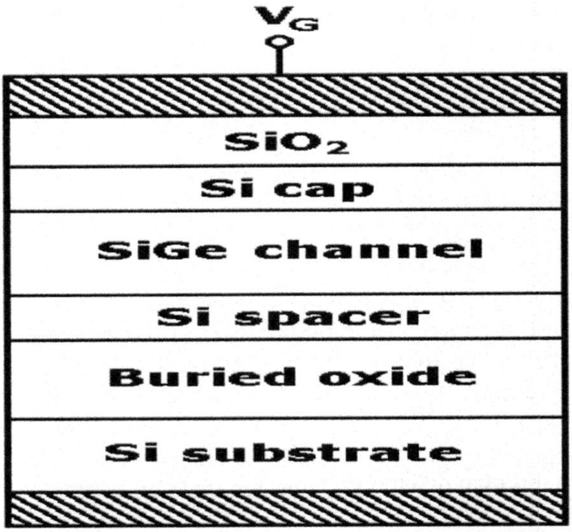

Fig. 2 SiGe MOS structure used in simulation

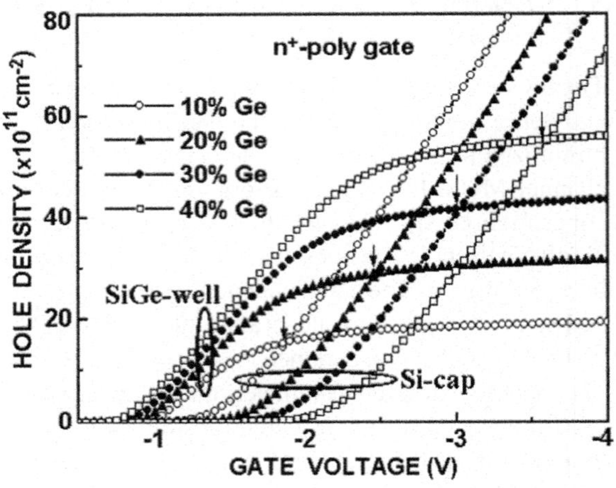

Fig. 3 Distribution of holes in the SiGe well and Si cap with the variation of Ge content (simulation result)

contains holes. With further decrease in gate voltage to −2.0 V for n^+ poly gate, the hole concentration is found to be larger in the parasitic Si cap in comparison to the SiGe quantum well. With further decrease in gate voltage, the hole concentration increases quickly in the Si cap layer whereas it remains almost constant in SiGe channel. The reason is that the holes at the Si/SiO_2 interface act as a shield between the gate and the SiGe channel. At crossover voltage, the value of hole density is equal in the SiGe channel and in the Si cap layer. The crossover voltage needs to be large negative gate bias for high hole density in the SiGe channel.

3.2 *Effect of Gate Oxide Thickness*

As found from simulation (Fig. 4), as the gate oxide thickness is increased, the hole confinement in the quantum well is improved. However, in order to obtain high current drive and high transconductance for short-channel MOSFET, the oxide thickness scaling is important. Because of these important requirements, the gate oxide thickness of a SiGe p-MOSFET will be controlled by device scaling. It is very difficult to achieve gate oxide of good quality directly on strained epitaxial SiGe by thermal oxidation due to presence of a pile-up of Ge atoms at the SiO_2/substrate interface. This may also lead to the formation of dislocation half-loops in the active region of a device. Si cap layer on SiGe overcomes the problem. Thus, low temperature oxidation directly on strained SiGe may be the right choice to deposit an oxide layer of good quality without affecting the strain at the interface.

3.3 *Effect of Ge Concentration*

The enhanced confinement and crossover voltage with increase in Ge concentration is shown in Fig. 3. The crossover voltage is more or less varies linearly with the Ge concentration. Confinement of hole in the quantum well is confirmed from the C–V (Capacitance–Voltage) measurement. The simulation results are shown in Fig. 3 with the four Ge concentrations (x = 0.1, 0.2, 0.3, 0.4). The plateau in the inversion capacitance indicates the confinement of hole in the SiGe quantum well (see Fig. 5). The result in Fig. 5 shows that, the plateau region becomes more noticeable with increase in Ge concentration. When inversion starts, the holes are present in the SiGe quantum well. This results in lower effective capacitance because the oxide and Si cap capacitances form a series combination. With an increase in the negative bias, the Si cap inversion layer is created and hence the capacitance

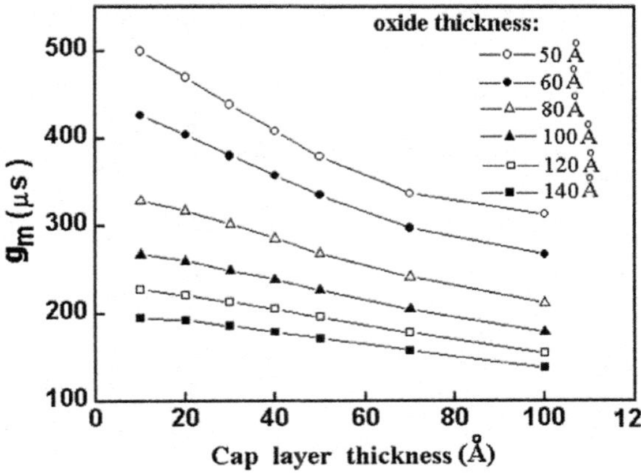

Fig. 4 Transconductance variation in the SiGe channel with cap layer thickess variation for different gate oxide thickness at a gate voltage ($V_g = -2$ V) (simulated result)

approaches the oxide capacitance. The strained epitaxial SiGe films should remain stable during the entire device fabrication process. This can be a limit on the thickness of SiGe as described by Matthews-Blakeslee stability criterion.

3.4 Effects of SiGe Well and Cap Layer Thickness

The effect of SiGe well thickness variation on the hole confinement is not much pronounced as shown in Fig. 6. For lower value of SiGe quantum well thickness, the crossover voltage and saturation value are slightly higher. The main limitation is the fact that the well thickness should be below critical thickness. The confinement of hole in terms of Si cap layer thickness is shown in Fig. 7. As the thickness of Si cap increases, the density of hole increases significantly. As the thickness of cap is increased for $V_g = -2$ V, there is a slight decrease in the density of holes in the channel. So in order to have the hole population minimum, the thickness of the cap should be as small as possible. So it becomes advantageous to keep the thickness of the cap small in order to have the population of hole in SiGe cap minimum as well as minimize the charge screening effect on the channel of SiGe. Also a high quality gate oxide should be grown using the Si cap. For a thin gate oxide (~5 nm), a 10 nm of silicon cap is adequate for the growth of a uniform oxide layer across the surface of wafer. Accordingly, the operation of the device should be performed at lower voltages applied across gate where the SiGe channel governs the electrical characteristics of the device. An optimization of the cap layer thickness is necessary to make it beneficial for future short-channel applications. The value of hole density is more for higher channel width at low gate voltage. However, this structure behaves like a surface channel device. Surface scattering and hot carrier may play an important role and may affect the device characteristics.

Fig. 5 Simulation results for low-frequency and high-frequency capacitance of a SiGe MOS structure with different Ge concentrations

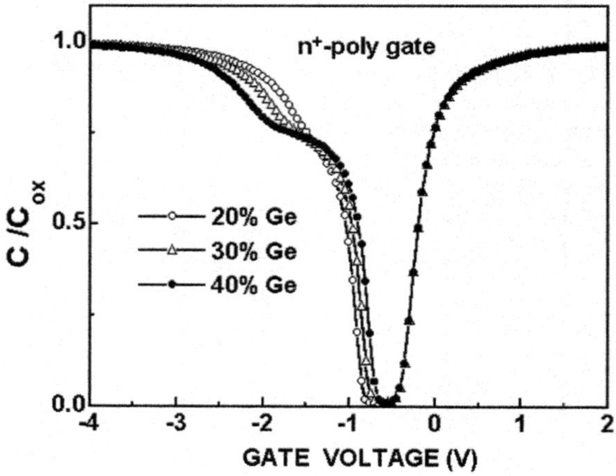

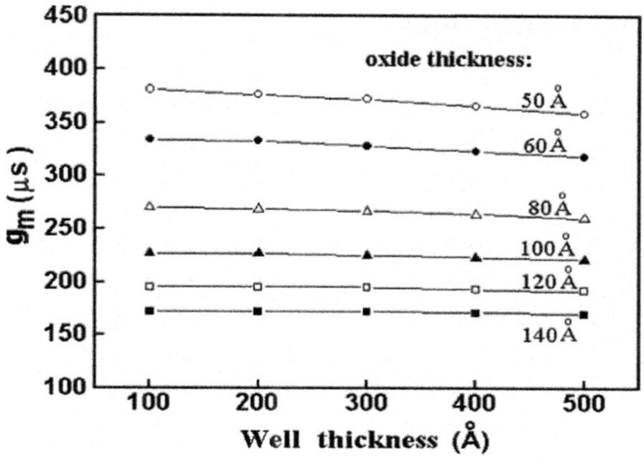

Fig. 6 Simulation results for transcondcutance in the SiGe channel with well thickness variation for different oxide layer thickness

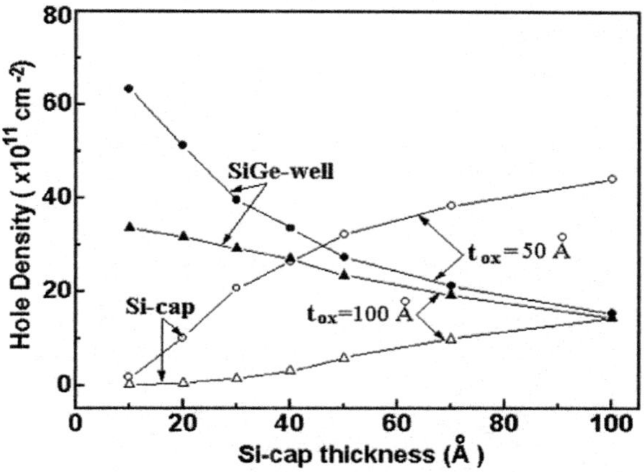

Fig. 7 Results for distribution of hole in the SiGe well as a function of Si cap layer thickness for a particular gate voltage ($V_g = -2$ V) from simulation

The main results obtained may be summarized as follows:

- As the Ge concentration increases, the hole concentration in the channel improves.
- The effect of increasing Si cap thickness is just the opposite. As the cap thickness increases, hole confinement in the parasitic channel increases. Therefore, small cap layer thicknesses are desirable.
- Increase in the gate oxide thickness improves the hole confinement.
- The thicknesses of the SiGe strained layer does not affect much the hole confinement.
- An n^+-poly gate is desirable over p^+-poly gate.

4 Conclusions

A critical review on the SiGe channel heterojunction p-MOSFET has been presented. The confinement of hole in the SiGe well is demonstrated by simulation and C–V measurements. The design trade-off for p-MOSFET devices with SiGe channel has been presented.

The device behaviour was studied using a 1-D Poisson Solver. The choice of material for gate, optimal selection of the SiGe channel width and profile, thicknesses oxide and Si cap, the method of threshold voltage adjustment have been addressed. It is shown that an optimally designed SiGe p-MOSFET should have a quantum-well 50 Å thick, a 40% Ge fraction in the quantum well, a Si cap layer 50 Å thick, a channel doping of about 2×10^{15} cm^{-3} and a gate oxide thickness about 50 Å.

References

1. D. Kahng, M.M. Atalla, Silicon-silicon dioxide field induced surface devices, in *Solid-State Device Research Conference*, Carnegie Institute of Technology, Pittsburgh, PA, 1960
2. Semiconductor Industry Association, *The International Technology Roadmap for Semiconductors*
3. T. Low, Y.T. Hou, M.F. Li, C. Zhu, A. Chin, G. Samudra, L. Chan, D.L. Kwong, investigation of performance limits of germanium double-gated MOSFETs, in *Proceedings of IEDM*, 2003, p. 691
4. A. Pethe, T. Krishnamohan, D. Kim, S. Oh, H.S.P. Wong, Y. Nishi, K.C. Saraswat, investigation of the performance limits of III-V Double-Gate n-MOSFETs, in *Proceedings of IEDM*, 2005, p. 605
5. C.K. Maiti, N.B. Chakrabarti, S.K. Ray, *Silicon Heterostructures: Materials and Devices* (Institute of Electrical Engineers (IEE), UK, 2001)
6. L.K. Bera, in Studies on applications of strained-Si for heterostructure field effects transistors. PhD Thesis, IIT Kharagpur, 1998
7. S.H. Li, J.M. Hinckley, J. Singh, P.K. Bhattacharya, Carrier velocity-field characteristics and alloy scattering potential in $Si_{1-x}Ge_x$/Si. Appl. Phys. Lett. **63**, 1393–1395 (1993)
8. D.K. Nayak, J.C.S. Woo, J.S. Park, K.L. Wang, K.P. MacWilliams, Enhancement-mode quantum-well Ge_xSi_{1-x} PMOS. IEEE Electron Dev. Lett. 1991;**EDL-12**:154–156
9. D.K. Nayak, J.C.S. Woo, G.K. Yabiku, K.P. MacWilliams, J.S. Park, K.L. Wang, High mobility GeSi PMOS on SIMOX. IEEE Electron Dev. Lett. **14**, 520–522 (1993)
10. S. Verdonckt-Vandebroek, E. Crabbe, B.S. Meyerson, D.L. Harame, P.J. Restle, J.M.C. Stork, A.C. Meydanis, C.L. Stanis, A.A. Bright, G.M.W. Kroesen, A.C. Warren, High-mobility modulation-doped grades SiGe-channel p-MOSFET's. IEEE Electron Dev. Lett. 1991;**EDL-12**, 447–449
11. S. Verdonckt-Vandebroek, E.F. Crabbe, B.S. Meyerson, D.L. Harame, P.J. Restle, J.M.C. Stork, J.B. Johnson, SiGe-channel heterojunction p-MOSFETs. IEEE Trans. Electron Dev. **41**, 90–101 (1994)
12. K. Bhaumik, Y.S. Diamand, J.P. Noel, J. Bevk, L.C. Feldman, 23 GHz f_T room temperature SiGe quantum-well p-MOSFETs, in *Proceedings of ISDRS*, 1993, p. 349–352

An Ultra Low Power CMOS RF Front-End-Based LNA and Mixer for GPS Application

Namrata Yadav, Deepak Prasad, Vijay Nath and Manish Kumar

Abstract In this research article, a 1.5-GHz low-noise amplifier and down-conversion double-balanced mixer have been designed for CMOS RF front receiver. It plays a vital role in Global Positioning System (GPS) for increasing the safety and efficiency of flight. Gilbert down-conversion topology has been adopted for the design of mixer, while single-differential topology with matching network has been implemented for low-noise amplifier. The conversion gain of the mixer is 16 dB, noise figure is 12 dB, IIP3 is −5.66 dBm, and 1-dB compression point is 1.369 dBm. The designed circuit is tested at 1.5 V, and the simulation has been carried out with the help of cadence analog design environment with UMC 90 nm technology.

Keywords Cascode transistor · Single-differential LNA
Double-balanced mixer · Gilbert cell · Noise figure

1 Introduction

The growing avionic industry has generated increasing interest in GPS. The new emerging technologies in GPS have increased data rates and capacity to reduce the power dissipation for longer operation time. Earlier GPS receivers were used in conventional systems like ship navigation which consists of several chips. With exploration and development of CMOS technology, we could have low-cost, small-size, and low-voltage circuitry promising to integrate whole system on single chip. The challenges are continuous and imply motivation in exploration of RF

N. Yadav · D. Prasad (✉) · V. Nath
VLSI Design Group, Department of Electronics & Communication Engineering,
Birla Institute of Technology, Mesra, Ranchi, India
e-mail: prasaddeepak007@gmail.com

M. Kumar
Department of Electronics & Communication Engineering, MMMUT,
Gorakhpur, Uttar Pradesh, India

V. Nath (ed.), *Proceedings of the International Conference on Microelectronics, Computing & Communication Systems*, Lecture Notes in Electrical Engineering 453,
https://doi.org/10.1007/978-981-10-5565-2_33

architectures. RF front end which is the most important block of receiver unit constitutes of low-noise amplifier (LNA) and mixer block. After antenna, the key unit of the receiver system is the low-noise amplifier (LNA) which has been used to amplify the very weak signal received by antenna [1]. In the designing of circuit, NMOS is generally preferred over PMOS due to its high gain and mobility property. This work is on RF front end which consists of differential gm stage, i.e., low-noise amplifier and down-conversion mixer. Low-noise amplifier is first key component of receivers. The block has simple topology consisting input and output matching networks. The main aim of the design is to have low-noise figure and high-voltage gain. LNA involves many design trade-offs such as voltage gain, linearity, stability, noise figure [2]. The next circuitry in RF front end is called down-conversion mixer. It is basically used for frequency translation from RF to IF.

2 Architecture

The architecture of the front end of receiver circuit which is to be followed while designing is given in Fig. 1.

The BPF1 is the band-selection filter. It is used to select the required range of frequency from the large bandwidth. It plays the role of eliminating and suppressing the undesired signals [3]. After the band-pass filter, LNA is placed as the desired signal containing noise elements with additive white gaussian noise (AWGN). In other words, it can be said that the output of BPF1 is amplified by LNA. BPF2 is an anti-image filter. It is used to filter out the image frequency which can degrade signal-to-noise ratio. The BPF is implanted before the mixer block because once the signal is being down converted, it cannot be rectified. Rather it can badly affect linearity of the mixer. It is constraint of receiver design as the bandwidth is too narrow. BPF3 is a channel filter. It works for the small channel of intermediate frequency. While performing channel filtering, the frequency band of this filter depends on channel spacing which is unique for particular standard. After BPF3, the signal follows IF amplifier for amplifying the mixer output around IF band.

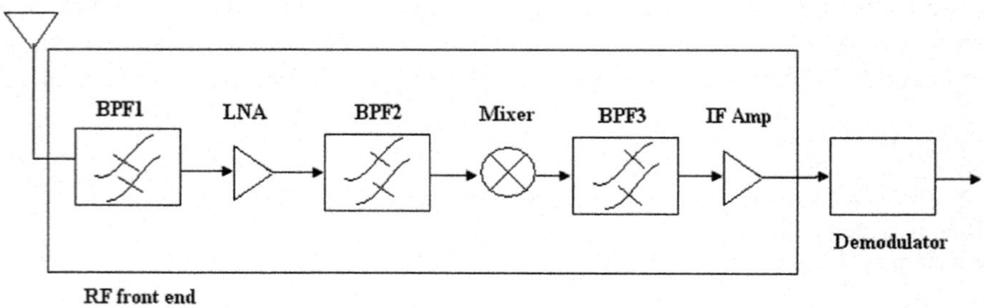

Fig. 1 Architecture of RF front end

3 Circuit Analysis

3.1 Single-Differential LNA

The differential circuit is operated at 1.5 V dc. A single-differential LNA is having a single RF input port and two differential outputs which helps to have good gain of the circuit. Input port is fed by sinusoidal signal of 1.575 GHz, and output port is connected to band-pass filter circuit for output, followed by next block that is mixer. At input of LNA, input matching is done at 50 Ω through capacitance (C1), gate inductor (Lg), source degeneration inductance (Ls). At differential output, capacitance (Cd) and inductor (Ld) are used as tuning circuit for output matching along with resistance (rd) [4]. *M*0 and *M*3 are the common source transistors used in differential circuitry having low gate resistance, high aspect ratio (W/L ~ 1000), and large gm. *M*0, *M*2 and *M*3, *M*2 are current mirror pairs providing biasing to CS amplifier *M*0 and *M*3. The aspect ratio of transistor *M*2 is not as high as of transistor *M*0/*M*3. The biasing is done so as to operate transistor in region 3, i.e., sub-threshold region. *M*1, *M*4 are a cascode transistor (CS-CG) which isolates the input from output (Miller capacitance) providing stability and reducing harmonics. At microwave frequencies, the parasitic capacitance of transistor becomes significant. Cgd is Miller capacitance at input and output terminal. This capacitance will create nonlinearity in the circuit which can be reduced by increasing fingers in transistor and implanting cascode transistor [5]. The method used to have maximum power transfer is done by inductive source degeneration technology which is done by Lg, Ls, and C1. Applying Kirchhoff's voltage law (KVL) at the input of small-signal model which is shown in Fig. 2 gives

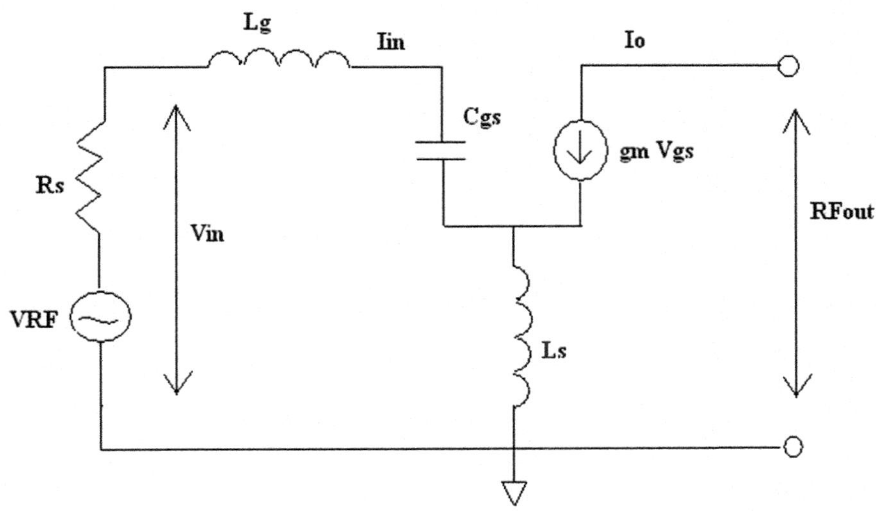

Fig. 2 Small-signal model of LNA

$$V_i = I_i j\omega L_g + \frac{I_i}{j\omega C_{gs}} + (I_i + g_m V_{gs}) j\omega L_S \quad (1)$$

For impedance matching, following condition should be met (Fig. 3)

$$\text{Re}[Z_{in}] = \frac{g_m L_s}{C} = R_s \quad (2)$$

$$\text{Im}[Z_{in}] = j\omega(L_g + L_s) + \frac{1}{j\omega C} = 0 \quad (3)$$

And for output LC tank

$$f_c = \frac{1}{2\pi\sqrt{(C_d + C_L + C_{db})L_d}} \quad (4)$$

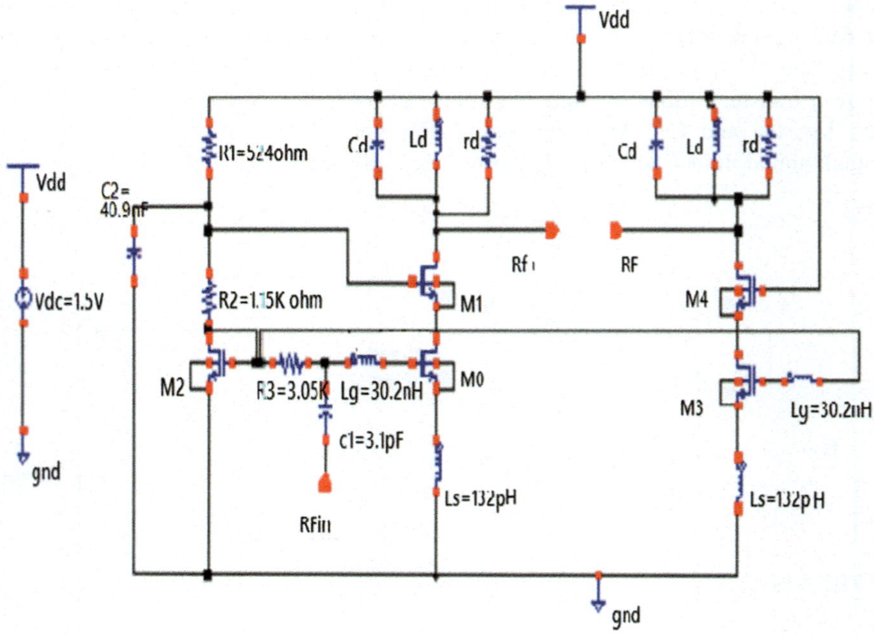

Fig. 3 Proposed single-differential LNA

4 Down-Conversion Double-Balanced Mixer

Mixers are commonly used to multiply signals of different frequencies in an effort to achieve frequency translation. It translates frequency from one band to another. This frequency translation or frequency conversion can be done in two ways, i.e., up-conversion mixer and down-conversion mixer. Up-conversion mixer is used in the transmitter circuit section where it multiplies the low-frequency message signal with a local oscillator signal to convert low-frequency message signal to high-frequency message signal [7]. On the other hand, down-conversion mixer is used in receiver circuit section where it multiplies a high-frequency signal with a local oscillator signal to obtain low-frequency signal (IF signal).

The Down-Conversion mixer which has been used in the RF receiver is basically featured with Gilbert Multiplier. There are two single-balanced mixers having four NMOS switches (M0–M3). These switches are driven with large Vgs else they are in cutoff region. Large square LO(±) signal (10 dBm, 1675 MHz) are cross-coupled multiplied with RF out of LNA to get required low-frequency

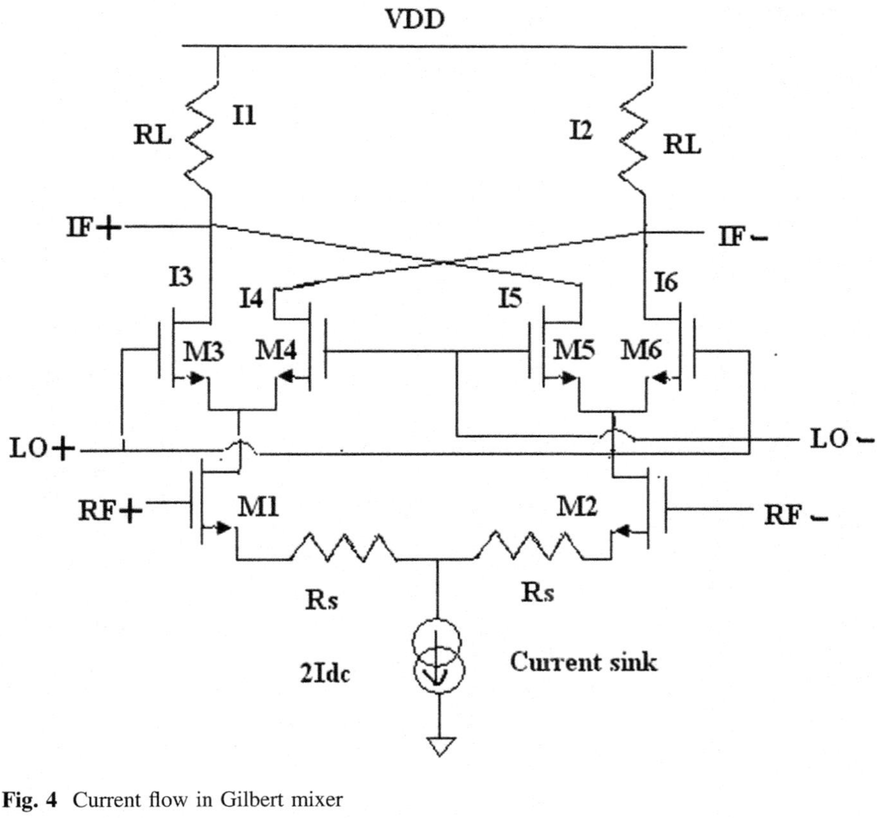

Fig. 4 Current flow in Gilbert mixer

component (IF signal) [8]. The local oscillator will have only odd harmonics because of differential circuitry. *M*0, *M*3 will conduct for LO+ and *M*1, *M*2 will conduct for LO. The duty cycle of LO can largely affect the gain of mixer block. There RF merits of a mixer include input third-order intercept point (IIP3), conversion gain, input 1-dB compression point, noise figure (Fig. 4).

Differential output of two single-balanced mixer is as follows:

$$I_3 - I_4 = \left(I_{dc} + I_{rf}\cos\left(\omega_{rf}t\right)\right)s(t) \tag{5}$$

$$I_5 - I_6 = \left(I_{dc} - I_{rf}\cos\left(\omega_{rf}t\right)\right) \tag{6}$$

Differential output of two single-balanced mixers is as follows:
Where $|s(t)| = 2/\pi$ for LO)

$$I_{od} = (I_3 - I_4) - (I_5 - I_6) \tag{7}$$

$$I_{od} = 2I_{rf}\cos\left(\omega_{rf}t\right)s(t) \tag{8}$$

So, conversion gain of double-balanced mixer is given as (Fig. 5)

$$G_C = \frac{\text{Output\_amplitude}}{\text{Input\_}R_f\text{amplitude}} = \frac{4I_{rf}R_l}{\pi 2V_{rf}} = \frac{2g_mR_l}{\pi} \tag{9}$$

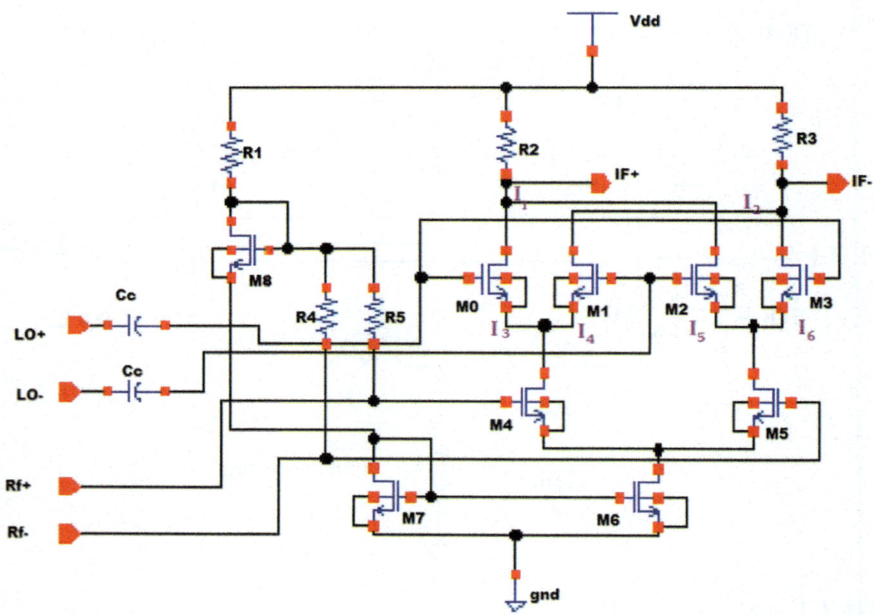

Fig. 5 Proposed down-conversion double-balanced mixer

5 Simulation Result and Discussion

The LNA is simulated and achieves parameters like Gain S21 (26.08 dB), input matching S11 (−8.59 dB), output matching S22 (−5.73 dB), reverse isolation S12 (−37.26 dB) which are shown in Fig. 6 along with NF (0.487 dB) in Fig. 7.

The IIP3 and 1-dB compression point of mixer are −5.66 and 1.369 dBm, which is shown in Figs. 8 and 9. Figures 10 and 11 show Noise Figure and conversion gain of mixer which is achieved to be 12 and 16 dB respectively. The simulation of IIP3, 1-dB compression point, and noise figure is depicted for 1 MHz as IF frequency. The performance summary of the proposed LNA and mixer where it is being compared to other recent paper is summarized in Table 1.

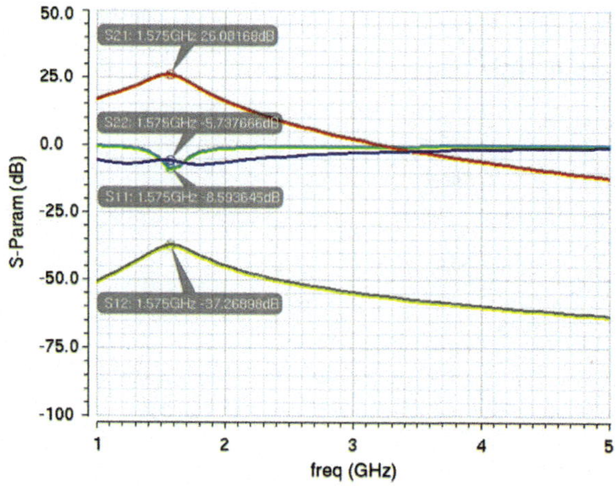

Fig. 6 S parameters of LNA simulation

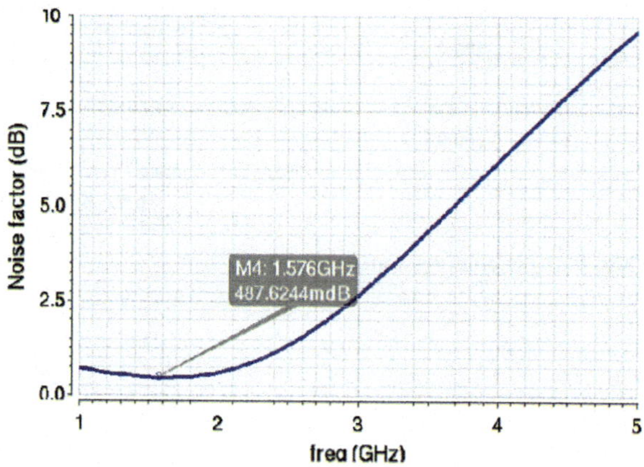

Fig. 7 Noise figure of LNA

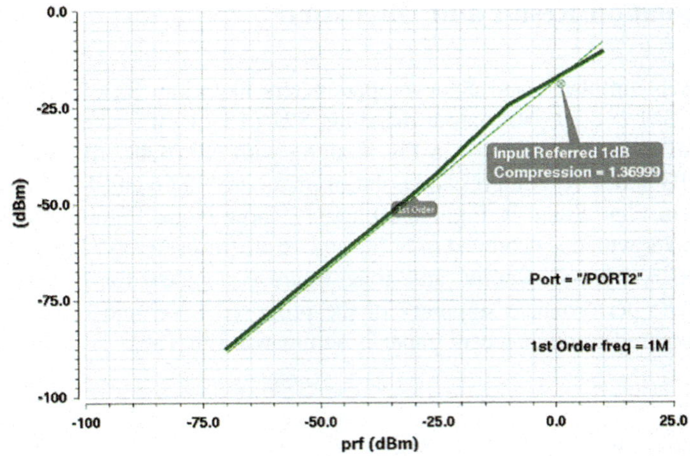

Fig. 8 IIP3

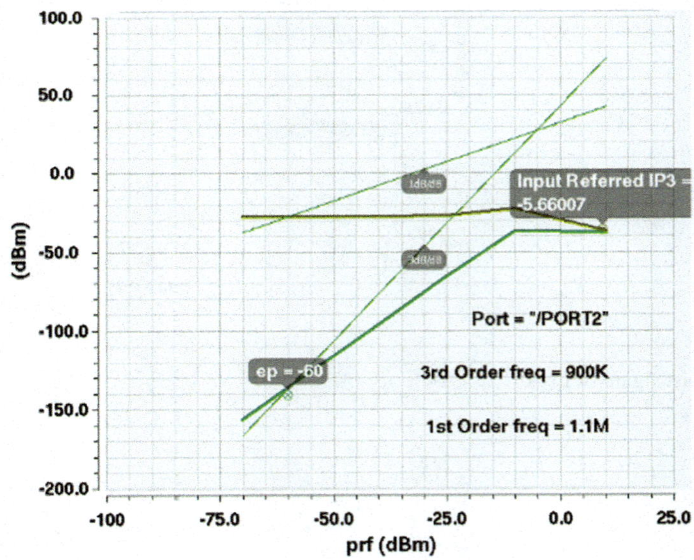

Fig. 9 1-dB compression point

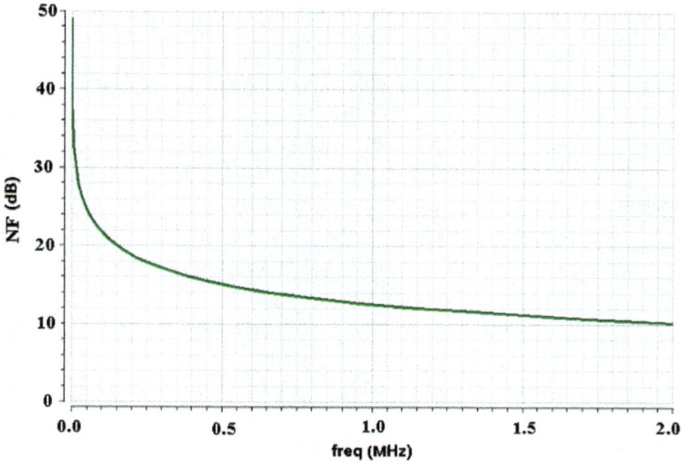

Fig. 10 Noise figure of mixer

Fig. 11 Conversion gain versus bulk voltage

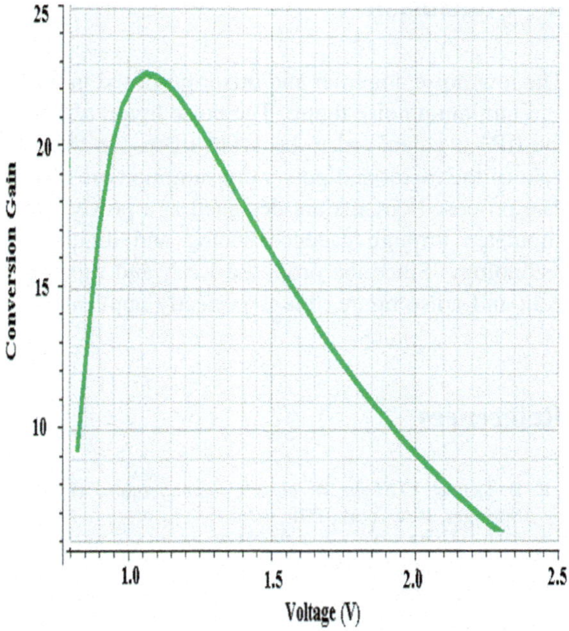

Table 1 Performance summary and comparison with recent paper of LNA and mixer

LNA				Mixer			
Parameter	[6]	[9]	This work	Parameter	[10]	[11]	This work
Technology (nm)	90	90	90	Technology (nm)	65	130	90
Operating frequency (GHz)	2.45	1.575	1.575	RF frequency (GHz)	1.9	2.4	1,575
S11 (dB)	−9.19	−16.34	−8.59	Noise figure (dB)	8.92	16.5	12
S12 (dB)	−38.03	−20	−37.26	IIP3 (dBm)	6	2.7	−5.66
S21 (dB)	31.53	14.64	26.08	1-dB compression point	−11.5	−10	1.369
IIP3 (dBm)	−5.70	1.655	1.114	Power conversion gain	12.42	13.1	16
Noise figure (dB)	2.34	.933	0.487	Power supply (V)	1.8	1.2	1.5
Power supply (V)	1.2	0.7	1.5	Power consumption	2	2.4	2.14

6 Conclusion

The low-noise amplifier and mixer proposed in this paper are relying with high gain and minimum noise figure. The noise figure of LNA is found to be 0.487 dB, while the IIP3 is calculated to be 1.114 dBm. On the other hand, the power conversion gain of the mixer and power consumption are observed to be 16 dB and 2.14 mW. The proposed circuits are designed at a power supply of 1.5 V. It is designed and simulated through cadence analog and digital system design tools of UMC90 technology. Since the proposed LNA and mixer give agreeable performance with gain and noise figure, this is perfectly applicable in GPS.

References

1. B. Leung, *VLSI for Wireless Communications* (Prentice Hall Electronics and VLSI series, Springer, New York, 2011)
2. T.H. Lee, *The Design of CMOS Radio Frequency Integrated Circuits* (Cambridge University Press, Cambridge, 1998)
3. B. Razavi, CMOS technology characterization for analog and RF design. IEEE J. Solid-State Circ. **34**, 268–276 (1999)
4. R. Kumar, Design and noise optimization of RF low noise amplifier for IEEE Standard 802.11a WLAN. Int. J. VLSI Design Commun. Syst. **2**, 165 (2012)
5. N. Baig, D.S Chandu, B Satish, Design and analysis of a CMOS 0.7 V low noise amplifier for GPS L1 band. Int. J. Eng. Innov. Technol. **2**, 272–276 (2012)
6. R. Kundu, A. Pandey, V. Nath, A CMOS low noise amplifier based on common source technique for ISM band application. Microsyst. Technol. doi:10.1007/s00542–015-2550

7. P.C. Pattewar, Performance comparison of various low noise-high speed amplifier topologies for GPS applications. Int. J. Curr. Eng. Technol. **4**(1), 426–429 (2014)
8. V. Venkatesan, A 3–14 GHz low noise amplifier for ultra wide band applications. Int. J. VLSI Design Commun. Syst. (VLSICS) **3**(1), 137 (2012)
9. M.T. Hsu, Y.C. Chang, Y.Z. Huang, Design of low power UWB LNA based on common source topology with current-reused technique. J. Microelectron. **44**, 1223–1230 (2013). doi:10.1016/j.mejo.2013.08.008
10. R. Mahmou, K. Faitah, Designing of RF single balanced mixer with a 65 nm CMOS technology dedicated to low power consumption wireless applications. Int. J. Comput. Sci. (IJCSI) **9**(3), 358–363 (2012)
11. H. Rashtian, A. Hossein, M.i Shirazi, S. Mirabbasi "On the use of body biasing to improve linearity in low LO-power CMOS active mixers". Microelectron. J. **45**(8), 1026–1032 (2014)

Author Index

V. Nath (ed.), *Proceedings of the International Conference on Microelectronics, Computing & Communication Systems*, Lecture Notes in Electrical Engineering 453,
https://doi.org/10.1007/978-981-10-5565-2

The manufacturer's authorised representative in the EU is Springer Nature Customer Service Centre GmbH, Europaplatz 3, 69115 Heidelberg, Germany. If you have any concerns regarding our products, please contact ProductSafety@springernature.com

Printed and bound by CPI Group (UK) Ltd, Croydon, CR0 4YY
01/10/2026
02225871-0011